Geohazards in Indonesia: Earth Science for Disaster Risk Reduction

Geological Society books refereeing procedures

The Society makes every effort to ensure that the scientific and production quality of its books matches that of its journals. Since 1997, all book proposals have been refereed by specialist reviewers as well as by the Society's Books Editorial Committee. If the referees identify weaknesses in the proposal, these must be addressed before the proposal is accepted.

Once the book is accepted, the Society Book Editors ensure that the volume editors follow strict guidelines on refereeing and quality control. We insist that individual papers can only be accepted after satisfactory review by two independent referees. The questions on the review forms are similar to those for *Journal of the Geological Society*. The referees' forms and comments must be available to the Society's Book Editors on request.

Although many of the books result from meetings, the editors are expected to commission papers that were not presented at the meeting to ensure that the book provides a balanced coverage of the subject. Being accepted for presentation at the meeting does not guarantee inclusion in the book.

More information about submitting a proposal and producing a book for the Society can be found on its website: www.geolsoc.org.uk.

It is recommended that reference to all or part of this book should be made in one of the following ways:

CUMMINS, P. R. & MEILANO, I. (eds) 2017. *Geohazards in Indonesia: Earth Science for Disaster Risk Reduction.* Geological Society, London, Special Publications, **441**.

WATKINSON, I. M. & HALL, R. 2017. Fault systems of the eastern Indonesian triple junction: evaluation of Quaternary activity and implications for seismic hazards. *In*: CUMMINS, P. R. & MEILANO, I. (eds) *Geohazards in Indonesia: Earth Science for Disaster Risk Reduction.* Geological Society, London, Special Publications, **441**, 71–120. First published online December 19, 2016, updated April 19, 2017, https://doi.org/10.1144/SP441.8

GEOLOGICAL SOCIETY SPECIAL PUBLICATION NO. 441

Geohazards in Indonesia: Earth Science for Disaster Risk Reduction

EDITED BY

P. R. CUMMINS
Australian National University, Australia

and

I. MEILANO
Bandung Institute of Technology, Indonesia

2017
Published by
The Geological Society
London

THE GEOLOGICAL SOCIETY

The Geological Society of London (GSL) was founded in 1807. It is the oldest national geological society in the world and the largest in Europe. It was incorporated under Royal Charter in 1825 and is Registered Charity 210161.

The Society is the UK national learned and professional society for geology with a worldwide Fellowship (FGS) of over 10 000. The Society has the power to confer Chartered status on suitably qualified Fellows, and about 2000 of the Fellowship carry the title (CGeol). Chartered Geologists may also obtain the equivalent European title, European Geologist (EurGeol). One fifth of the Society's fellowship resides outside the UK. To find out more about the Society, log on to www.geolsoc.org.uk.

The Geological Society Publishing House (Bath, UK) produces the Society's international journals and books, and acts as European distributor for selected publications of the American Association of Petroleum Geologists (AAPG), the Indonesian Petroleum Association (IPA), the Geological Society of America (GSA), the Society for Sedimentary Geology (SEPM) and the Geologists' Association (GA). Joint marketing agreements ensure that GSL Fellows may purchase these societies' publications at a discount. The Society's online bookshop (accessible from www.geolsoc.org.uk) offers secure book purchasing with your credit or debit card.

To find out about joining the Society and benefiting from substantial discounts on publications of GSL and other societies worldwide, consult www.geolsoc.org.uk, or contact the Fellowship Department at: The Geological Society, Burlington House, Piccadilly, London W1J 0BG: Tel. +44 (0)20 7434 9944; Fax +44 (0)20 7439 8975; E-mail: enquiries@geolsoc.org.uk.

For information about the Society's meetings, consult *Events* on www.geolsoc.org.uk. To find out more about the Society's Corporate Affiliates Scheme, write to enquiries@geolsoc.org.uk.

Published by The Geological Society from:
The Geological Society Publishing House, Unit 7, Brassmill Enterprise Centre, Brassmill Lane, Bath BA1 3JN, UK

The Lyell Collection: www.lyellcollection.org
Online bookshop: www.geolsoc.org.uk/bookshop
Orders: Tel. +44 (0)1225 445046, Fax +44 (0)1225 442836

British Library Cataloguing in Publication Data

A catalogue record for this book is available from the British Library.
ISBN 978-1-86239-966-2
ISSN 0305-8719

Distributors
For details of international agents and distributors see:
www.geolsoc.org.uk/agentsdistributors

Typeset by Nova Techset Private Limited, Bengaluru & Chennai, India
Printed and bound by CPI Group (UK) Ltd, Croydon CR0 4YY

Contents

Geohazards in Indonesia: Earth science for disaster risk reduction – introduction

PHIL R. CUMMINS

Research School of Earth Sciences, Australian National University, Canberra 2604, Australia
Correspondence: phil.cummins@anu.edu.au

Indonesia encompasses one of the most active tectonic regions on Earth. Its territory includes over 18 000 km of major tectonic plate boundary, more than twice that of Japan or Papua New Guinea (Fig. 1). This includes: the Great Sumatran Fault, at 1900 km length about 50 per cent longer than the San Andreas and North Anatolian faults; a 6000 km length of convergent plate margin, stretching from off Sumatra to the Banda Sea that has experienced the world's second largest megathrust earthquake (Stein & Okal 2005) and its largest intraslab earthquake (Osada & Abe 1981); and one of the highest-curvature subduction zones in the world, with the Banda Arc curving through 180° in 2200 km of its length. Indonesia is home to at least 127 active volcanoes, more than any other country in the Asia-Pacific. These eruptions include three of the largest and deadliest in human history: 1257 Rinjani (VEI (Volcanic Explosivity Index) 7, fatalities unknown; Lavigne *et al.* 2013), 1815 Tambora (VEI 7, 71 000 fatalities; Oppenheimer 2003), 1883 Krakatau (VEI 6, 36 000 fatalities; Self & Rampino 1981). Other remarkable features of Indonesia's tectonic environment are the divergent double subduction of the Molucca Sea Plate resulting in arc–arc collision (Widiwijayanti *et al.* 2003); the subduction of continental lithosphere in the Banda Sea arc–continent collision (Fichtner *et al.* 2010); one of the world's youngest orogenies along the Timor–Tanimbar section of the Banda Arc (Kaneko *et al.* 2007; Keep & Haig 2010); and areas of pronounced mud volcano activity (Milkov 2000).

Indonesia's pronounced tectonic activity poses a grave threat to its large and often vulnerable population. By population Indonesia is the fourth largest country in the world, with a population expected to grow from 238.5 to 305.7 million between 2010 and 2035 (Indonesian Bureau of Statistics 2014). Almost 50 per cent of Indonesians live on the island of Java, the world's most densely populated island. Java accounted for over 60 per cent of Indonesia's urban land use in 2010 and is expected to accommodate 80 per cent of Indonensia's urban expansion between 2000 and 2030 (Muis *et al.* 2015). As noted for buildings in Banda Aceh following the 2004 Great Sumatra Earthquake (Saatcioglu *et al.* 2006), urban residential construction in Indonesia often comprises low-rise, non-engineered structures that are unable to withstand the loads that might be placed on them by a major event such as an earthquake, tsunami or volcanic eruption. From the combination of increased population, greater urbanization and poor construction standards, we can expect that Indonesia's vulnerability to natural hazards is high, much higher than it has ever been in the past.

Because geological hazards in Indonesia comprise a potent threat to a large and vulnerable population, it is important that decisions made by disaster managers be informed by the best available Earth science. Considerable progress has been made in recent years in providing the Earth science that can support disaster risk reduction, and it is the purpose of this Geological Society of London Special Publication to document some of this progress.

Tsunamis

The vulnerability of Indonesia to tsunamis was starkly illustrated by the 2004 Indian Ocean Tsunami (IOT), which killed over 165 000 people along the coast of northern Sumatra. This single event, the most lethal tsunami disaster in human history, accounts for almost 70 per cent of the recorded deaths due to natural disasters in Indonesia since 1900 (Fig. 2). While it may be tempting to regard this event as unique and therefore unrepresentative of the type of tsunami impact Indonesia will experience in the future, it is important to remember that the 1883 eruption of Krakatau generated a tsunami that killed tens of thousands. Also, in this issue **Harris & Major (2016)** use the historical catalogue of Wichmann (1918) to highlight the many destructive tsunamis that have occurred historically in eastern Indonesia during the colonial period (seventeenth to nineteenth centuries). They point out that the past 160 years has been a time of relative quiescence, but that if the previous level of activity resumes it will have a much greater potential for disaster now than in the past because of the explosive growth in population during the twentieth century.

From: Cummins, P. R. & Meilano, I. (eds) 2017. *Geohazards in Indonesia: Earth Science for Disaster Risk Reduction*. Geological Society, London, Special Publications, **441**, 1–7.
First published online September 6, 2017, updated September 7, 2017, https://doi.org/10.1144/SP441.11

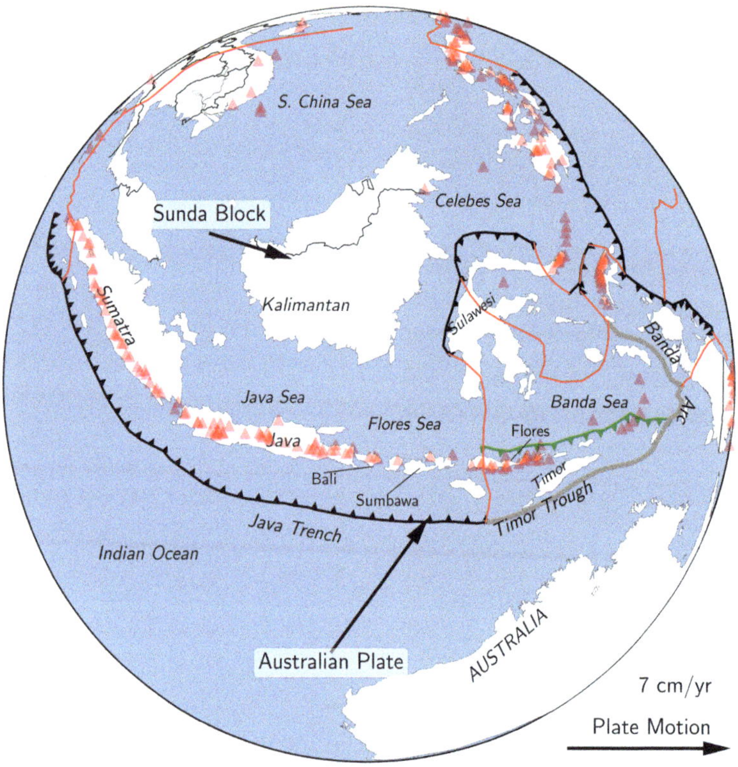

Fig. 1. Active tectonics of the Indonesian Region. Major plate boundaries are indicated: subduction zones and back-arc thrusts by the black and green 'toothed' curves, respectively; plate boundaries with uncertain relative movement by grey curves; and other transform boundaries by red curves. Also indicated are Holocene volcanoes (red triangles) and the absolute motion of the Australian Plate and the Sunda Arc. Plate boundaries are from Bird (2003); volcanoes from Global Volcanism Program (2013).

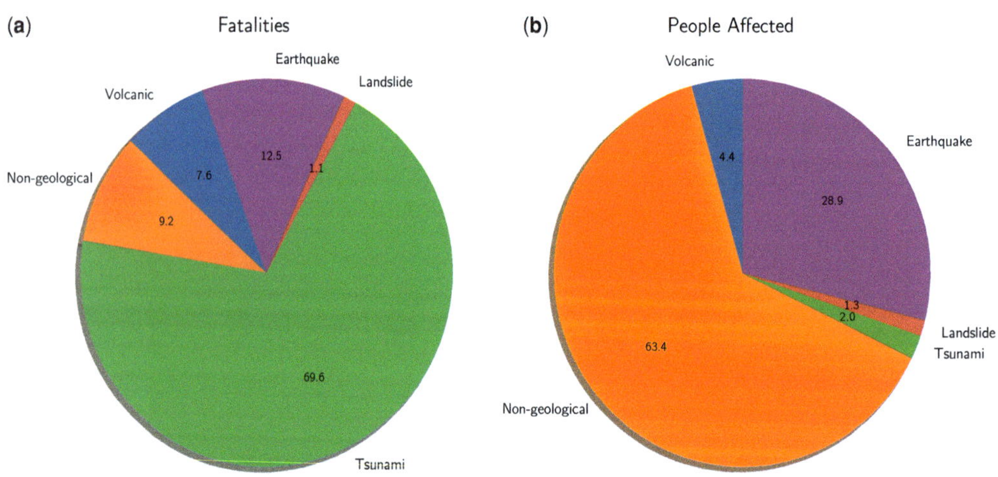

Fig. 2. Impacts of natural disasters occurring in Indonesia from 1900 to 2016 in terms of (**a**) fatalities and (**b**) people affected. Data from the EM-DAT database (Guha-Sapir *et al.* 2017).

Understanding just how large the impacts of future tsunami events might be requires careful modelling of tsunami inundation that takes into account the wide variety of rupture styles of tsunamigenic earthquakes. **Griffin *et al.* (2016)** have undertaken such a study using the highly probable scenario of a tsunamigenic earthquake off Padang, Sumatra, as a case study. In this case there are considerable palaeogeodetic data that can be used to constrain the range of plausible rupture scenarios, and **Griffin *et al.* (2016)** show that sampling a range of such scenarios yields more robust estimates of which areas may be inundated than does considering any single scenario. The challenge in future tsunami hazard assessments will be to determine how a range of plausible scenarios can be considered when there are few if any data to constrain them.

Earthquakes

Figure 2a shows that, following the massive fatalities caused by the 2004 IOT, earthquakes are the second largest source of natural disaster fatalities, accounting for 12.5 per cent of the total (about 40 per cent if the IOT is discounted). The sequence of large earthquakes that ruptured the Sumatran subduction zone since 2004 has called attention to the potential for damaging earthquakes to occur in Indonesia. Historical records prior to 1900 indicate that this level of activity is not unusual, and that while the largest Sumatran events of the 1700–1800s have largely repeated themselves since 2004, historical events in Java and eastern Indonesia have not, that is, the post-1900 occurrence of large earthquakes in Java and eastern Indonesia has been low by historical standards (Nguyen *et al.* 2015). As was true for tsunamis, the exponential growth in population in the late twentieth century means that the impact of these events when they recur will far exceed their historical levels. Therefore, there is an urgent need to improve our understanding of earthquake hazard and risk in Indonesia. The studies published in this issue deal with several aspects of earthquake risk: earthquake sources, ground motion and site response, fatality models and probabilistic seismic hazard assessment.

Sources of earthquakes

Most of the known damaging Indonesian earthquakes have occurred in the pre-instrumental, historical past. The difficulty in assessing the threat they pose to present-day populations is exacerbated by the fact that most of them cannot be unequivocally associated with specific faults. Indeed, very few of the many known crustal faults in Indonesia have been conclusively categorized as active. In this Special Publication, **Harris & Major (2016)** have made important progress in identifying the faults associated with some of the most damaging earthquakes known to have occurred during the colonial period (seventeenth to nineteenth centuries). In another paper **Watkinson & Hall (2017)** have taken a more systematic approach by using geological methods to identify active faults in eastern Indonesia. Of the 27 faults they studied, 16 show rapid and three show low tectonic activity. **Watkinson & Hall (2017)** call particular attention to the Palu-Koro Fault in Sulawesi, which they characterized as having a high slip rate with fault strands proximate to a large urban population in the city of Palu.

Probabilistic seismic hazard assessment

In order to inform engineering requirements for seismically resilient construction, a hazard assessment needs to specify exceedance levels of ground motion for different return periods and frequencies of wave motion. The only widely accepted approach for estimating these is probabilistic seismic hazard assessment (PSHA), which has been used to support the Indonesian building code since at least 2002 (SNI-1726). A revision to the national-scale bedrock PSHA was undertaken in 2010, and another is currently underway. These more recent national PSHAs generally involve much higher levels of hazard due to the inclusion of active crustal faults, but the earthquake potential for these faults is often poorly determined. **Omang *et al.* (2016)** in this Special Publication show how this uncertainty, specifically in the estimated fault slip rate and locking width, can be directly incorporated into PSHA so that hazard assessment results more robustly reflect the quality of information used as input, although this methodology has yet to be incorporated in any Indonesian PSHA.

The Indonesian Geological Agency has also developed province-scale PSHAs that include site amplification, and **Cipta *et al.* (2016)** in this Special Publication present an example of this work for the six provinces of Sulawesi. As suggested by **Watkinson & Hall (2017)**, the hazard levels in the city of Palu are alarmingly high: the 10 per cent in 50 years exceedance levels of ground motion are about 1 g for Peak Ground Acceleration (PGA); 2.5 g for 0.2 s Response Spectral Acceleration (RSA); and about 1.5 g for 1.0 s RSA. **Cipta *et al.* (2016)** compare two widely used proxy-based approaches for estimating site class with site classes determined at selected sites using *in situ* measurements of the average shear velocity in the top 30 m of the soil profile (Vs_{30}). This comparison shows that these three methods of determining site class can often give inconsistent results. While this suggests considerable uncertainty in the estimated ground motion exceedance levels, the extremely high levels of hazard argue for much more work to be done in Palu in order to help prioritize the improvement and to enforce the building codes of that city.

Earthquake site response and ground motion

Site response is an important consideration, not only in Palu but throughout Indonesia, due to the pervasiveness of poorly consolidated sediments in the surface geology, as these can act to greatly amplify earthquake ground motion. In this Special Publication **Ridwan *et al.* (2016)** report on an extensive survey of the shallow shear-wave velocity structure of the Jakarta Basin, home to Indonesia's largest city and capital, Jakarta. With a population that has recently topped 10 million, the city of Jakarta itself now ranks as a megacity. However, the combined population of Jakarta and its neighbours Bogor, Depok, Tangerang and Bekasi (popularly known as 'Jabodetabek') is over 28 million, making it one of the world's great urban agglomerations. Most of these people live in the Jakarta Basin, with a surface geology of alluvial sediments, beach ridge deposits and volcanic tuff. **Ridwan *et al.* (2016)** show that these materials have near-surface shear-wave velocities of 200–400 m s^{-1}, corresponding mainly to National Earthquake Hazard Reduction Program (NEHRP) site classes D and E, and that material with shear velocity <750 m s^{-1} extends to considerable depth, from 350 m in the south to 725 m in the north. Such low-velocity sediments near the surface combined with the considerable depth of the sediment column suggests a complicated resonance structure that may affect high-rise buildings as well as low-rise residential structures.

Nothing contributes more to the understanding of the ground motions that may be experienced in future Indonesia earthquakes than observations of actual ground motions recorded following past earthquakes. Such recordings have only recently become available in Indonesia, thanks to the deployment of an extensive network of strong ground motion recording instruments as part of the massive investment in earthquake monitoring infrastructure that followed the 2004 Great Sumatra Earthquake. In this Special Publication, **Pramono *et al.* (2016)** describe the implementation of real-time 'ShakeMaps' within Indonesia's Agency of Meteorology, Climatology and Geophysics (BMKG). These are intended to support rapid response following the occurrence of a large, damaging earthquake, but **Pramono *et al.* (2016)** also show that recorded ground motions are generally consistent with theoretical models.

Earthquake fatality model

Ultimately, in order to make informed decisions regarding earthquake mitigation, public officials need to understand not only the hazard – that is, the potential earthquake ground motion levels – but also the impact earthquakes can cause. Natural hazard impacts can be measured in a variety of ways, but one of the most important is in terms of fatalities. While other models have been developed for earthquake fatalities in the Indonesian region (e.g. Jaiswal & Wald 2010), very little consideration has hitherto been given to the assessment of model uncertainty. In this Special Publication, **Sengara *et al.* (2017)** address this issue by using a Bayesian approach to develop a fatality model that estimates both uncertainty and the expected number of fatalities for a given earthquake. The authors expect this model to be useful in rapid assessments of earthquake impact that are needed for disaster response.

Volcanic eruptions

Volcanoes are a major concern in Indonesia, with about 100 000 lives having been lost to volcanic disasters in the last 200 years. A statistical analysis of volcanic eruptions in Indonesia by Simpson *et al.* (2011) showed that volcanic disasters having an impact upon populations of 100 000 or more can be expected at least once every decade in Indonesia, and a 'catastrophic' event affecting over 1 per cent of the population (over 2 million based on current population) can be expected every 500 years. While fatalities have been greatly reduced due to the effective warning system established by Indonesia's Geological Agency, not all of the 77 historically active volcanoes are monitored, nor are the over 50 volcanoes that have had no historical eruptions but are thought to have erupted in the Holocene. The potential for such volcanoes to resume activity after a long hiatus was highlighted by the recent eruption of Mt Sinabung, which had been dormant for the past 400 years.

Volcanic ash is a hazard associated with volcanic eruptions that threatens communities living in the vicinity of active volcanoes. In this Special Publication, **Bear-Crozier *et al.* (2016)** describe how they have adapted an existing open-source volcanic ash dispersion model for use in Indonesia. They used Gunung Ciremai in West Java as a case study, and considered a complete range of possible wind conditions for both dry and rainy seasons, showing that communities on the western side of Gunung Ciremai are susceptible to volcanic ash loading in any season, while communities on the eastern side are more susceptible during the rainy season. The authors suggest that disaster management officials can use this kind of detailed information on volcanic hazard to better focus mitigation efforts.

Mud volcanoes

While mud volcanoes are present in several tectonically active regions in the world (Milkov 2000), they are normally regarded as a secondary hazard associated with earthquakes, and it is rare for them to cause a significant societal impact. The Lusi mudflow disaster in East Java is unique in this respect (Lusi is a contraction of *Lumpur Sidoarjo,* where *lumpur* is the Indonesian word for 'mud' and Sidoarjo is the name of the regency in East Java where Lusi is located). The volcano began erupting on 29 May 2006, two days after the Mw 6.3 Yogyakarta earthquake about 275 km away. To date the Lusi eruption has displaced about 40 000 people and resulted in about US$2.7 billion in economic loss. Since its initial eruption, the issue of whether it was triggered by the Yogyakarta earthquake or by drilling activity of the Lapindo oil company has been hotly debated (see, e.g. Tingay *et al.* 2015).

In this Special Publication, **Andreas *et al.* (2017)** analyse the ground deformation signatures of the Lusi mud volcano to help resolve this controversy. These ground motion signatures include GPS and Interferemetric Synthetic Aperture Radar (InSAR) measurements of ground deformation as well as observations of crack orientations and geophysical electric field measurements. The authors show that there is no discernible pattern in the ground motion signatures that would indicate fault reactivation, and that these signatures are instead consistent with the triggering of Lusi by the Lapindo drilling activity.

Conclusion

While it was the intention that the articles of this Geological Society Special Publication would span a wide range of geological hazards, it has not been particularly successful at doing so. The editors were unable to solicit any contribution on landslide hazards, which account for 1.1 per cent of fatalities and 1.3 per cent of people affected by natural hazards and are therefore significant (Fig. 2). Tsunamis were covered by only two papers, and volcanoes by only one paper, when clearly their importance in terms of impact (as indicated in Fig. 2) merits more attention. Instead, the Special Publication focuses strongly on earthquake hazard, which is the subject of six of the ten contributions.

Is the emphasis on earthquake hazard warranted? Does this reflect only that the science for earthquakes is well developed relative to the other hazards, or that the editors were more involved in the earthquake community than in others? We believe it is the former, and that in fact the communities of Indonesian scientists specializing in the other geological hazards – tsunamis, volcanoes and landslides – are actually quite small, despite the significant threat these hazards pose to Indonesian society.

Earthquake hazard in Indonesia is very important. As indicated in this Special Publication and elsewhere, historical records of pre-instrumental earthquake activity suggest that many more earthquakes occurred in the seventeenth to nineteenth centuries in areas of Indonesia that are now densely populated than have occurred in living memory. Given the high population density, especially in Java, and the poor construction standards, it can be surmised that the likelihood of Indonesia experiencing a major earthquake disaster in the twenty-first century is reasonably high. Furthermore, there is something we can do about it. With rapid urbanization and increasing population, as well as increasing incomes, new construction is taking place at a fast pace. If building codes can be better informed by reliable hazard estimates, and if these codes are followed, fatalities due to future earthquakes may be substantially reduced. There is every reason to believe a strong effort at better earthquake hazard assessment and mitigation will bring enormous benefits to society.

However, the editors cannot help but feel that other geological hazards, not to mention non-geological hazards, need more investment if Indonesia is to develop into a society with an acceptable level of disaster resilience. Even after the 2004 IOT killed over 160 000 Indonesians, there are still very few faculty staff at Indonesian universities who could be described as 'tsunami scientists'. Regarding volcanoes, despite the huge social cost of evacuating large populations from the vicinity of active volcanoes on a yearly basis, and despite the potential for cataclysmic eruptions that might affect the whole region, it is difficult to identify even one Indonesian university scientist who specializes in volcanoes as a hazard (i.e. as opposed to specializing in igneous rocks). While the situation may be slightly improved in government technical agencies, we would argue that progress in the science of geohazards is best achieved by a community of academic and government scientists working together, as is the case for earthquake science in Indonesia. We therefore hope that our introduction to the Geological Society of London Special Publication on geohazards in Indonesia will help to call attention not only to the excellent work that has been done to improve our knowledge of these threats to the well-being of all Indonesians, but also to the need for more research investment in all geohazards.

Correction notice: The catchline of the original version was incorrect.

References

ANDREAS, H., ABIDIN, H.Z., SIDIQ, T.P., GUMILAR, I., AOKI, Y., HAKIM, A.L. & SUMINTADIREDJA, P. 2017. Understanding the trigger of the LUSI Mud Volcano from ground deformation signatures. *In*: CUMMINS, P.R. & MEILANO, I. (eds) *Geohazards in Indonesia: Earth Science for Disaster Risk Reduction*. Geological Society, London, Special Publications, **441**. First published online July 10, 2017, https://doi.org/10.1144/SP441.10

BEAR-CROZIER, A.N., KARTADINATA, N., HERIWASESO, A. & NIELSEN, O. 2016. Evaluating a volcanic ash ground-loading hazard at Gunung Ciremai, West Java, Indonesia using PF3D. *In*: CUMMINS, P.R. & MEILANO, I. (eds) *Geohazards in Indonesia: Earth Science for Disaster Risk Reduction*. Geological Society, London, Special Publications, **441**. First published online March 24, 2016, https://doi.org/10.1144/SP441.1

BIRD, P. 2003. An updated digital model of plate boundaries. *Geochemistry Geophysics Geosystems*, **4**, 1027, https://doi.org/10.1029/2001GC000252

CIPTA, A., ROBIANA, R., GRIFFIN, J.D., HORSPOOL, N., HIDAYATI, S. & CUMMINS, P. 2016. A probabilistic seismic hazard assessment for Sulawesi, Indonesia. *In*: CUMMINS, P.R. & MEILANO, I. (eds) *Geohazards in Indonesia: Earth Science for Disaster Risk Reduction*. Geological Society, London, Special Publications, **441**. First published online April 26, 2016, https://doi.org/10.1144/SP441.6

FICHTNER, A., DE WIT, M. & VAN BERGEN, M. 2010. Subduction of continental lithosphere in the Banda Sea region: Combining evidence from full waveform tomography and isotope ratios. *Earth and Planetary Science Letters*, **297**, 405–412.

GLOBAL VOLCANISM PROGRAM 2013. Volcanoes of the world, v. 4.5.5. VENZKE, E. (ed.) Smithsonian Institution [ONLINE]. https://doi.org/10.5479/si.GVP.VOTW4-2013 [last accessed 1 May 2017].

GRIFFIN, J.D., PRANANTYO, I.R. *ET AL*. 2016. Assessing tsunami hazard using heterogeneous slip models in the Mentawai Islands, Indonesia. *In*: CUMMINS, P.R. & MEILANO, I. (eds) *Geohazards in Indonesia: Earth Science for Disaster Risk Reduction*. Geological Society, London, Special Publications, **441**. First published online June 2, 2016, https://doi.org/10.1144/SP441.3

GUHA-SAPIR, D., BELOW, R. & HOYOIS, PH. 2017. *EM-DAT: International Disaster Database*. Université Catholique de Louvain, Brussels, Belgium [ONLINE], http://www.emdat.be [last accessed 1 May 2017].

HARRIS, R. & MAJOR, J. 2016. Waves of destruction in the East Indies: the Wichmann catalogue of earthquakes and tsunami in the Indonesian region from 1538 to 1877. *In*: CUMMINS, P.R. & MEILANO, I. (eds) *Geohazards in Indonesia: Earth Science for Disaster Risk Reduction*. Geological Society, London, Special Publications, **441**. First published online May 24, 2016, https://doi.org/10.1144/SP441.2

INDONESIAN BUREAU OF STATISTICS 2014. Proyeksi Penduduk menurut Provinsi, 2010–2035 (Ribuan) [ONLINE], https://www.bps.go.id/linkTabelStatis/view/id/1274 [last accessed 1 May 2017].

JAISWAL, K. & WALD, D. 2010. An empirical model for global earthquake fatality estimation. *Earthquake Spectra*, **26**, 1017–1037.

KANEKO, Y., MARUYAMA, S. *ET AL*. 2007. On-going orogeny in the outer-arc of the Timor-Tanimbar region, eastern Indonesia. *Gondwana Research*, **11**, 218–233.

KEEP, M. & HAIG, D.W. 2010. Deformation and exhumation in Timor: distinct stages of a young orogeny. *Tectonophysics*, **483**, 93–111.

LAVIGNE, F., DEGEAI, J.P. *ET AL*. 2013. Source of the great AD 1257 mystery eruption unveiled, Samalas volcano, Rinjani Volcanic Complex, Indonesia. *Proceedings of the National Academy of Sciences*, **110**, 16742–16747.

MILKOV, A.V. 2000. Worldwide distribution of submarine mud volcanoes and associated gas hydrates. *Marine Geology*, **167**, 29–42.

MUIS, S., GÜNERALP, B., JONGMAN, B., AERTS, J.C.J.H. & WARD, P.J. 2015. Flood risk and adaptation strategies under climate change and urban expansion: a probabilistic analysis using global data. *Science of the Total Environment*, **538**, 445–457.

NGUYEN, N., GRIFFIN, J., CIPTA, A. & CUMMINS, P.R. 2015. *Indonesia's Historical Earthquakes: Modelled examples for improving the national hazard map*. Record 2015/23. Geoscience Australia, Canberra, https://doi.org/10.11636/Record.2015.023

OMANG, A., CUMMINS, P., ROBINSON, D. & HIDAYATI, S. 2016. Sensitivity analysis for probabilistic seismic hazard analysis (PSHA) in the Aceh Fault Segment, Indonesia. *In*: CUMMINS, P.R. & MEILANO, I. (eds) *Geohazards in Indonesia: Earth Science for Disaster Risk Reduction*. Geological Society, London, Special Publications, **441**. First published online June 9, 2016, https://doi.org/10.1144/SP441.5

OPPENHEIMER, C. 2003. Climatic, environmental and human consequences of the largest known historic eruption: Tambora volcano (Indonesia) 1815. *Progress in Physical Geography*, **27**, 230–259.

OSADA, M. & ABE, K. 1981. Mechanism and tectonic implications of the great Banda Sea earthquake of November 4, 1963. *Physics of the Earth and Planetary Interiors*, **25**, 129–139.

PRAMONO, S., ALLEN, T.I. *ET AL*. 2016. Towards real-time earthquake impact alerting in Indonesia. *In*: CUMMINS, P.R. & MEILANO, I. (eds) *Geohazards in Indonesia: Earth Science for Disaster Risk Reduction*. Geological Society, London, Special Publications, **441**. First published online May 18, 2016, https://doi.org/10.1144/SP441.4

RIDWAN, M., WIDIYANTORO, S., IRSYAM, M., AFNIMAR & YAMANAKA, H. 2016. Development of an engineering bedrock map beneath Jakarta based on microtremor array measurements. *In*: CUMMINS, P.R. & MEILANO, I. (eds) *Geohazards in Indonesia: Earth Science for Disaster Risk Reduction*. Geological Society, London, Special Publications, **441**. First published online July 28, 2016, https://doi.org/10.1144/SP441.7

SAATCIOGLU, M., GHOBARAH, A. & NISTOR, I. 2006. Performance of structures in Indonesia during the December 2004 Great Sumatra Earthquake and Indian Ocean Tsunami. *Earthquake Spectra*, **22**, 295–319.

SELF, S. & RAMPINO, M.R. 1981. The 1883 eruption of Krakatau. *Nature*, **294**, 699–704.

Sengara, I.W., Suarjana, M., Yulman, M.A., Gashemi, H. & Ryu, H. 2017. Empirical fatality model for Indonesia based on a Bayesian approach. *In*: Cummins, P.R. & Meilano, I. (eds) *Geohazards in Indonesia: Earth Science for Disaster Risk Reduction*. Geological Society, London, Special Publications, **441**. First published online April 24, 2017, https://doi.org/10.1144/SP441.9

Simpson, A., Johnson, R.W. & Cummins, P. 2011. Volcanic threat in developing countries of the Asia-Pacific region: probabilistic hazard assessment, population risks, and information gaps. *Natural Hazards*, **57**, 151–165.

Stein, S. & Okal, E.A. 2005. Seismology: speed and size of the Sumatra earthquake. *Nature*, **434**, 581–582.

Tingay, M.R.P., Rudolph, M.L., Manga, M., Davies, R.J. & Wang, C.Y. 2015. Initiation of the Lusi mudflow disaster. *Nature Geoscience*, **8**, 493–494.

Watkinson, I.M. & Hall, R. 2017. Fault systems of the eastern Indonesian triple junction: evaluation of Quaternary activity and implications for seismic hazards. *In*: Cummins, P.R. & Meilano, I. (eds) *Geohazards in Indonesia: Earth Science for Disaster Risk Reduction*. Geological Society, London, Special Publications, **441**. First published online December 19, 2016, updated April 19, 2017, https://doi.org/10.1144/SP441.8

Wichmann, A. 1918. Die Erdbeben des Indischen Archipels bis zum Jahre 1857. *Verhandelingen Der Koninklijke Akademie van Wetenschappen Te Amsterdam*, **20**, 193.

Widiwijayanti, C., Mikhailov, V., Diament, M., Deplus, C., Louat, R., Tikhotsky, S. & Gvishiani, A. 2003. Structure and evolution of the Molucca Sea area: constraints based on interpretation of a combined sea-surface and satellite gravity dataset. *Earth and Planetary Science Letters*, **215**, 135–150.

Waves of destruction in the East Indies: the Wichmann catalogue of earthquakes and tsunami in the Indonesian region from 1538 to 1877

RON HARRIS[1]* & JONATHAN MAJOR[1,2]

[1]*Department of Geological Sciences, Brigham Young University, Provo, UT 84602–4606, USA*

[2]*Present address: Bureau of Economic Geology, The University of Texas at Austin, Austin, TX 78758, USA*

**Correspondence: rharris@byu.edu*

Abstract: The two volumes of Arthur Wichmann's *Die Erdbeben Des Indischen Archipels* [*The Earthquakes of the Indian Archipelago*] (1918 and 1922) document 61 regional earthquakes and 36 tsunamis between 1538 and 1877 in the Indonesian region. The largest and best documented are the events of 1770 and 1859 in the Molucca Sea region, of 1629, 1774 and 1852 in the Banda Sea region, the 1820 event in Makassar, the 1857 event in Dili, Timor, the 1815 event in Bali and Lombok, the events of 1699, 1771, 1780, 1815, 1848 and 1852 in Java, and the events of 1797, 1818, 1833 and 1861 in Sumatra. Most of these events caused damage over a broad region, and are associated with years of temporal and spatial clustering of earthquakes. The earthquakes left many cities in 'rubble heaps'. Some events spawned tsunamis with run-up heights >15 m that swept many coastal villages away.

2004 marked the recurrence of some of these events in western Indonesia. However, there has not been a major shallow earthquake ($M \geq 8$) in Java and eastern Indonesia for the past 160 years. During this time of relative quiescence, enough tectonic strain energy has accumulated across several active faults to cause major earthquake and tsunami events, such as those documented in the historical records presented here. The disaster potential of these events is much greater now than in the past due to exponential growth in population and urbanization in areas destroyed by past events.

Supplementary material: Translation of the catalogues into English, scanned PDFs of the original catalogues and geographical locations of most place names found in the catalogue (as a KMZ file) are available at https://dx.doi.org/10.6084/m9.figshare.c.2860405.v1

Recent earthquake catastrophes in Japan (2011), Haiti (2010) and Chengdu, China (2008) all occurred in regions mapped as relatively 'low' seismic hazard (Stein *et al.* 2012). One factor contributing to this forecasting failure is the over-reliance on the relatively short (50 year) instrumental earthquake record. Although historical records of centuries of seismic events exist in these regions, they were rarely used in hazard assessments. Reliable data from historical accounts of earthquake and tsunami events can help constrain long-term seismic potential in areas with little to no earthquakes during the past 50 years (Musson & Jimenez 2008).

An example of earthquake and tsunami forecasting success using historical records is the reconstruction by Newcomb & McCann (1987) of mega-thrust events along the Sumatran subduction zone. They primarily used *Die Erdbeben Des Indischen Archipels* [*The Earthquakes of the Indian Archipelago*] by Arthur Wichmann (1918, 1922) to demonstrate that a series of megathrust earthquakes occurred during the nineteenth century along various segments of the subduction zone. Some of these events have now reoccurred, such as the 2005 northern Sumatra earthquake near Nias Island, which was nearly of the same magnitude and ruptured close to the same area as that estimated by Newcomb & McCann (1987) from descriptions in the Wichmann catalogue of an 1861 earthquake and tsunami (see Supplementary material).

Arthur Wichmann was a German geologist and professor in geology at Utrecht University in the Netherlands. Much of Wichmann's professional career was spent studying the geology of Indonesia, including several expeditions to eastern Indonesia. He wrote the catalogue in old German, but included passages in Dutch, English, French and Latin. It is evident that Wichmann tried to preserve and use the original accounts as much as possible. However, he was careful to note conflicting reports, errors in dates

From: CUMMINS, P. R. & MEILANO, I. (eds) 2017. *Geohazards in Indonesia: Earth Science for Disaster Risk Reduction*. Geological Society, London, Special Publications, **441**, 9–46.
First published online May 24, 2016, https://doi.org/10.1144/SP441.2

and other inconsistencies. The catalogue includes 350 years of observations of earthquakes and tsunamis, and some volcanic eruptions for the entire Indonesian region (see Supplementary material).

The observations were mostly compiled from Dutch records kept by the Dutch East India Company of Indonesia (Vereenigde Oost-Indische Compagnie: VOC). Seismic events are included that reach west to east from the Cocos Islands to New Guinea, and north to south from Bangladesh to Timor. Although the catalogue is cited in some tsunami and earthquake databases (i.e. Soloviev & Go 1974), it remains largely unknown to the scientific community (i.e. Hamzah *et al.* 2000; Rastogi & Jaiswal 2006). For this reason, we have translated the catalogue into English and supplied scanned PDFs of the original; these are included in the Supplementary material to this paper.

For ease of comparison and consistency, the English translation is formatted and paginated in the same way as the original document. Footnotes with abbreviations used and works commonly cited are included with the translation. The Index, containing mostly locations, was not translated and so readers should refer to the original catalogue. Volume II also contains a list of additions and corrections for the previous volume. We have also included geographical locations of nearly all the place names found in the catalogue as a KMZ file (see Supplementary material).

The number of entries in the catalogue increases towards the end (1877), which reflects, to some extent, the increasing number of observation posts. However, many of the events in the later part of the catalogue are also the result of temporal and spatial clustering from a few extreme events, such as megathrust earthquakes in Sumatra, Java and eastern Indonesia. Other clusters may be associated with eruptions of Tambora and precursor events to the eruption of Krakatoa, which happened shortly after the end date of the catalogue. Events associated with the collapse of buildings, tsunamis (flood waves) and fatalities are highlighted by bold text in the catalogue.

Most of the locations listed in the catalogue where major earthquakes or tsunamis occurred have experienced a dramatic increase in population and urbanization (see the maps of Padang in Natawidjaja *et al.* 2006). To make the catalogue accessible to those actually in harm's way, we have worked with Indonesian geoscientists to translate large parts of it into Indonesian. For example, the sections of the catalogue for the Maluku and Nusa Tenggara regions have already been translated into Indonesian, and updated to include earthquakes and tsunamis to 2010. In November 2013, the Indonesian version of the catalogue was distributed throughout some of the most vulnerable areas of eastern Indonesia by 'In Harm's Way', a non-profit organization for natural disaster prevention (inharmswayhelp.org).

During the time covered by the catalogue (1538–1877), the population of Indonesia doubled from 15 million to 30 million. Since 1877, the population has increased by nearly an order of magnitude and is projected to reach 300 million by 2030. At least 94% of the population of Indonesia lives in active seismic and volcanic regions, or in tsunami inundation zones. However, from our experience of conducting informal surveys throughout Java and eastern Indonesia over the past 25 years, for most Indonesians, their sense of history, especially regarding natural disasters, is limited to the living memory of the oldest residents of each area. One factor that is likely to contribute to the lack of awareness is a distinct lull in seismic and tsunami activity in the region between 1883, when Krakatoa erupted, and the recent past (Ghose & Oike 1988). Another problem is that historical records of past events in Indonesia are not available to those in harm's way (Harris *et al.* 1997; Harris & Prasetyadi 2002). These problems contributed to the disaster potential of recent earthquakes such as the Banda Aceh event and other earthquakes during the past 10 years.

Many of the events documented in the Wichmann catalogue have reoccurred in western Indonesia during the past decade and account for around 200 000 fatalities. However, in eastern Indonesia, seismic quiescence continues to mask the fact that significant amounts of elastic strain energy is accumulating on many active faults dangerously close to burgeoning urban centres. As in the past, the strain will eventually be unleashed and may result in a cluster of large events like those documented throughout eastern Indonesia in the Wichmann catalogue during 1629–99, 1754–80 and 1814–77, and like the cluster of events that initiated in 2004 in western Indonesia.

The recurrence of temporal and spatially clustered seismic events in Sumatra and Java also demonstrate a significant increase in the disaster potential of Indonesia. More death and destruction is occurring from events that caused few losses in the past. The reasons behind the greater losses to nature are multifaceted, but the one main contributing factor is non-resilient urbanization in hazardous regions. For example, some of the most seismically active regions of Indonesia, from a historical perspective, are transitioning from traditional woody building materials to unreinforced masonry structures built in hazardous areas. With limited resources to address these issues, it is imperative that mitigation efforts focus on the regions at highest risk, which are proximal to active faults. Our hope is that this paper can help better identify these regions by extending the earthquake records back far enough to reveal likely areas of high seismic potential and to better constrain the locations of active fault zones in Indonesia.

Summary of significant events and possible source regions

The purpose of this section is to augment what we know about earthquake source regions in Indonesia (Fig. 1) with data from the Wichmann catalogue. We first summarize as many characteristics as possible about each significant event in the catalogue in Tables 1–7. By 'significant event' we mean those that caused damage or a tsunami. A summary of all of the events felt regionally is provided in Table 1. Summaries of significant events for various regions of similar tectonic setting are listed in Tables 2–7. Besides the obvious characteristics of place and time, the catalogue also notes direction of ground motion and documents possible foreshocks or aftershocks. Other characteristics of note include fatalities (F), regional extent (R), tsunami (T) with the height in metres and volcanic activity (V). Distinguishing between volcanic and tectonic earthquakes can be problematic in the catalogue. Many times volcanic eruptions are noted in association with strong ground shaking, but it is not clear whether the eruption or its precursors caused the earthquakes or a tectonic earthquake caused the eruption.

One way of addressing this problem is to determine the regional extent of each earthquake. Columns in the tables to the right of 'Other' list several of the observation stations in each region and how many reported each event. We have divided Indonesia into six regions based mainly on tectonic setting (Table 1). In the tables for each region (Tables 2–7), the locations of individual reporting stations are listed. An upper case 'X' refers to strongly felt (>Mercalli scale 5), while a lower case 'x' refers to weakly felt (<Mercalli scale 5). 'Mercalli' always refers to the Modified Mercalli Intensity (MMI). In many cases, earthquakes were felt over several sub-regions, but records documenting the event in the sub-regions are lacking. Other locations where the earthquake was felt that are not listed in columns are provided in the last column on the right of each table.

The majority of the earthquakes listed in the catalogue are probably aftershocks from major events. For example, in Volume I, a detailed list of aftershocks from each of the Uliasser Islands is reported after the 1754 earthquake in Ambon. In Volume II, most of the events are low-intensity earthquakes following larger events in Sumatra (1861), Java (1865), northern Timor (1857) and in the eastern Banda Arc (1852). The number of reliable reporting stations increased dramatically during the late nineteenth century, which provides a better estimate of the location and parameters of earthquake sources.

Since more is now known about the structure and tectonics of earthquake and tsunami source regions in Indonesia than when Wichmann compiled the catalogue, we include here a summary of active faults for each region.

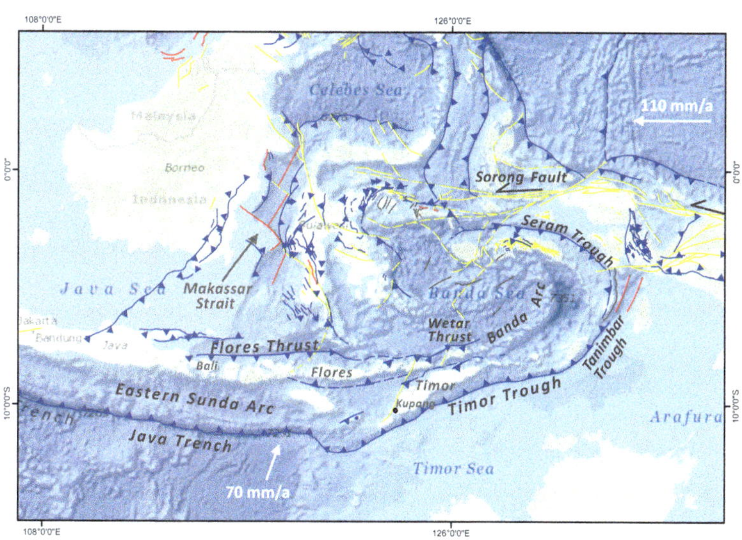

Fig. 1. Location map of active faults through Java and eastern Indonesia.

Table 1. *Significant earthquakes and tsunamis throughout Indonesia, 1584–1877*

Primary reporting location	Year	Date (month/day)	Time	Intensity (MMI)	Aftershocks reported	Other	Molucca Sea	Banda Arc	Lesser Sunda	Java	Sumatra	New Guinea	Other locations
Java	1584–86			VIII		T?				x			
Ternate	1608	7/18	Night	VIII		V	X						
Banda Islands	1629	8/1	9:30 p.m.	XI	9 years	T (15 m)		X					
Amboina (Ambon)	1644	5/12	Morning	VIII–IX	2 weeks			X					
Molukken (Makjan) Island	1646	7/19		VIII–IX		F, V, T (1.5 m)	X						
Fort Henricus, Solor	1648	2/2		IX–X	3.5 months	F			X				
Saparua (Ambon)	1671	10/17	Evening	X–XI	2 months			X					
Ambon	1672	?		X				X					
Ternate	1673	8/12	10:00 p.m.	IX	1 month	T	X						
Ambon	1675	2/17	7:30 p.m.	XI	3 months	F, T (2.5 m)		X					
Banda Islands	1683	10/16	8:00 a.m.	IX	1 year	V		X					
Batavia (Jakarta)	1699	1/5	1:30 a.m.	X	13 months	F				X	X		
Ambon	1705	10/?		IX	1 month			X					
Banda Islands	1710	3/6		IX	4 weeks	T		X					
Batavia (Jakarta)	1722	10/?	8:00 a.m.	IX		T				X			
Banda Islands	1743	?		X				X					
Ambon	1754	8/18	3:30 p.m.	X	1.5 months	F, T		X					
Bengkulen (west Sumatra)	1756	11/3		X	1 month						X		
Assam (India)	1762	4/3	5:00 a.m.										Burma, Bangladesh
Banda Islands	1763	9/12	5:00 p.m.	IX	4 years			X					
Ternate	1770	7/26		IX	5 years	V	X						
Bengkulen (west Sumatra)	1770	?		X		V, T					X		
Ambon	1777	3/30	9:00 a.m.	VIII	3 months			X					
Banda Islands	1778	4/2		IX		T?		X					
Batavia (Jakarta)	1780	1/22	2:39 p.m.	IX	12 months					X	X		
Kupang (Timor)	1793–95	?		IX					X				
Padang (west Sumatra)	1797	2/10	10:00 p.m.	IX	1 week	T					X		
Palembang (west Sumatra)	1799	?		IX		T (15 m)					X		
Ambon	1802	8/25		X		T		X					
Kupang (Timor)	1814	?		IX		T		X	X				

Sumbawa (Tambora)	1815	4/1	10:00 p.m.		1 year	T (4 m), V		X	X	x			1 year
Ambon	1815	4/11		VIII–IX		V, T		X					
Banda Islands	1816	10/8	7:00 a.m.	IX	3 months	F		X					
Bima (Sumbawa)	1818	?		X		T (4 m)			X				
Benkulen (SW Sumatra)	1818	3/18		IX	2 months	T					X		
Makassar (west Sulawesi)	1820	12/29	10:00 a.m.	X	3 months	F, T (29 m)			X				
Ambon	1830	3/28	10:00 a.m.	IX	4 months			X					
West Sumatra	1833	11/24	8:30 p.m.	X	1.5 years	T				x	X		Singapore
Batavia (Jakarta)	1834	10/10	5:30 a.m.	X						x	X		
Ambon	1835	11/1	3:00 a.m.	X	2–3 years	F		X					
Pajitan (central Java)	1840	1/4	1:15 p.m.	VIII	1 month	T				X			
Ternate	1840	2/14	10:00 a.m.	X	3 months	V	X						
Barus (NW Sumatra)	1843	1/5	11:30 p.m.	XI	3 months	T					X		
Menado, (north Sulawesi)	1845	2/8	3:30 p.m.	XIII	2 weeks	T	X						
Nicobar minor	1847	10/31	3:30 a.m.	X	3 weeks	T					X		
Cheribon (Java)	1847	11/16	12:55 p.m.	X	2 weeks					X	X		
Saparua (Ambon)	1849	5/30		VIII	1 week			X					
Banda Islands, Ambon	1852	11/26	7:50 a.m.	XI	3 + years	F, R, T (8 m)	x	X	x	x	x		
Solok (Sumatra)	1853	2/8	Night	VIII							X		
Ternate	1855	4/25	Morning	IX	1 year	F	X						
Ternate	1855	6/14	11:00 a.m.	IX			X						
Ternate	1855	7/14	4:00 p.m.	IX	1 year		X						
Gorontolo (north Sulawesi)	1856	8/6		VIII	16 months		X						
Astrolabe Bay (New Guinea)	1856	?	Night	IX		F, T						X	
Rook Island (New Guinea)	1857	4/17	Evening	VIII								X	
Dilli (Timor)	1857	5/13	10:30 a.m.	IX–X	1 month	F, T (3 m)		x	X				
Minahassa (north Sulawesi)	1858	11/13	4:00 p.m.	IX	Years	T	X		x				
Ternate	1859	7/29	3:15 p.m.	IX	Years	T (1 m)	X	x?					
Kau Territory (Halmahera)	1859	10/8	2:00 p.m.	VIII			X						

(Continued)

Table 1. *Significant earthquakes and tsunamis throughout Indonesia, 1584–1877* (*Continued*)

Primary reporting location	Year	Date (month/day)	Time	Intensity (MMI)	Aftershocks reported	Other	Molucca Sea	Banda Arc	Lesser Sunda	Java	Sumatra	New Guinea	Other locations
West Sumatra	1861	2/16	6:30 p.m.	X	4+ years	F, T (7 m)				x	X		Malaysia, Singapore
Geelvink Bayk (P.N.G.)	1864	5/23	12:30 a.m.	IX	1 week	F, T (3 m)						X	
Ambarawa (central Java)	1865	7/17	2:27 a.m.	VIII	15 months					X			
Djogjakarta (central Java)	1867	6/10	4:30 a.m.	IX	1 year	F				X			
Bengkulen (SW Sumatra)	1871	8/18	8:16 p.m.	VIII						x	X		
Astrolabe Bay (Papua New Guinea)	1873	?		IX		T						X	
Singkel (NW Sumatra)	1873	8/19	3:15 p.m.	IX	1 month						X		
Cheribon (West Java)	1875	10/25	5:50 a.m.	X	3 months	F					X		
	63 total significant events (MMI >VII)												

F, fatalities reported; V, during or followed by volcanism; T, tsunami (maximum run-up); X, strongly affected; x, weakly affected.

Table 2. *Molucca Sea (Maluku North) events*

Primary reporting location	Date			Intensity (MMI)	Aftershocks reported		Affected locations						
	Year	Month/day	Time			Other	Ternate	Halmahera	Menado	Makjan	Minahassa	Gorontalo	Other locations
Ternate	1538	4					X						
Ternate	1538	9					X						
Ternate	1546	9/29					X						
North Halmahera	1550	11 or 12				V	X	X					
NW Halmahera	1564					V	X	X					
Molukken (Makjan)	1608	7/1				T				X			
Molukken (Makjan)	1646	7/20		IX		F, T, V				X			
North Halmahera	1673	5/20		VI		T, V	x	X					
Ternate	1673	8/12	10:00 p.m.	IX	1 month	T	X						
Ternate	1686	10/26	11:00 a.m.	VI			X						
Ternate	1770	7/27		X	5 years	T, V	X						
Ternate	1770	8/2	9:00 a.m.	VIII									
Ternate	1773	2/21, 10/21		VIII		V	X						
Ternate	1775	7/4	8:00 a.m.	VI			X						
Ternate	1840	2/15	10:00 a.m.	X	3 months	V	X						
Menado, (north Sulawesi)	1845	2/9	3:30 p.m.	XIII	2 weeks	T			X		X		
Ternate	1854	11/24	1:30 p.m.	VIII									
Ternate (east–west)	1855	4/25	Morning	IX	1 year	F	X						
Ternate	1855	6/14	11:00 a.m.	IX	1 year	T							
Ternate	1855	7/14	4:00 p.m.	IX	1 year	F, T							
Gorontolo (north Sulawesi)	1856	8/7		VIII	16 months							X	
Ternate	1858	6/4	7:00 a.m.	VII			X						
Minahassa (north Sulawesi)	1858	12/13	4:00 p.m.	IX	Years	F, T	X		X		X	X	
Ternate (NW–SE)	1859	6/28	8:07 p.m.	IX		R, T (2 m)	X	T (9 m)	x		T, (east–west)		
Ternate	1859	7/29	3:15 p.m.	IX		R, T (>1 m)	X	X	X	x	T (2 m)	X	T, Banggai
NE Halmahera	1859	10/8	2:00 p.m.	VIII			X	X					
Minahassa (north Sulawesi)	1859	12/17	8:00 p.m.	VII	Aftershock?	T			x		X	X	
Menado (north Sulawesi)	1859	12/26	Early morning	?	Aftershock?	T			T (>2 m)				
Halmahera	1860	10/6	12:00 a.m.	VIII		T		X					

F, fatalities reported; V, during or followed by volcanism; T, tsunami (maximum run-up); X, strongly affected; x, weakly affected.

Table 3. *Banda Arc (Maluku South) events*

Primary reporting location	Date			Intensity (MMI)	Aftershocks reported	Other	Affected locations							
	Year	Month/day	Time				Ambon	Saparua	Haruku	Buru	Banda Islands	Kei Islands	Seram	Other locations
Ambon	1612			V			X							
Banda Islands	1618			VII							X			
Banda Islands	1629	8/2	9:30 p.m.	XI	9 years	F, T (15 m)	X				X			
Ambon	1644	5/12 5/13	Morning	IX	2 weeks		X						x	
Ambon	1644	5/17		IX	Aftershock		X							
Ambon	1648	2/29				T	X							
Kei Islands	1649			VIII		New island						X		
Teon Island	1659	11/9		VII	2 years	V								
Buru	1659	1/13		VII		T	x			X				
South Seram	1664	1/8		V									X	
Saparua[a]	1671	10/17–10/18	Evening	XI	2 months		x	X	x				x	
Ambon	1672			X			X							
Ambon	1674	2/17	7:30 p.m.	XI	3 months	F, T (80 m)	X	X	X	X	X	X	x – west	
Banda Islands	1683	10/16–10/30	8:00 a.m.	IX	1 year	V					X			
Serua SW (Islands)	1683			IX		F								
Ambon	1687	2/19–4/4	6:30 a.m.	IV	3 months		X							
Banda Islands	1691			V		V					X			
Banda Islands	1699	5/11	2:00 p.m.	V							X			
Ambon	1705	10/?		IX	1 month	F	X	x	x				x – west	
Ambon	1708	11/28	10–11 p.m.			T	x	x	x					
Ambon	1710	2/15–2/17		IX		T	X				X			
Banda Islands	1710	3/6		IX	4 weeks									
Ambon	1711	9/5		VI		T	X	x	X		x			
Leti (SW Islands)	1714	1/12		VI	2 years									

Leti (SW Islands) (south–north)	1716	12/20		VII										
Banda Islands	1743			IX							X			
Banda Islands	1750			VI							X			
Amboina	1754	8/19	3:30 p.m.	X	1.5 months	F, T	X	X	X	X				Felt in Java?
Banda Islands	1763	9/13	5:00 p.m.	IX	4 years						X			
Ambon (SW–NE)	1775	4/19	1:00 a.m.	VII			X							
Ambon (NW–SE)	1777	3/31	9:00 a.m.	VIII	3 months		X							
Ambon	1777	4/28		VIII	Aftershock		X							
Ambon	1781			VIII			X							
Amboina (south–north)	1802	8/26		X		R, T	X	X	X	x	x	x	x	East Sunda
Banda Islands	1811			IX	6 years						X			
Sumbawa (Tambora)	1815	4/1	10:00 p.m.		1 year	T (4 m), V	x				x			Java, Ambon, Ternate
Amboina	1815	4/11–4/12		VIII–IX		T	X		X					SE Borneo
Banda Neira	1816	10/8–10/11	7:00 a.m.	IX	3 months	F					X			
Saparua, Ambon	1817	11/11	3:00 a.m.	VI				X	x					
Banda Neira (east–west)	1824	5/26		VII		T, V					X			
Ambon	1830	3/28	10:00 a.m.	IX	4 months		X							
Ambon	1835	11/2	3:00 a.m.	X	2–3 years	F	X	X	X		x			
Ambon	1836	9/16		VI			X							
Saparua, Ambon	1837	1/21	9:00 p.m.	VIII	1 month		x	X	X					
Ambon	1841	12/16	2:00 a.m.	VI	2 years	T (1.7 m)	x			T(>2m)				
Buru	1848	8/7		VI	12 days					X				
Saparua, Ambon	1849	5/30		VIII	1 week		x	X						
Ambon	1851	11/20	11:55 p.m.	VIII			X	x	x					
Banda Islands (NW–SE)	1852	11/26	7:50 a.m.	VII	10 years	F, R, T (8 m)	X	X	X	X	X	X	X	Java, Ternate
Ambon	1852	11/26	7:30 a.m.	VIII	10 years		T (1.8 m)	T (5 m)	T	T				
Haruku and Saparua	1854	1/2		VIII	Aftershock	T	x	X	X					
Banda Islands (SW–NE)	1857	4/6	1:30 a.m.	VII	Aftershock									
Banda Islands	1858	11/9	5:30 a.m.	VIII	Aftershock									
Banda Islands	1859	9/25	Evening	VII	Aftershock	T					X			

F, fatalities reported; V, during or followed by volcanism; T, tsunami (maximum run-up); X, strongly affected; x, weakly affected.

Table 4. *Lesser Sunda Island events*

Primary reprting location	Date			Intensity (MMI)	Aftershocks reported	Other	Affected locations						
	Year	Month/day	Time				Sumbawa	Flores	Makassar	Timor	Solor	Kisar	Other locations
Timor	1638?			IX		V?				X			
Solor Island	1648	2/3		X	4 months						X		
Makassar	1690	12/1		VI					X				
Kupang (West Timor)	1793–95			VIII						X			
Mampawa (west Borneo)	1803			VI									
Kupang (West Timor)	1814			IX		T, new island				X		X	
Tambora (Sumbawa)	1815	4/10	10:00 p.m.	VIII	4 days	V, T (4 m)	X	x	x	x			Java, Banda, Ternate?
Bima (Sumbawa)	1818			X		T (4 m)	X						
Makassar	1820	12/30	10:00 a.m.	X	3 months	F, R, V, T (29 m)	X	X	X				Java
Kisar	1823	6/1		V								X	
West Timor	1829	2		VIII			X			X			
Bima (Sumbawa)	1836	3/5		VIII		T	X						
Kisar	1836	10		VIII	3 months								
Bima (Sumbawa)	1836	11/28	10:30 a.m.	IX	1 month	T							
Timor	1854	8/21								X			
Flores	1855	5/14	09:00 p.m.			T		T					
Ocussi (NW Timor)	1857	4/26	05:30 p.m.	VII						X			
Dilli (Timor Leste) (south–north)	1857	5/14	10:30 a.m.	IX–X	1 month	F, T (3 m)				X			Ambon
Atapupu (NE–SW, north–south)	1858–69			VI	Nine aftershocks					X			
West Timor (east–west, north–south)	1866			VI	Two aftershocks					X			Savu
Roti Island (west–east)	1866			VI	Aftershock					X			
Sumba (NE–SW)	1869	8–9		VI	Three aftershocks								

F, fatalities reported; V, during or followed by volcanism; T, tsunami (maximum run-up); X, strongly affected; x, weakly affected.

Table 5. *Java and Bali events*

Primary reporting location	Date			Intensity (MMI)	Aftershocks reported	Other	Affected locations						
	Year	Month/day	Time				Jakarta	Buitenzorg	Cheribon	West Java	Central Java	East Java	Other locations
Java	1500					V							
Java	1584			VII		R							
Java	1638			VIII						x	X	x	
Batavia (Jakarta)	1681	11/2		V			X						
Batavia	1699	1/6	1:30 a.m.	X	13 months	F	X	X		X			South Sumatra
Batavia	1722	10/?	8:00 a.m.	IX		T	X						
Batavia	1737			V			X						
Batavia	1739	7/22		VII			X						
Batavia	1757	8/24	2:00 a.m.	VII	9 months		X						
Batavia	1765	1/10		VIII			X						
Java	1771			X			x	x	x	x	x	x	
Batavia	1772	5/10		V?			X						
Semarang (central Java)	1773	4/21	12:00 a.m.	VII							X		
Batavia	1780	1/23	2:39 p.m.	IX	12 months		X	X		X	X	x	South Sumatra
Batavia	1807						X						
Bali	1808			VII								X	Bali
Batavia	1810	12/5	7:30 p.m.	VII			X						
Buitenzorg	1812	10/14	11:00 p.m.	VII			X	X					
Batavia	1814	3/24	12:00 a.m.	VIII			X	X					
Bali	1815	11/22	10:00 p.m.	IX		F, T						X	Lombok, Sumbawa
Surabaya (Java)	1815	11/22	11:00 p.m.	VII	Aftershock							X	
Jakarta	1818	3/29	2:30 p.m.	VII			X						NW–SE
Buitenzorg	1818	10/2	1:30 p.m.	VI			x	X					
Pasuaran, East Java	1818	11/8	11:15 p.m.	VIII			x	x		x	x	X	NNW–SSE
Galunggung, Java	1822	10/8	1:30 p.m.	VIII		F, V				X			
Buitenzorg	1823	9/9	8:00 a.m.	VIII	4 months	T		X	X				
Batavia (Jakarta)	1833	1/28	12:00 p.m.	VII			X						
Batavia (Jakarta)	1834	10/11	5:30 a.m.	X			X	X		X			South Sumatra
Jakarta	1836	3/22	3:30 a.m.	VIII			South–north	X		X	x		
Pajitan, (central Java)	1840	1/5	1:15 p.m.		1 month	T					X		
Cheribon (Java)	1847	11/16	10:18 a.m.	X	2 weeks	R	ESE–WNW	X	SE–NW	X	X		Sumatra

(*Continued*)

Table 5. *Java and Bali events (Continued)*

Primary reporting location	Date			Intensity (MMI)	Aftershocks reported	Other	Affected locations						
	Year	Month/day	Time				Jakarta	Buitenzorg	Cheribon	West Java	Central Java	East Java	Other locations
Java	1848	1/7	4:00 a.m.	X	1 month	R	x			North–south	North–south	North–south	
East Java, north Bali	1848	2/17	10:00 a.m.	VII	8 years	R, V						X	Bali
Sunda Strait	1848	6/4	11:00 a.m.	V	12 years	V	x	x	x	x			South Sumatra, Krakatoa?
Sunda Strait	1851	5/4	3:00 p.m.	VI		V?, T (1.2 m)	x						South Sumatra, Krakatoa?
Batavia and Sunda Strait	1852	1/9	6:00 p.m.	VI		T	North–south	SSW–NNE	East–west	X	x		South Sumatra
Central Java	1856	1/19	6:00 a.m.							x	N–S	x	
Patjitan (central Java)	1859	10/20	5:30 p.m.	VII		F, T							
Central and East Java	1865	5/17	7:00 p.m.	VIII	15 months		x	x	x		X	X	Madura, Bali
Ambarawa (central Java)	1865	7/18	2:27 a.m.	VII	Aftershock						X		
Ambarawa (C Java)	1866	4/22	6:30 p.m.	VII	Aftershock						X		
Central and East Java	1866	9/30	9:18 a.m.	VI	Aftershock						X	X	Madura, Borneo
Djogjakarta (central Java)	1867	6/11	4:30 a.m.	IX	1 year	F				x	X	x	
Cheribon (West Java)	1875	10/26	5:50 a.m.	X	3 months	F	x		X		X		

F, fatalities reported; V, during or followed by volcanism; T, tsunami (maximum run-up); X, strongly affected; x, weakly affected.

Table 6. *Sumatra events*

Primary reporting location	Date			Intensity (MMI)	Aftershocks	Other	Affected locations					
	Year	Month/day	Time				South Sumatra	Central Sumatra	North Sumatra	Andaman Islands	Nicobar Islands	Other locations
Ache (north Sumatra)	1621	3/7	1:00 a.m.		3 years	V?			X			Sumatra Fault?
West Sumatra	1681	12/11		VI			x	x	x			
Bengkulen, (west Sumatra)	1755	11/3		X	1 month	V	X	x?	x?			
Nias (west Sumatra)	1763			X		F		X				
Bengkulen (west Sumatra)	1770			X		F, T, V						
Padang (west Sumatra)	1795	2/11	10 p.m.	IX	1 week	T	x	X	x			1797?
Palembang (west Sumatra)	1799			IX		T (15 m)	New land	x				
Benkulen	1818	3/18		IX	2 months	T	X					
Bengkulen	1818	5		VII	Aftershock							
Padang (west Sumatra)	1824	4		VI								
Bengkulen, Padang	1833	11/25	8:30 p.m.	XI	24 years	F, R, T, V	X	X	x			SSW–NNE, Java, Singapore
Padang (west Sumatra)	1835	8/26	9:00 p.m.	VIII	Aftershock	R		North–south				Singapore, Java, Malaysia
Ache (north Sumatra)	1837	9/30		VIII		T, V						
Barus (NW Sumatra)	1843	1/5 to 1/6	11:30 p.m.	XI	3 months	T		x	West–east, north–south			Singapore, Malaysia
Nicobar Minor	1847	11/1	3:30 a.m.	X	3 weeks	T					X	
Padang Mountains	1852			V	1 year	F, V?		X				Sumatra Fault?
Solok, Sumatra	1853	2/9	Night	VIII		F				X		
Sumatra	1861	2/16	6:30 p.m.	X	4+ years	F, T (7 m)	SE–NW	North–south	X	X		Malaysia, Singapore, Java
NW Sumatra	1861	4/26	6:30 a.m.	VII	Aftershock	T (>2 m)			X			
Bengkulen	1871	8/19	8:16 p.m.	VIII			X					West Java
Singkel (NW Sumatra)	1873	8/19	3:15 p.m.	IX	1 month			x	X			

F, fatalities reported; V, during or followed by volcanism; T, tsunami (maximum run-up); X, strongly affected; x, weakly affected.

Table 7. *New Guinea events*

Primary reporting location	Date			Intensity (MMI)	Aftershocks reported	Other	Affected locations		
	Year	Month/day	Time				Geelvink Bay	Astrolabe Bay	Rook Island
Astrolabe Bay (New Guinea)	1856–58	No date	Night	IX		F, T		X	
Rook Island (New Guinea)	1857	4/18	Evening	VIII		F, T			North–south
Geelvink Bayk (New Guinea)	1864	5/24	12:30 a.m.	IX	1 week	F, T (2.6 m)	X		
Astrolabe Bay	1873	No date		IX		T		X	

F, fatalities reported; V, during or followed by volcanism; T, tsunami (maximum run-up); X, strongly affected; x, weakly affected.

The Molucca Sea region (north Maluku: Table 2), including Ternate, Makjan, Minahassa (north Sulawesi) and Halmahera

The first authenticated earthquake in the Wichmann catalogue is from 1538 in Ternate (Fig. 2, Supplementary material). At that time, the volcanic island of Ternate was one of the world's major producers of cloves. Ternate is part of the Halmahera intra-oceanic arc, which forms above the eastwards-subducting Molucca Sea oceanic slab (Hamilton 1979; Hall 2011). The same slab is also subducting westwards beneath the Sangihe Arc. The accretionary wedges of these opposing subduction zones are colliding near the middle of the Molucca Sea to form one of the youngest submarine collisional mountain belts in the world (Fig. 2).

Earthquakes and tsunamis in this region could have several different sources. In the northern part of the region, the major active faults include the southern Philippine Trench and the Cotabato Fault, which overlap with the Molucca collision complex (Fig. 1). Benioff zones beneath the Sangihe Arc to the west and the Halmahera Arc to the east dip in

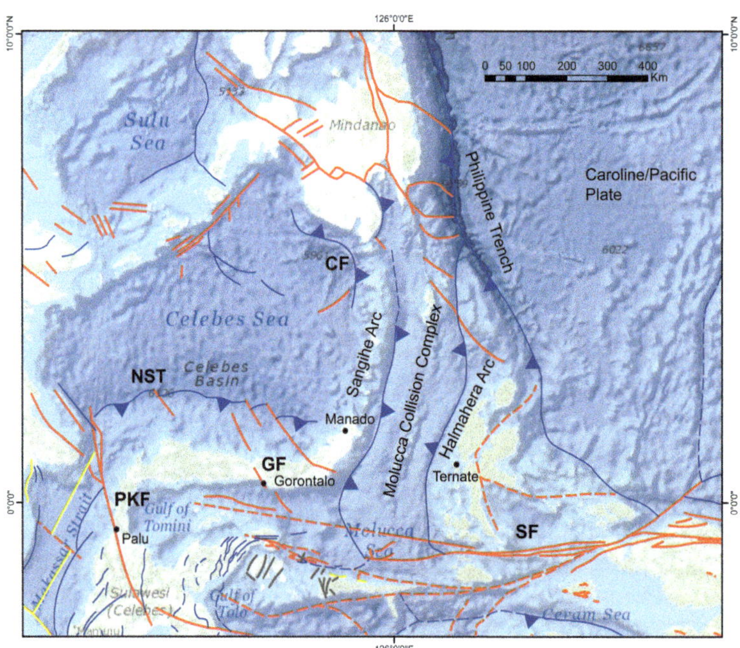

Fig. 2. Location map and active faults of the Molucca Sea region. Fault colours: blue, convergence; red, transvergence; yellow, divergence; grey, uncertain motion. Fault abbreviations: CF, Catabato Fault; GF, Gorontalo Fault; NST, North Sulawesi Trench; PKF, Palu-Koro Fault; SF, Sorong Fault. For the locations of earthquake events in the catalogue use the KMZ file in the Supplementary material.

opposite directions, which indicates that the Molucca Sea Plate in the middle is almost completely consumed by the opposing subduction zones. Most earthquakes occur along the subduction interfaces of these colliding arcs, from internal deformation within the opposing accretionary wedges, within the Molucca Sea Plate and from explosive volcanic eruptions. However, most of the active faults in the region, like nearly all of Indonesia, are submarine. Marine geophysical studies (Silver & Moore 1978; Bader & Pubellier 2000; Watkinson *et al.* 2011) document surface ruptures in several places throughout the Moluccas, but many areas where earthquakes are common remain unexplored. Local explosive volcanic events may also account for a few of the earthquakes included in the catalogue, such as those in 1646, 1673, 1770 and 1840.

Active thrust faults are found within the accretionary wedge collision complex near the Talaud Islands, rising up in the middle of the Molucca Sea (Silver & Moore 1978; Bader & Pubellier 2000). Other major thrust faults involve backthrusting of the accretionary wedges over associated forearc basins (Hamilton 1979). Along the East Sangihe Fault, the accretionary wedge collision complex is thrust westwards over the Sangihe Forearc (Fig. 2). A similar relationship is inferred on the east side of the accretionary wedge collision complex, with top-to-the-east thrusting over the Halmahera Forearc (Fig. 2). To the west of the Molucca collision complex, the Minahassa Trench forms from SSE subduction of the Celebes seafloor. The volcanic arc above this subduction complex forms the eastern part of the northern arm of Sulawesi. The Gorontalo Fault slices through the middle of the northern arm of Sulawesi (Fig. 2).

The southern part of the Molucca Sea region hosts many active faults associated with the westwards translation of the Caroline–Pacific Plate against continental crust of the northern Australian Plate (Silver *et al.* 1983*a*, *b*; Garrard *et al.* 1988; Smith & Silver 1991). The Sorong fault system forms the southern boundary of the Molucca Sea Plate, which may reach as far west as the Gorontalo Basin (Watkinson *et al.* 2011). The Sula and Matano faults form the southern-most strike-slip faults of the Sorong fault system, with associated pop-up and pull-apart structures (Watkinson *et al.* 2011).

Relative to the Sunda Craton, GPS measurements indicate 89 mm a^{-1} of westwards convergence across the Sangihe arc–arc collision, 22–46 mm a^{-1} of NNW convergence across the Minahassa Trench and as much as 53 mm a^{-1} of left-lateral oblique WNW convergence along the Sula–Sorong fault system (Rangin *et al.* 1999). Models combining these GPS data with patterns of seismicity over the past 50 years yield very high shear and dilatational strain rates in the Molucca Sea region (Kreemer *et al.* 2000). Yet, several earthquakes documented in the Wichmann catalogue are larger than anything recorded instrumentally. For example, between 1855 and 1860, there were several significant earthquakes and at least six tsunami that caused damage on both sides of the Molucca Sea, including Ternate, Tidore, Makjan and Halmahera on the west, and Minahassa, Gorontolo and Kema (north Sulawesi) on the east (Supplementary material). One tsunami, 9 m high, crashed into the west coast of Halmahera and was also noted on other islands (Table 2). The largest earthquake in the Harvard catalogue for the region (post-1976) is a M_w 7.9 subduction interface event in 1996. However, this event caused few casualties and only a local tsunami (Pelinovsky *et al.* 1997).

The regional nature of the events documented in the Wichmann catalogue, and the years of aftershocks that followed, indicate they were probably megathrust earthquakes. In a recent paper on possible megathrust earthquake sources (Heuret *et al.* 2011), the Moluccas and the Banda Arc (see below) are not included. This omission highlights the importance of examining historical records to validate models based on the relatively short instrumental catalogue and other geophysical criteria.

The Banda Sea region (south Maluku: Table 3), including Seram Island, Ambon, Buru, Amblau, Saparua, Haruku, Nusalaut (the Uliassers), Banda Neira, Lonthor and Ai (Banda Islands)

The majority of earthquakes and tsunami documented in the Wichmann catalogue are from the southern Maluku region (Fig. 3). However, just the opposite is true for the earthquake and tsunami rate during the past 100 years. Even though the post-1970 seismic rate in Maluku is the second highest in the Indonesia (United States Geological Survey: USGS), it is much lower than the rate of major earthquake and tsunami events between 1629 and 1876 (Tables 3). For example, from 1612 to 1859, the earthquake rate was one major event (>MMI 6) every 3 years (this catalogue) compared with a rate of one major event every 7 years since 1859 (Hamzah *et al.* 2000).

The abundance of observations in the region is partly attributed to the many Dutch settlements in the 'Spice Islands' found there (Hanna 1978; Milton 1999). Most of the observations in the catalogue from the Maluku region are from Ambon (Figs 4 & 5), the neighbouring islands to the Uliassers, and the Banda Islands 200 km to the SW (Fig. 3, Supplementary material). These island groups are part of the northern section of the Banda Arc. The

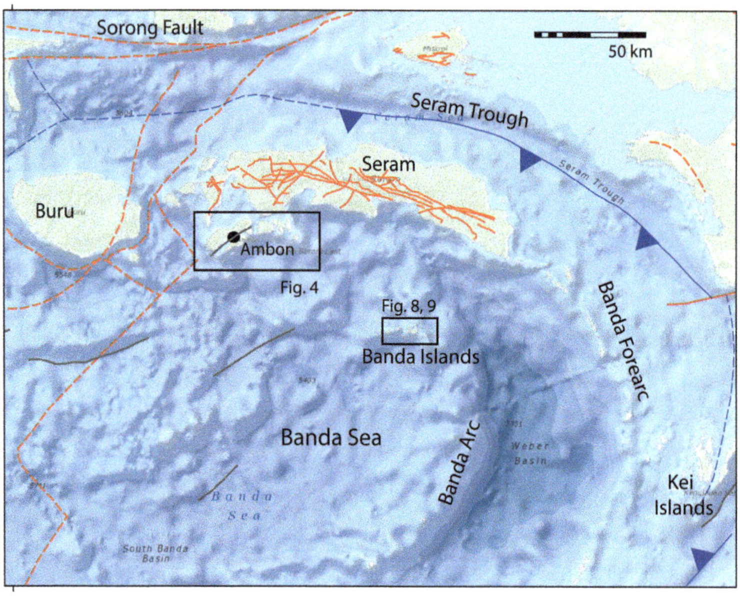

Fig. 3. Location map and active faults of the southern Maluku region. Fault colours: blue, convergence; red, transvergence; grey, uncertain motion.

Banda Arc formed in the Late Miocene above an intra-oceanic subduction system. At around 16 Ma, the arc began to collide with the NW continental margin of Australia in the Seram region (Berry & McDougall 1986; Linthout *et al.* 1996; Pownall *et al.* 2014). The current arc–continent collision zone makes a 180° bend from the Southwest Islands east of Timor to Ambon (Fig. 1). The collision is more advanced in the Uliasser Islands, where the volcanic arc is extinct and forearc basement is exposed.

Volcanism still persists in the Banda Islands and other volcanoes of the Banda Arc, although they are contaminated by subducted continental crust. Notwithstanding the active volcanism, only one of the earthquakes cited in the Wichmann catalogue, the disputed eruption of Banda Api in 1824, is likely to have been associated with a local volcanic eruption.

On multiple occasions, earthquakes and tsunami destroyed most of the buildings in Ambon or Banda Neira, including the regional earthquakes of 1648, 1754 and 1802. The colonial-era Fort Victoria, built in the early seventeenth century, is the only structure from this time that remains standing in Ambon. The Wichmann catalogue documents that

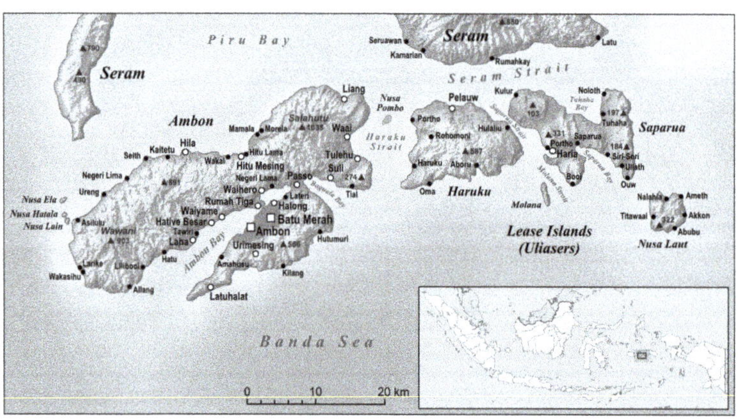

Fig. 4. Map of Ambon and the Lease Islands showing locations of many of the events mentioned in the text.

Fig. 5. Ambon City was destroyed by earthquakes in 1644, 1675, 1754, 1781, 1835 and 1841. Multiple tsunamis are also documented, inundating what is now the major urban centre of the Maluku region.

the walls of the fort were damaged by earthquakes in 1644, 1673 and 1705, and inundated by a tsunami in 1648 (Fig. 6).

Around 1.5 million people now live in the southern Maluku region, with more than a third of a million crowded into the low-lying coastal plain that has been inundated with past tsunamis. The city has now burgeoned up onto landslide-prone hills (Fig. 4). There are several accounts of major landslides in and around Ambon City throughout the Wichmann catalogue that are similar to the fatal slides that occurred in 2012 and 2013. Development is also increasing in coastal areas throughout the region that were probably inundated by the 15 regional tsunamis that occurred between 1629 and 1877.

The active fault sources for the earthquakes and tsunami in the Maluku region are poorly constrained

Fig. 6. Conjugate fractures in the walls of Fort Victoria from earthquake damage. The fort was damaged four different times by earthquakes and inundated by a tsunami. Surrounding the fort now is the urban centre of Ambon City.

(a)

Fig. 7. (**a**) Fort Amsterdam on the north coast of Ambon., which was built in the fifteenth century and is described in the catalogue as completely inundated by the 1675 tsunami. The central part of the fort has been reconstructed. (**b**) The 1675 tsunami poured over this wall and killed most of the inhabitants who went to the fort for shelter.

owing to a lack of marine geophysical surveys. One of the most dangerous aspects of this active arc–continent collision is the distribution of strain away from the deformation front in the Tanimbar and Seram troughs into other parts of the upper plate, such as the populated backarc region of Ambon and the surrounding islands (Harris 1991, 2011). It is likely that there are active faults near Ambon, such as in Ambon Bay, (Hamilton 1979) and offshore to the north and south associated with accretion of the Banda Arc onto Australian continental margin. Earthquakes felt in Ambon may also be sourced from the Sorong strike-slip fault system to the north (Fig. 3).

Intermediate to deep earthquakes in the Benioff zones surrounding this region on three sides are broadly felt, but have weak intensities and cause little to no damage. The largest instrumentally recorded event (M_w 8.5) occurred in 1938. The epicentre was located at the rear of the accretionary wedge south of the Watubela Island of Tioor (Okal & Reymond 2003). The depth is estimated at 60 km, which is consistent with the lack of damage reported (MMI 6) and the small tsunami run-up (1.5 m). This event was felt as far south as Darwin, Australia and in eastern New Guinea. The focal mechanism is an east–west-striking thrust with some left-lateral oblique slip. A similar focal mechanism is also found for the M_w 8.3 earthquake in 1963 in a similar tectonic position north of Babar Island (Welc & Lay 1987). These events are the largest along a NNE–SSW line of mostly intermediate-depth earthquakes of M_s 6.5–7.1 beneath the southern Weber Basin. They are interpreted as deformation of the subducting slab, with slip in the NNE–SSW direction of plate convergence (Okal & Reymond 2003).

One of the most damaging historical earthquakes occurred in 1675. It flattened the entire Chinese district and many other stone buildings in Ambon. Landslides also caused at least 2500 fatalities. A coastal landslide on the north shore of Ambon generated a tsunami with a run-up height of up to 100 m, based on the coastal vegetation trim line that it left (Soloviev & Go 1974). The wave flowed over the walls of Fort Amsterdam (Fig. 7) and destroyed most of the homes inside the fort, causing many fatalities.

The most damaging instrumentally recorded earthquake was in 1899, a M_s 7.8 event in Seram. It generated a 10 m tsunami that was probaby assisted by a landslide. The tsunami killed around 3500 people (Brune *et al.* 2010). No earthquake of this magnitude has occurred since this event. With elastic strain energy accumulating across the Seram Trough at rates of up to 7 m/century (Bock *et al.* 2003), it is likely that earthquake recurrence intervals are relatively short (*c.* 100 years).

Banda Neira, which is nestled in the Banda Islands (Fig. 8), has been destroyed several times by earthquakes and inundated by some of the most massive waves recorded in Indonesia. Two earthquakes in particular, one in 1629 and the other in 1852, showed clear characteristics of megathrust seismic activity. Both of these events were felt over a large region, generated large tsunamis of at least 15 and 8 m, respectively, at Banda Neira (Fig. 9), and are associated with at least a decade of aftershock-like seismic activity. The accounts of both events provide key details, such as the duration of shaking and the travel time of the tsunami after the shaking stopped:

> 1629, August 1 9:30 p.m. Banda-Islands. A half hour after the termination of a violent seismic shock there formed in the sound ... a high mountain of water. The...wave rolled westward ... against Fort Nassau ... where it achieved a height of 9 fathoms [15.3 m] above the springtide stand ... Houses lying on the beach were swept away, while others were laid to rubble ... Eastwards the tidal wave crashed on the west beach of the island Lonthor, where ... it achieved ... a height of 13 feet [4 m].

(b)

Fig. 7. *Continued.*

These particulars make it possible to inversely model the tsunami, and to test likely source areas and earthquake magnitudes. Models of the 1629 tsunami by Liu & Harris (2013) required a megathrust earthquake (M_w 8.2–8.8) along the Seram Trough. The Seram Trough accumulates strain from NNW convergence relative to the Sunda Craton at a rate of around 40 mm a^{-1} (Socquet *et al.* 2006). At these strain rates, enough elastic strain energy has already accumulated along the Seram Trough to cause an earthquake of similar magnitude to what occurred in 1629.

Archaeological excavations in Banda Neira (Lape 2000) found evidence of several destructive events throughout the colonial era. We excavated a trench adjacent to site BN4 of Lape (2000) and found three likely tsunami deposits, which we are currently analysing. The stratigraphic ages of the two largest deposits, based on the archaeological ages, correspond to the 1629 and 1852 tsunamis (Fig. 10).

The 1852 event was recorded at 20 different locations from Sumatra to Ternate. From these accounts and from modelling of the tsunami that was recorded in several locations, the source region is most likely to have been in the Tanimbar Trough (Fisher & Harris, in press). This collisional plate boundary accumulates strain from highly oblique NNE convergence of the Australian continental margin beneath the Banda Arc at a rate of around 70 mm a^{-1}. The event, like another one previously noted in the catalogue in 1649, caused the rise of new mud volcano islands in the Kei archipelago (Fig. 3).

We are concerned about the hazard potential of this very active tectonic region. During the past century, population and urbanization in the region has increased 10-fold, particularly in coastal area susceptible to tsunami inundation. The tsunami

Fig. 8. Panorama of Banda Islands. Photograph looking NE from Gunung Api. Banda Neira is the island in the middle. Lonthor is the crescent-shaped island behind it. The Lonthor Strait, to the left of Lonthor Island (near the rainbow), is where the tsunamis of 1629 and 1852 approached Banda Neira.

Fig. 9. Detail of Figure 8 showing the city of Banda Neira. The Wichmann catalogue documents four tsunamis that inundated Fort Nassau. The dashed lines are the approximate run-up heights of the 1629, 1841 and 1852 tsunamis.

potential also threatens large cities outside the region. For example, modelling of the 1629 earthquake and tsunami on the southern Seram Trough (Liu & Harris 2013) predicts run-up heights of 5.3 m along the north coast of Tanimbar, >3 m in Dili and >1 m in Darwin, Australia.

The Nusa Tenggara and South Sulawesi region (Table 4), including Lombok, Sumbawa, Flores, Sumba, Timor and Wetar

Nusa Tenggara and Sulawesi form a transition zone between the Greater Sunda Islands of Java and Bali,

Fig. 10. Trench near Fort Nassau of Banda Neira exposing tsunami deposits from earthquakes most likely in 1629 (lower white layer) and 1852 (upper white layer). The darkest black layers are volcanic ash. The trench site corresponds to archeological site BN4 of Lape (2000). See Figure 9 for the location of trench site.

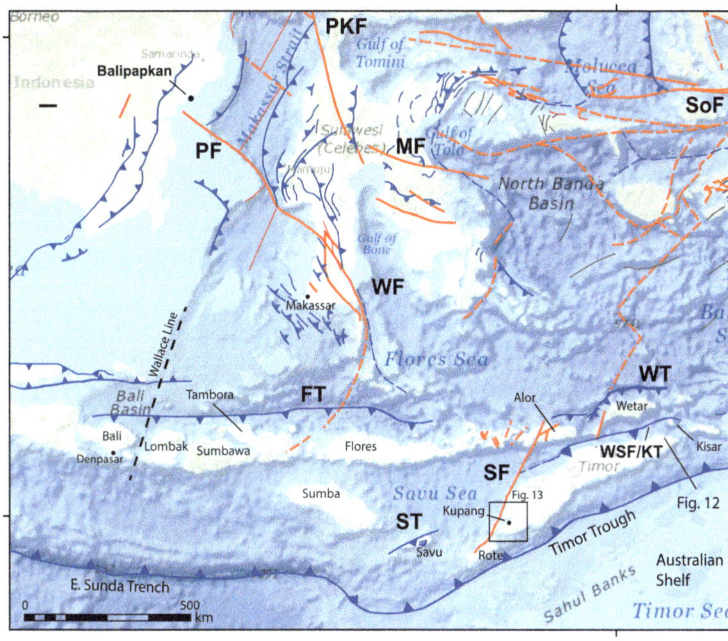

Fig. 11. Location map and active faults of the Nusa Tenggara and south Sulawesi region. FT, Flores Thrust; MF, Matano Fault; PKF, Palu-Koro Fault; PF, Pasternoster Fault; SF, Semau Fault; SoF, Sorong Fault; ST, Savu Thrust, WF, Walanae Fault; WST/KS, Wetar Strait Fault/Kisar Thrust; WT, Wetar Thrust. The locations of Figures 12 and 13 are shown.

which are part of the Sunda Craton, and the Banda Sea marginal basin. Sulawesi and various continental fragments in the Banda Sea basin were rifted from the Sunda Craton by intra-arc extension (Harris 2006). The intra-oceanic Banda Arc forms along the outer edge of the young oceanic crust of the Banda Sea basin.

The western boundary of Nusa Tenggara is the famous Wallace Line through the deep straight between Bali and the Lombok Islands (Fig. 11). The eastern boundary is disputed, but here is defined as the eastern edges of Wetar and the Timor Islands. The Timor region, which is similar to the Seram and the Ambon portion of the Banda Arc to the east, represents an evolving arc–continent collision that propagates westwards from central Timor at around 110 km myr^{-1} (Harris 1991). The collision initiation point is south of the island of Sumba, with progressively more mature parts of the collision represented to the east on the island of Timor (Harris 2011).

Also included in the Nusa Tenggara region is the Makassar Strait, which has one of the highest rates of tsunami in Indonesia since 1900 (Prasetya *et al.* 2001). However, only a few events are recorded in the Wichmann catalogue during the previous three centuries. How much of this relative quiescence is from a lack of reporting or a lack of seismic events is unknown. Between 1600 and 1900, only three earthquakes are recorded and one tsunami. Since 1820, there have been at least 18 tsunamis (Baeda 2011).

The most active faults in the Makassar Strait (Fig. 11) are found by geophysical investigation (e.g. Guntoro 1999; Hall *et al.* 2009), which includes: (1) thrust faults along the east and west edges of the northern Makassar Basin, with the eastern edge forming the Makassar Trench and Majene fold–thrust belt (Bergman *et al.* 1996); (2) the Palu Koro Fault (Katili 1978) that forms the northern boundary of the northern Makassar Basin; (3) the Mentano Fault of central Sulawesi; (4) the Paternoster Fault (Guntoro 1999) that forms the southern boundary of the northern Makassar Basin; and (5) the Walanae Fault (van Leeuwen 1981), forming the Salajar Trough offshore and the western edge of the Bone Mountains onshore, which is 85 km west of Makassar City. Makassar City is one of the largest urban areas in eastern Indonesia, with a population of 1.4 million (Fig. 11). The region surrounding the city is identified as a seismic gap (Baeda 2011) owing to the fact that it has a history of earthquakes, but very little seismicity, over the past 50 years.

The GPS velocity field in Sulawesi (Socquet *et al.* 2006) shows between 8 and 21 mm a^{-1} of strain accumulation across the Makassar Thrust.

Earthquakes near the convergent boundary have fault-plane solutions consistent with low-angle, east-dipping thrust faults beneath the western coast of Sulawesi, where most of the population is concentrated.

Five significant tsunamigenic earthquakes caused major damage in Nusa Tenggara in 1814, 1818, 1820, 1836 and 1856. A tsunami was also generated in 1815 from the colossal eruption of Tambora (Fig. 11).

The Tambora eruption (Volcanic Explosivity Index (VEI) 7) is the largest of recorded history. It is also one of the most notable extreme global geohazards of the Holocene (Wood 2014). The colossal eruption caused at least 100 000 deaths locally and years of global climate upheaval that accounted for, perhaps, as many as 6 million deaths worldwide (Oppenheimer 2011). Months prior to the eruption, there was an earthquake near Kupang, West Timor that created a new island and caused a tsunami. Shortly after the Tambora eruption, there were significant regional earthquakes in Maluku in 1815, 1816 and 1817, and a major event in the southern Makassar Basin in 1820. The spatial and temporal clustering of these events is a common feature throughout the Indonesian region and many other plate boundaries (Jagla & Kolton 2010).

Accounts of the 1820 Makassar (Belekomba) earthquake in the Wichmann catalogue are consistent with a Walanae Fault source. For example, the strongest ground motion (>1*g*, cannons hopping) and the longest time of shaking (4–5 min) are recorded from Bulukomba, which is only 20 km from the fault. The Dutch Fort there was also inundated by an 18–26 m-high tsunami. It is possible, from the account, that the tsunami was from an earthquake-induced submarine landslide near, but west of, Bulukomba:

> Fort Bulekomba fluctuated to and fro. The six-pounders ... hopped from their mountings. After the 4–5 minute long quake, shots were believed to be heard in the west, coming from the sea. Barely had the sent envoy returned with the news that ships were nowhere to be seen, than did the sea, under both a whistling and thunder-like rumble, come in, formed as a 60–80 foot high wall, and flooded everything. The barracks in the fort were destroyed, likewise, as a result of the flood wave that penetrated 400–500 feet inland ... 400–500 persons drown.

The earthquake 'violently' shook Flores and Sumbawa for 2 min, and was felt as far west as Madura (800 km from Belekomba). Each of these locations also experienced tsunamis. In Sumbawa, the tsunami, 'flung anchored ships far inland'.

Investigations of the GPS velocity field in Nusa Tenggara reveals a pattern of increased coupling with the NNE-subducting Australian Plate towards the eastern edge of the province (Genrich *et al.* 1996; Nugroho *et al.* 2009). It is likely that this pattern of strain is associated with increased amounts of subducted continental crust towards East Timor (Harris 1991). Between Sumba Island, where the arc–continent collision initiates, and the island of Rote (Fig. 11), 70–80% of the convergence is taken up at the deformation front (Timor Trough), with the bulk of the remainder occurring along the Flores backarc thrust (Nugroho *et al.* 2009). This changes to only 35% east of the 121° shear zone off the coast of West Timor (Harris *et al.* 2009). East of the shear zone, an additional 30% of the NNE motion of the Australian Plate is transferred to the Wetar backarc thrust system (Nugroho *et al.* 2009).

Recent earthquakes along the Flores and Wetar thrust systems demonstrate just how vulnerable this rapidly growing region is to even moderate-sized seismic hazards (McCaffrey & Nabelek 1984). In 1992, a M_w 7.8 earthquake on the Flores Thrust struck the area around Maumere and generated a tsunami with submarine landslide-assisted wave heights of up to 26 m (Budiono *et al.* 1995). The earthquake destroyed 90% of the buildings in the region, including around 800 schools. Half of the deaths were caused by the tsunami.

The Wetar thrust system, which is further to the east and along strike of the Flores Thrust, is also very seismically active and caused a M_w 7.5 earthquake and tsunami in November 2004. Around 800 homes were destroyed on Alor Island (USGS). Damage from the earthquake and tsunami also reached Dili, the capital of East Timor (with a population of *c.* 200 000).

Little onshore historical seismicity is recorded for Timor Leste. However, several thrust events have occurred offshore, with hypocentres on or close to the décollement at depths of 18–30 km below the northern flank of the Timor Trough. Immediately north of the Timor Leste capital city of Dili, a thrust event was recorded at approximately 30 km, along with extensional earthquakes at 16–20 km.

The Wichmann catalogue documents that, between 1793 and 1836, several tsunamigenic earthquakes struck the Nusa Tenggara region and were felt more widely than both the 1992 and 2004 events. One of the most notable earthquakes occurred along the north coast of East Timor in 1857. The earthquake caused a tsunami consisting of four waves, the highest of which was >3 m. The tsunami inundated the north coast around Dili, and completely flooded the villages of Liquica and Hera.

It is likely that the source of this earthquake/tsunami event was the Wetar Strait Thrust first inferred by Carter *et al.* (1976), and later documented as an active fault by Breen *et al.* (1989). This south-dipping thrust system accommodates the northwards motion of the Timor orogenic wedge over the forearc basin of the Banda Arc (Fig. 11). The thrust system can be traced just offshore the north coast of Timor eastwards to the islands of Kisar and Leti, where it was imaged breaching the seafloor by the BIRPS seismic reflection profile (Snyder *et al.* 1996). Mapping and age analysis of uplifted and warped coral terraces on the north coast of Timor (Fig. 12) and Kisar document co-seismic uplift, tsunami deposits and continued shortening along the Wetar Strait and the Kisar Thrust (Cox 2009; Major *et al.* 2013).

Tsunamigenic earthquakes have also struck the area around Kupang, West Timor (Fig. 13). Kupang is one of the main commercial centres of Nusa Tenggara, with a population of around 0.5 million. There are four known active faults in the region that may have sourced these events. The most significant is the Timor Trough to the south. Recent GPS measurements determined a strain rate across the Timor Trough of 22 mm a^{-1} (Nugroho *et al.* 2009). As noted earlier, there are several earthquakes with low-angle thrust fault-plane solutions along the projection of the Timor Trough beneath Timor. It is possible that there may have been a major earthquake along the Timor Trough some time between 1793 and 1814, which immediately preceeded the 1815 eruption of Tambora.

Another possible source of major earthquakes much closer to Kupang is the Semau Fault (Fig. 13) or 121° discontinuity shear zone (Harris *et al.* 2009). This left-lateral transpressional fault system is mapped on Rote Island (Roosmawati & Harris 2009) and expressed along the west coast of Timor as a linear zone of mud volcano islands (Barber *et al.* 1986). The fault is identified in seismic reflection profiles just offshore of Kupang (Karig *et al.* 1987) and is associated with uplift of several flights of coral terraces (Merritts *et al.* 1998). GPS measurements show 11 mm a^{-1} of strain accumulation across the fault (Nugroho *et al.* 2009).

Known active faults onshore in West Timor are indicated by earthquakes with generally transtensional fault-plane solutions, and include strike-slip and normal earthquakes with T-axes orientated east–west (McCaffrey 1988, 1989). A surface-rupturing M_W 5.7 strike-slip earthquake in westernmost Timor in 1975 (Fig. 13) showed both dextral and sinistral slip along vertical faults striking mostly NW–SE (Tjia 1983; Duffy *et al.* 2013).

The other active fault found in the Savu Sea is the Savu thrust system (Fig. 11). This north-verging thrust forms the rear of the Sumba–Savu accretionary wedge (Harris *et al.* 2009), much like the Wetar Strait Fault north of Timor. Thrust faults are imaged breaking the seafloor on seismic reflection profiles (Harris *et al.* 2009). It is likely that the March 1866 earthquake felt in Kupang, but strongly felt in Savu, was along the Savu Thrust. GPS measurements indicate 6–8 mm a^{-1} of strain across this plate boundary segment (Nugroho *et al.* 2009). At this rate of strain, at least 1 m of strain may have accumulated along the Savu Thrust since 1866. Uplifted coral terraces along the north coast of Savu attest to the long-term activity of the thrust fault.

Fig. 12. Photograph looking south from the Wetar Strait at flights of uplifted coral terraces along the north coast of Timor Leste. The uplift is mostly attributed to movement along the Wetar Strait Thrust (Cox 2009).

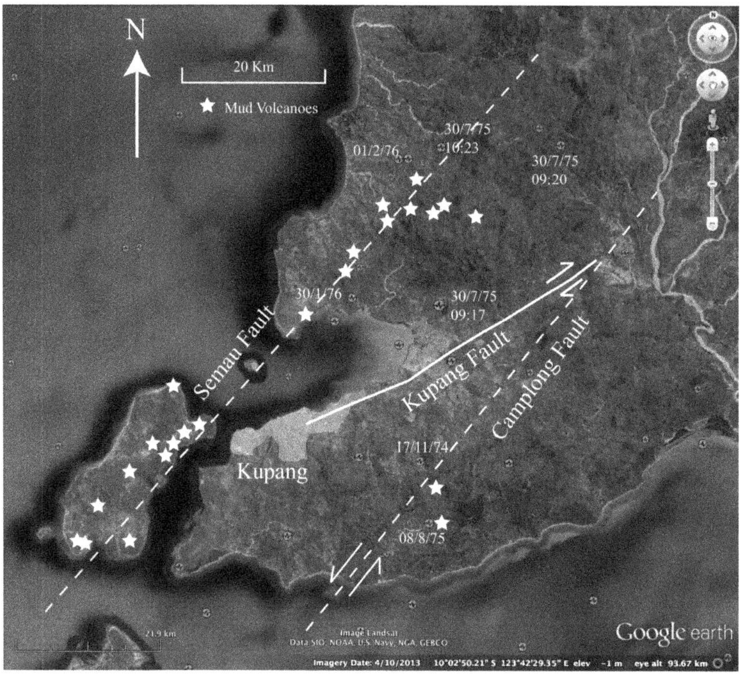

Fig. 13. Active faults of the Kupang region of West Timor. The Semau Fault is inferred from lines of mud volcanoes (white stars). The mud volcanoes and Kupang Fault are taken from Rosidi *et al.* (1979). The Camplong Fault ruptured on 30 July 1975 at 09:17 GMT. The dates and times of the associated earthquake cluster are shown. The urban centre of Kupang is shaded.

The earthquake and tsunami disaster risk of the Nusa Tenggara region has increased almost exponentially since the events documented in the Wichmann catalogue occurred. The only major event along the subduction zone to the south of the region is the 1977 M_w 8.0 Sumba earthquake, which generated a 15 m-high tsunami. However, this event was caused by normal faulting on the downgoing oceanic slab. There is no evidence of major earthquakes along the subduction interface from Java to Timor in over 400 years, notwithstanding the 70 mm a^{-1} of convergence along this section of the eastern Sunda subduction zone. It is possible that up to 28 m of strain has accumulated along the subduction interface.

The population of Nusa Tenggara region (Nusa Tenggara) is currently around 10 million. Another 2 million people live along the coast of the Makassar Strait. Most of the people now live in cities, most of the buildings in the cities are of unreinforced masonry construction and many of the cities are within historical tsunami inundation zones. The western shores of the Makassar Strait are near active faults and coastal cities, such as Balikpapan and Bontang, are vulnerable to tsunami hazards from events such as the one in 1820. Other parts of Kalimantan, which account for <6% of Indonesia's population, are considered highly unlikely to experience earthquakes. No entries of significant damage due to earthquakes in Kalimantan are found in the Wichmann catalogue.

Java and Bali (Table 5)

The islands of Java and Bali mostly consist of products of the Sunda Arc, which is mounted on the southern edge of the relatively young Sunda Craton (Hamilton 1979). Jurassic-age seafloor with >1 km of sediment cover is subducting beneath the edge of the craton at the Java Trench. Between the trench and the southern coast of Java and Bali is a well-developed accretionary ridge and forearc basin. Orthogonal convergence is estimated at a rate of around 75 mm a^{-1} (Nugroho *et al.* 2009). GPS measurements indicate little or no internal shortening across Java (Bock *et al.* 2003), which is interpreted as evidence for loose coupling across the subduction interface. However, there is a broad seismic zone across Java that is mostly the result of intermediate-depth events in the Benioff zone and more localized events associated with deformation of the volcanic arc (Ghose & Oike 1988). The Benioff zone continues to a depth of 650 km and produces deep earthquakes that are felt over long distances, but cause little or no damage.

Some of the most widely felt and damaging events in this region were in 1586, 1699, 1722, 1757, 1780, 1815, 1834, 1840, 1847, 1859, 1862, 1865, 1867 and 1875. It is difficult from the historical records to constrain which of these events are from inter- v. intra-plate earthquakes or large intermediate to deep Benioff zone earthquakes. Most of the known deep events are characterized by broad zones of small intensity, such as the 2007 M_w 7.5 event at a depth of 290 km that was felt throughout Java and as far away as Malaysia, but did little or no damage. Descriptions in the Wichmann catalogue of the event of 2 January 1780 fit the characteristics of a strong, but deep event.

It is also possible that some earthquakes, where little damage is observed (MMI <5) but a tsunami was generated, may have been slow-rupture events, such as the M_w 7.8 and 7.7 inter-plate earthquakes on the Java Trench in 1994 and 2006, respectively. These earthquakes were hardly felt on shore, but produced larger than expected tsunamis that claimed hundreds of lives (Kerr 2006; Kato *et al.* 2007).

The most damaging events in the historical catalogue, and therefore the ones that are likely not to be slow slip or deep earthquakes, were those in 1699, 1780, 1834, 1840, 1847, 1865, 1867 and 1875. Some of these events were likely to have been sourced from onshore faults accommodating deformation in and around the Sunda Arc (Fig. 14). However, inferring source regions is hampered by a lack of data for active faults in Java.

Active faults in Java have been described by Van Bemmelen (1949) and Tjia (1977). These include: (1) the left-lateral Lembang Fault, north of the densely populated urban centre of Bandung; (2) the Cimandiri normal fault that has sourced several destructive earthquakes within the NE-striking Cimandiri River Valley; (3) the Bency normal fault, which sourced the surface-rupturing Sukabumi earthquake, 1982 (Kertapati & Koesoemadinata

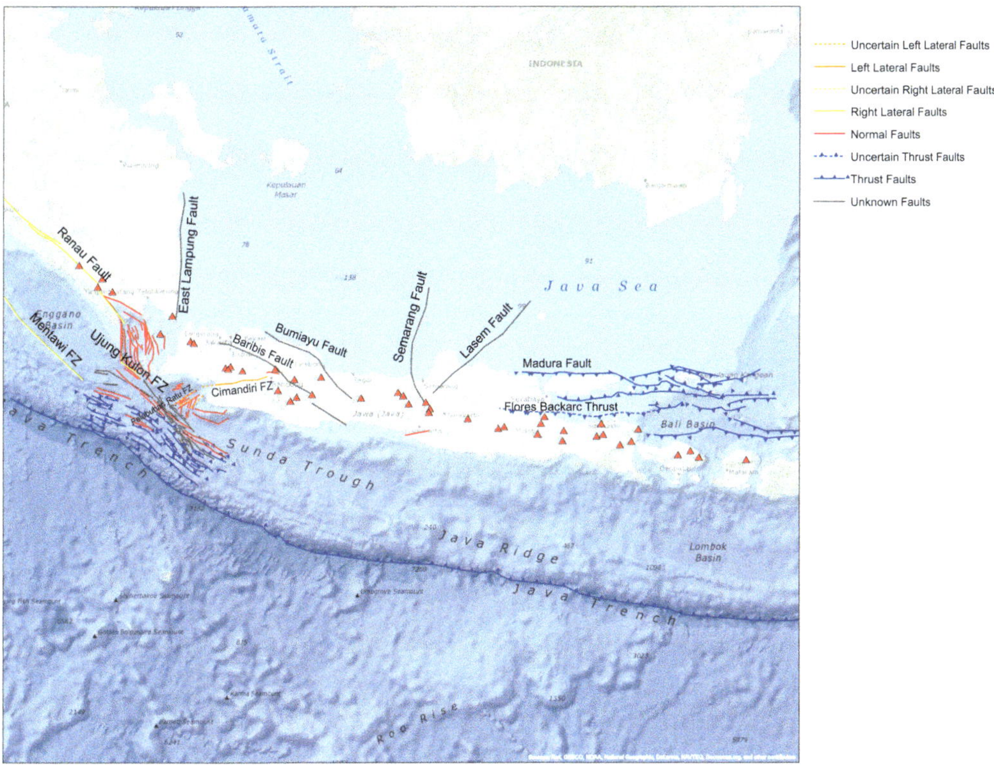

Fig. 14. Active faults and volcanoes of the Java and Bali region. Blue, thrust; red, normal; yellow, strike slip; grey, oblique slip; red triangles, volcanoes. The Lembang Fault is the small red line south of the Bandis Fault. The Opak Fault is the red line near the south central coast. Marine geophysical mapping off the SW coast of Java (Kopp *et al.* 2001) provides a glimpse of the structural complexity in the forearc ridge south of Java. Map modified from Irsyam *et al.* (2008).

1983); (4) the left-lateral Opak Fault, south of densely populated Yogyakarta; and (5) fault zones in the Citanduy Valley near Banjarsari, which has sourced many destructive earthquakes (Simandjuntak 1993).

The general trend for Indonesia of increasing casualties per event is clearly demonstrated in the Wichmann catalogue as well as in more recent events in Java. In 2006, a shallow M_w 5.9 earthquake on a previously unknown fault near Yogyakarta caused around 6000 casualties and damaged around 451 000 homes (Tsuji *et al.* 2009). Much of the damage from the earthquake resulted from the poor soil conditions upon which much of the built environment rests. The relatively small extent over which this event was felt and caused damage was not significant enough to have made the above list of most damaging earthquakes. Five million people now live within 100 km of this fault and the nearby Opak Fault, which is expressed geomorphically (Fig. 15).

The question of whether a megathrust earthquake event can occur on the Java Trench is still debated (McCaffrey 2007; Stein & Okal 2007). The well-developed forearc basin and accretionary wedge, and subduction of a relatively thick (>1 km) package of sediment, are characteristic of subduction zones that produce megathrust earthquakes $>M_w$ 8.5 (Ruff & Kanamori 1980).

There is circumstantial evidence from multiple sources that a megathrust event may have happened during the period of 1584–87, when the whole island shook and several volcanoes erupted that were previously, and have since been, dormant. During this time, the two largest temple complexes in central Java (Prambanan and Borobudur) were also severely damaged by an earthquake (Fig. 15). A tsunami deposit along the south coast of Java (Fig. 16) also dates to around this time (Eko Yulianto pers. comm.). If the Java Trench is capable of causing a megathrust earthquake and the last one was 428 years ago, then up to 30 m of potential slip could have accumulated along the trench since the last megathrust event. The densely populated areas that would be affected by the strong ground motions of this event are mostly constructed of unreinforced masonry buildings that would most probably experience a high rate of structural failure. Tsunami hazards are also high along much of the southern Java coast, where higher than expected run-up heights (Fig. 17) were observed from the July 2006 Tsunami (Fritz *et al.* 2007).

Sumatra (Table 6)

Much attention has been currently focused on the Sumatra region since the megathrust event occurred on 26 December 2004, and the spatial and temporal clustering of events in the region since 2004. Publications describing the history and potential of megathrust earthquakes along the Sunda Trench offshore Sumatra relied heavily on the Wichmann

Fig. 15. The Opak Fault scarp (ridge on the horizon) as seen from the Prambanan Temple site. The Prambanan Temple was severely damaged in the ?1584–86 regional earthquake and again in 2006 by a local M_w 5.9 earthquake on a previously unknown fault associated with the Opak Fault.

Fig. 16. Tsunami deposit near Pangandaran. It is most likely from a 1584–86? earthquake that damaged the Prambanan and Borobudur temples of central Java. The tsunami deposit is the 10 cm-thick dark layer of coarse sand, near the top of the shovel handle. The tsunami deposit (light grey layer by shovel handle) overlies a palaeosol, which overlies peat (dark grey). A tsunami deposit from the 2006 M_w 7.7 event overlies the surface.

catalogue (e.g. Newcomb & McCann 1987). Some of the earthquakes reconstructed from the catalogue have reoccurred. For example, the 2005 northern Sumatra earthquake near Nias Island was nearly of the same magnitude and ruptured close to the same area as that which was proposed by Newcomb & McCann (1987) from descriptions in the Wichmann catalogue of an earthquake in 1861.

The catalogue documents tsunamis that may have been generated by megathrust earthquakes in 1797, 1799, 1818, 1833 and 1861. Investigations of microatolls of the forearc islands offshore of Sumatra extend the record of abrupt land-level changes that are likely from co-seismic events back to over 1 kyr (Natawidjaja *et al.* 2004; Meltzner *et al.* 2010).

Besides the inter-plate seismic events along the Sunda Trench, there are three other sources of intra-plate earthquakes in Sumatra. These include thrusting within the accretionary wedge, and strike-slip faulting within the arc and forearc. The strike-slip faults are known as the Sumatran and Mentawai faults, respectively.

The Sumatran Fault is a dextral-slip shear zone that accommodates the oblique convergence of the subducting Australian Plate. The fault is marked by a linear topographical depression filled with offset river valley segments that extend over the entire

Fig. 17. Damage in Pangandaran from the 2006 earthquake and tsunami. The wave destroyed most of the buildings along the beach and reached the second floor of the lone surviving home in the distance.

length of the island (Sieh & Natawidjaja 2000). At least 25 km of offset of granitic intrusions has been documented since the Late Miocene. Some Jurassic units may be offset by as much as 180 km (Tjia 1977). At least 18 different fault segments have been identified that have caused more than 20 destructive earthquakes over the past century. Some of the earthquakes in the Wichmann catalogue share characteristics with these events.

The Papua region (Table 7)

Several large earthquake events were recorded in the 1800s in the Papua region, which included two destructive tsunamis. However, the rate of seismicity in this region has been much higher over the past 50 years than in the sparse accounts found in the Wichmann catalogue. This discrepancy is likely to be due to a lack of reporting. The source of most large earthquakes in this region is thrusting along the New Guinea Trench and oblique-slip deformation associated with the Sorong fault system, and other strike-slip structures accommodating the mostly westwards motion of the Pacific Plate relative to the Australian Plate (Fig. 1).

Significant earthquake and Tsunami events in Indonesia from 1600 to 1877

The following summaries are taken from the Wichmann catalogue. We have speculated on the likely source of each event from known active faults. The complete account of each event and many others is found in the translation of the catalogue (Supplementary material). Summaries of the events for each region are provided in Tables 2–7.

1584–86, Java: An earthquake spread over the entire island and caused damage to both the Prambanan and Borobudur Temple complexes [Fig. 15]. Tsunami deposit in Pangandaran may document that this event was a megathrust earthquake along the Java Trench [Fig. 16].

1608, 1 July, Makjan (Makian): A large wave wrecked ships. No earthquake was felt, suggesting a different source, but there was also no storm at the time.
Likely source: Volcanic eruption or distant earthquake.

1608, 18 July, Ternate: Earthquake at night during 'the first watch' 1 year after the Dutch took control of the island. The earthquake shook 9 or 10 cannons loose from their mountings and was accompanied by a volcanic eruption that emitted smoke and steam. No tsunami was noted.
Likely source: Nearby volcanic eruption possibly triggered by a seismic event that generated the tsunami on 1 July.

1629, 1 August, Banda Islands: At 9:30 p.m., a violent shock struck the Banda Islands, followed in half an hour by a large tsunami. The earthquake was also felt in Ambon, but no tsunami is mentioned. In the Banda Islands, the tsunami entered the sound between Lonthor and Banda Niera from the NE. It swept the village of Banda Neira away and damaged many stone structures. The tsunami inundated Fort Nassau and reached a height of 15.3 m above springtide [Fig. 9]. The quake was also felt on Ambon, but no tsunami is mentioned. At least 9 years of aftershocks followed the event.

Likely source: Megathrust earthquake $>M_w$ 8.2–8.8 on the Seram Trough (Liu & Harris 2013).

1644, 12–13 May, Ambon Island: The Portuguese had first arrived on the island in 1513, and the Dutch had occupied it for over 50 years when 'a terrible earthquake never before seen in this province' struck. Fort Victoria's walls were heavily cracked [Fig. 6] and some collapsed. Many buildings were damaged such that their collapse was feared. Fissures formed in the ground in Leitimor (peninsula), but no loss of life was mentioned. The ground remained in motion for many days. A strong aftershock struck on 17 May, causing the near-total collapse of the governor's residence, and was especially damaging on the island of Haruku.

Likely source: Unknown active fault near Haruku.

1646, 19–21 July, Makjan (Makian): A very violent earthquake accompanied a powerful eruption of the volcano. The earthquake was so terrible that residents thought the area would have to be abandoned. Several locals were killed by fire (or lava?).

Likely source: Nearby volcanic eruption.

1648, 2 February, Solor (Lesser Sunda): Earthquake caused wall collapse in Fort Henricus on the north side of Solor. Gaping fissures formed in the mountains and in Larantuka on the east coast of Flores. A probable aftershock cracked walls again in May of the same year.

Likely source: Flores Thrust.

1648, 29 February, Ambon Island: A flood wave formed and crashed against Fort Victoria [Fig. 6], but not causing notable damage.

Likely source: Distant earthquake or volcanic eruption.

1659, 11 November, Ambon Island: Earthquakes followed by a volcanic eruption of Teon in the southwest Banda Arc sent locals to neighbouring islands. The eruption was heard on Ambon and the Banda Islands a few hundred kilometres away. A 1–1.5 m-high tsunami was observed in the Bay of Ambon.

Likely source: Eruption on Teon Island.

1671, 17–18 October, Saparua Island: Earthquake causing destruction of the entrenchment Velsen and heavy damage of Fort Hollandia. The beach at Hatuwana sank by 0.3 m and the reef before Paparu village even more. Ground fissures were so deep that coco palms sank to the crown. The earthquake was also felt on Haruku, Ambon and the south coast of Seram. Aftershocks kept the ground in motion for a month and isolated shocks were felt for a year.

Likely source: Haruka–Saparua Fault

1672, date unknown, Ambon Island: Very strong earthquake that triggered landslides. Villages were buried under the rubble, and fissures 35–55 m deep formed.

Likely source: Ambon Fault Zone.

1673, 20 May, Halmahera: An eruption of Gamkanora on the NW coast of Halmahera was accompanied by an earthquake that was felt in Ternate 60 km away. A tsunami followed that flooded villages.

Likely source: Unknown.

1673, 12 August, Ternate: A violent earthquake that caused Mount Gamalama to split open on the south side from bottom to top, but no eruption was reported. Rocks sank away and houses collapsed. A tsunami crashed against the beach with such force that shipwrecks were feared.

Likely source: Unknown, but perhaps related to the event on 20 May of the same year.

1675, 17 February, Ambon and the Uliasser Islands; Buru; Ceram; Banda Islands; Nusa Tello Islands: The most violent earthquake experienced in living memory in the Ambon region. The quake was felt, and a tsunami was observed on Ambon, Haruku, Buru, Amblau, Manipa, Keland, Buano, the Banda Islands and the western peninsula of Ceram.

Many buildings collapsed and claimed victims. Fissures opened belching bluish sludge 6–7 m upwards. Landslides buried villages and dammed streams.

On the north coast of Ambon, whole villages slid into the sea causing a major tsunami that engulfed the entire region. The smallest of the three Nusa Tello islands off the NW coast of Ambon was fully flooded over by the wave. Trees were washed away as high as 90–100 m on some slopes near Cyst. At least 2243 persons perished from the tsunami. At Fort Amsterdam, it spilled over the walls and killed most of those who fled to the fort for shelter [Fig. 7].

Likely source: Ambon Bay Fault.

1681, 11 and 17 December, west coast, Sumatra: Sky-high mountains were shaken and shifted. Seaquakes reported by several ships.
Likely source: Subduction interface near Mentawai Islands (perhaps the same event as the *c.* 1685 one documented by Sieh *et al.* 2008).

1683, 16–30 October, Banda Islands: A violent earthquake with 2 weeks of strong aftershocks. Cracks and fissures appeared on the mountains. Most of the houses became heaps of rubble and the walls of Fort Revengie were damaged.
Likely source: Banda Api or intermediate-depth event on Banda Trench.

1699, 5 January, Jakarta: Powerful earthquake in West Java and southern Sumatra. Several homes collapsed and landslides dammed rivers. An increase in earthquake frequency, including two strong events in the next 10 years, likely to represent aftershocks or stress contagion onto other faults.
Likely source: Source unknown.

1705, October, Ambon Island: Earthquakes were felt over the entire month, especially on Hitu peninsula and west Ceram. A particular 2 day quake opened fissures, some which damaged houses. Fort Victoria's walls were rent asunder in several places and many were surprised that all the houses did not collapse.
Likely source: Aftershock from the likely stress contagion from the 1674 event.

1708, 28 November, Ambon Island: A tsunami penetrated the Bay of Ambon up to Batu Merah, three waves struck in total, and they were also observed on other coasts.
Likely source: Unknown.

1711, 5 September, Ambon and Uliasser Islands: A similar event to the 1708 tsunami, but three waves went up the Bay of Ambon within an hour. A violent earthquake was felt in Haruku and a weak shock was felt on Nusalaut and Saparua, 13 or 14 flood waves were observed on the beaches of these islands. The earthquake also felt in Banda Islands.
Likely source: Haruka–Saparua Fault.

1754, 18 August, Ambon and the Uliasser Islands: Earthquake caused the market hall sitting on 60 stone piers to collapse, several other buildings were destroyed, and houses were cracked. Fissures a few centimetres wide formed with one of them in Batu Merah ejecting bluish sludge with water and sand like a fountain. A tsunami struck the south coast of Ambon at Hutumuri.
Likely source: South of Ambon.

1755, 3 November, Mana, Bengkulu, SW Sumatra: Powerful shock with 12 aftershocks felt until 3 December. The walls of the Cumberland House, the administration building in Fort Marlborough in Bengkulen, and other buildings, suffered heavy damage. The same was the case with houses in Lais, Mana, as well as on a sugar plantation located 5–6 miles away from Bengkulu. In the area of Mana several villages were also destroyed. On the Bengkulu River [Ajer Bangkulu] or, more precisely, at the delta and also further upriver, the ground opened and a great amount of water gushed forth with 'sulphureous earth that spread an unbearable stench'. At the same time on Pablo Point [Cape of Si Abung], numerous cracks and fissures formed.
Likely source: Sumatra Fault.

1763, (no date), Nias Island, west Sumatra: Earthquake, by which a village was buried.
Likely source: Sunda Trench or accretionary wedge fault.

1763, 12 November, Banda Islands: Violent earthquake was followed by 16 shocks in the evening and night. 75% of the houses on Banda Neira were reduced to rubble. Great boulders crashed down from the mountains.
Likely source: Volcanic quake (Banda Api) or nearby fault zone.

1770, no date, District of Mana, Bengkulu, west Sumatra: Very violent earthquake that destroyed a village as the ground rose up and a fissure 400 m in length, 4 m wide and 8 m deep formed. Bituminous material gushed from the fissure. At the same time, near the delta of the Padang Gutji, likewise located in Mana, a piece of land was laid dry. Finally, a flood wave occurred simultaneously with this quake.
Likely source: Forearc thrust fault.

1793–95, date unknown, Kupang, Timor: The church and several buildings were destroyed in a violent earthquake.
Likely source: Timor Trough or Semau Fault.

1797 *(1795 in Wichmann catalogue)*, **10 February, west coast of Sumatra**: The first vibration lasted a minute, following which a flood wave immediately arose, which penetrated into the River of Padang with great force, such that the city was flooded. Afterwards, the water ran back out so far that even the riverbed was left dry. The process repeated itself three times. The village on the beach, Ajer Manis, was flooded and at the same time many huts were washed away. In Padang, proper fissures of 7–8 cm in breadth formed and closed again. Most buildings were damaged. During the entire night, and the next day the ground was in movement. Every 15–20 minutes a powerful vibration occurred, which persisted for a week, with the pauses between events gradually became longer.

The earthquake spread over two degrees north and south of the equator. Moreover, the flood wave it produced had also struck the Batu Islands. [See Natawidjaja *et al.* 2006 for more details.]

Likely source: Sunda megathrust.

1799, (no date), Sumatra: Earthquake and seaquake associated with a flood wave that achieved a height of 15 m above the usual water level. A reef was uplifted, such that it became a danger for sea navigation. The shock caused significant damage at Palembang.

Likely source: Aftershock or stress contagion to another segment of the Sunda Trench after 1795 event.

1818, 18 March, Bengkulu, SW Sumatra: Earthquake observed at Fort Marlborough where people were thrown out of bed and walls collapsed. The sea withdrew before a tsunami drove everything with great force back on shore. On the ships *Northumberland* and *Sunbury*, the shock was so strong that the sailors were thrown from their berths. The water flooded up over the bridge inland. Aftershocks were noted until at least 8 April.

Likely source: Sunda megathrust or accretionary wedge fault.

1833, 24 November, Bengkulu, Sumatra: Violent earthquake that was felt as far away as Singapore and Java. The shaking in Singapore was felt for 1 minute, 5 minutes after it was noted in Bengkulu. The report also notes that it was the first earthquake since Singapore has been in English possession [1819]. Ships also felt a seaquake.

In Bengkulu, the earthquake lasted for 5 minutes and caused building collapse. A tsunami stormed the coast destroying the port dams and houses in the vicinity, two schooners and several smaller vessels were thrown onto land.

In Padang the earthquake lasted for 3 minutes with several days of aftershocks. Direction SSW–NNE. Buildings were damaged, and water and sulphurous steam rose from cracks. A tsunami caused considerable damage.

In Indrapura and Pulu Tjingko, the tsunami caused considerable damage and killed many people. Two local volcanoes in this area were also affected. Gunung Singalang cracked and Gunung Merapi erupted.

In Pariaman, an extremely powerful shock caused ground fissures up to 1 m in width. The sea retreated and then returned in the form of a massive flood wave that tore all of the ships from their anchors. The vibrations continued for multiple days on end.

On the volcano Bukit Kaba, Ajer Lang a landslide dammed a river that was subsequently destroyed by an aftershock. The flooding destroyed many villages downstream. [See Natawidjaja *et al.* 2006 for more details.]

Likely source: Sunda megathrust.

1840, 14 February, Ternate: A violent eruption of Gamalama started on 2 February, followed by powerful earthquakes several days later that 'nearly transformed the city into a rubble heap'. The strongest quake on 14 February caused 500 houses to collapse along with heavy damage to Fort Oranje. Shocks continued through March.

1841, 16 November, Banda Islands: A weak shock was felt, followed by a wave 15 minutes later that struck the south coast at a height of 2.5–3 m. Water reached the door of Fort Nassau, even though it was low tide at the time [Fig. 9].

Likely source: Active fault zone south of Banda Islands.

1841, 16 December, Ambon, Amblau and Buru Islands: A not particularly strong earthquake was felt on Ambon, which was followed 15 minutes later by a wave *c.* 1.5 m above the highest water level that crashed against the beach in Ambon Bay. Several huts were washed away west of Ambon City [Figs 4 & 5]. The quake was much stronger on Buru and Amblau Islands. On Amblau, the wave washed away huts and mosques of beach villages.

Likely source: Sula or Matano strike-slip faults.

1843, 5 January, Nias Island and Baros, NW Sumatra: Violent quake causing damage to buildings and numerous cracks to form in Sumatra.

On the Island of Nias, shocks came from the west then continued to the north. The shaking was initially weak, but increased in intensity gradually for 9 minutes. Part of Mount Horifa fell into the chasm, and part of the ramparts of the fort at Gunungsitoli sank. With the exception of the barracks and the apartment of the commandant, all the houses collapsed. After the passage of 9 minutes, a 'sky-high' flood wave coming from the SE crashed against the NE beach and flooded away Kampong Mego, an hour south of Gunungsitoli. The flood phenomena lasted for 5 hours, after which a yet more violent earthquake occurred that lasted about 6 minutes. The shock waves came 'from the west and roamed towards the north'. Many days later, the vibrations were still perceptible, though to a weaker degree.

In Barus a flood wave stormed against the coast at 12.30 am and cast boats 600 m on land.
The earthquake was felt in Penang and Singapore.
Likely source: Sunda megathrust or accretionary wedge fault.

1845, 8 February, Menado, north Sulawesi: A violent earthquake spread over all of Minahassa peninsula, north Celebes. At Menado, it was barely possible to stand upright and objects were thrown about in houses. The walls of Fort Amsterdam formed gaping cracks. Many houses collapsed resulting in 56 fatalities and more injured. Many of the walls of the fort in Amurang collapsed. Many fissures formed in the ground and landslides occurred. The water in the roadstead (harbour) emptied, leaving fish to gather from the seafloor.
Likely source: Northern Sulawesi Trench

1846, 15 January, Ternate and neighbouring coasts; Menado, north Sulawesi: A non-violent, but long-duration earthquake was followed by a 1 m flood wave at Ternate. The water level rose in wells. The wave was also observed on the coasts of neighbouring islands and as far away as Menado, north Celebes.
Likely source: Molucca Sea subduction zones.

1847, 31 October, Nicobar Minor: A series of earthquakes and tsunami waves are documented for 19 days. On the small islands located in the St George Channel, the shocks were most powerful and caused large boulders to fall, collapse of houses and formation of fissures that spewed cold and salty water, and flooding of coastal areas tsunami waves.
Likely source: Sunda Megatrench.

1852, 26 November, Banda Islands; Ambon and Uliasser Islands: At 7:50 a.m., perhaps the largest earthquake ever documented in the Banda Sea region, similar to the 1629 event. It was felt from S. Sumatra to Australia to Ternate. New islands fitting the description of mud volcanoes appeared in the Kei Islands region [Fig. 3]. In Banda Neira 'the majority of the residences … were transformed to a rubble heap … and those that remained standing were uninhabitable'. 15 minutes after the shaking stopped the bay emptied to a narrow river. Three waves followed that came through Lonthor Sound and 'reached the hill on which Fort Belgica is built' [Fig. 9] and the foot of the mountains on Lonthor. The difference between the high and low of the waves was 8.2 m. The tsunami was not observed on the north coast of Bandaneira or on the south coast of Lonthor. The islands of Rosengain and Ai were also affected by the earthquake and the tsunami, which reached around 1 m above the usual sea level on Ai. Gunung Api did not show volcanic activity.
Violent shaking was experienced in and around Ambon where a wave > 1 m was observed. Aftershocks were noted for at least a decade.
Likely source: Megathrust on the eastern Seram and northern Tanimbar troughs (Fisher & Harris, in press).

1854, 2 January, Haruku and Saparua Islands (Uliassers): Powerful shocks were followed by a 'not particularly strong flood wave' also noticed on Ambon.
Likely source: Haruka–Saparua Fault.

1855, 25 April, Ternate; Halmahera: Shocks were felt from the morning until 2 p.m.. Multiple buildings were damaged and the shocks were felt aboard ships. The quake was also felt on Halmahera and the fort on Dodingga 'was damaged so badly that it had to be abandoned'.
Likely source: Volcano or Molucca collision.

1855, 4 May, Nanga Rama, south Flores: A series of waves struck the south coast. The first penetrated into a schooner, the second broke the anchor chain and set it on the beach, and the following wave turned it into wreckage.
Likely source: Java Trench, Timor Trough, Savu Thrust or thrust in Savu Sea.

1855, 14 July, Ternate and Tidore Islands; Halmahera: A very violent quake struck at 4 p.m. and lasted 2 minutes, badly damaging all buildings. The earthquake was stronger on Tidore Island and caused 25 houses to collapse, killing 24. 32 more died from rockfalls from Mt Dojado. The quake was felt with equal force at Dodingga, Halmahera. Mount Gamalama on Ternate showed no activity at the time.
Likely source: Halmahera Trench.

1856, 2 March, Sangihe Island (north of Sulawesi): An eruption of Mount Awa, north of Celebes on Sangihe Island, was accompanied by a powerful shock. A tsunami penetrated far inland.

1856, 25 July, Ampenan, Lombok: A quake was felt and a 'heavy surge' noticed on the beach at Ampenan.
Likely source: Fault between Lombok and Bali, West Flores Thrust or Java Trench.

1856, 6–7 August, Gorontolo, Minahassa peninsula, north Sulawesi: Powerful shocks heavily damaged the walls of the fort.
Likely source: Gorontolo Fault or North Sulawesi Subduction Zone.

1857, 13 May, northern East Timor: A violent earthquake threw people to the ground. Part of the walls of the fort collapsed and ground subsidence was observed. A tsunami with four waves struck the bay and reached 3.1 m high. Fissures were created on the beach. The quake was also 'very strong' in Hera, Laclo, Lautem, Lale and Batugade. A mud volcano erupted near Viqueque. At Liquica village, a tsunami flooded almost the entire village. A hill sank on Palau Kambing (north of Dili) killing more than 36 persons. The quake was felt in Kupang (285 km SW of Dili) and Ambon (600 km to NE). Aftershocks were felt through June when the record stops.
Likely source: Wetar Strait Thrust or Wetar Thrust.

1857, 17 October, Kema and Manado, Minahassa peninsula, north Sulawesi; Ternate: At 4:30 a.m., a shock was felt in Kema and Menado followed by a tsunami that washed away multiple huts and trees near Kema. Other shocks occurred at 6 and 9 a.m., and another tsunami accompanied the second shock at Kema. Shocks were still felt in Menado in March and later.
At Ternate, on the other side of the Molucca Sea, a weak shock was also felt at around 4:30 a.m. and the following groundswell (tsunami?) reached a height that was unknown even to the oldest residents and caused devastations.
Likely source: Sangihe megathrust.

1858, 13 November, Minahassa, north Sulawesi; Ternate: 'An extraordinarily strong and long lasting earthquake' in all of Celebes. Effects in Tondano were strongest with 15 huts collapsing and landslides. According to ships in the area, a tsunami appeared along the entire east coast of Celebes to the Banggai Islands, destroying villages and causing other damage.
A powerful quake was also felt in Ternate at 'noon', but no volcanic eruption occurred. It was also strongly felt on Batjan Island (Bacan) to the south. A tsunami also may have appeared.
Likely source: Sangihe megathrust.

1859, 28 May, Sidangola Bay, Ternate; east Sulawesi: At around 8:30 p.m. and midnight shocks were accompanied by waves that threw a government schooner onto the beach at Kema, north Celebes.
In Ternate also, a strong quake was felt and ships began to roll in the roadstead. Two small boats were thrown on the beach. At the Bay of Sidangoli on the west coast of Halmahera the tsunami reached over 10 m.
Likely source: Sangihe megathrust.

1859, 29 July, Minahassa, Kema, Gorontalo, Sulawesi; Banggai Islands; Ternate, Tidore and Makjan Islands: At Minhassa at 1:30 p.m. powerful shocks lasted 5 minutes 'which threatened to knock everything down'. Shocks repeated at 4:30, and through the evening and night. A tsunami stuck Kema and washed away a shed. The quake was also felt in Gorontalo, Mondono and Banggai Island. A tsunami also occurred on Banggai Island.
At 3:15 p.m. on Ternate, a quake struck creating fissures that opened and closed. Directly after, a flood wave came in 1 m above flood level, even though it was at low tide. The earthquake was also observed on Tidore, Makjan and other islands.
Likely source: Sangihe megathrust.

1859, 25 September, Banda Islands: A powerful shock occurred in the evening accompanied by a flood wave 'with great force' that struck the southern coast.
Likely source: Banda Sea Fault like the 1975 event.

1859, 17 December, Belang, Minahassa peninsula, north Sulawesi: A shock was felt accompanied by a tsunami.
Likely source: Sangihe megathrust (aftershock of earlier events)

1859, 25 December, Kema, north Sulawesi: A tsunami reached the roof of a coal shed and partly beat away palisades (protective fence) along the gangway.
Likely source: Sangihe megathrust (perhaps aftershock of earlier events).

1860, 6 October, south Halmahera: A tsunami was observed not far from the south tip of the island but cause no damage. 10–12 more waves came in before 'the sea became again smooth as glass'.
Likely source: Sangihe megathrust (aftershock of earlier events).

1861, 16 February, 6.30 pm, Nias Island and west Sumatra: Thought at the time to be the most widespread quake that ever struck Sumatra, which was also felt on the Malay peninsula and in Java. The earthquake was most strongly manifest in Nias Island, and the area between the Batu Islands to the north and Bengkulu to the south. The shaking lasted for 4 minutes around Nias Island and the immediately adjacent parts of the Sumatran coastline, as noted in Sibolga, Natal and Tapanuli where numerous buildings collapsed. The duration of shaking and damage from shaking was less to the north and south of this region. Shaking lasted for 1 minute in Bengkulu. The shaking in Singapore lasted for about 2 minutes and was in a SW–NE direction. On the west coast of Nias, coral reefs were uplifted that changed the coastline. Within 15 minutes of the earthquake, a 7 m tsunami arrived at Langundi on the south coast of Nias where it destroyed most of the buildings and claimed 50 lives. In Gunungsitoli on the NE coast of Nias, the sea retreated 32 m, then inundated multiple beach villages bringing ships onto the beach destroying nearly everything. On Simuk Island, which is south of Nias, most of the villages were destroyed by the tsunami, which claimed 675 lives. Tsunami waves also struck the west coast of Sumatra where the sea withdrew for some time then flooded shorelines from Singkil to the north to Bengkulu to the south, a distance of around 5 degrees of latitude. Even on the north coast of Java, 1.5 hours after the earthquake, a 1.6 m tsunami was noted in the Tji River, which tossed watercraft around and tore them from their anchors. Several ships also reported a seaquake.

Likely source: Sunda megathrust similar to the M_W 8.6 Nias earthquake in 2005.

1864, 23 May, Cenderawasih Bay, Papua: A strong earthquake struck in Dore territory, which collapsed buildings and generated a 2.5–3 m tsunami. The tsunami washed away the huts and claimed 250 lives.

Likely source: New Guinea Trench or Yapen Fault Zone.

1871, 2 March, Tagulandang Island (north of Sulawesi): An eruption of Ruang volcano caused a landslide that triggered a large tsunami. The wave hit the west beach, penetrating 188 m inland with a run-up height of 26 m in the village of Buhias. The village was entirely destroyed and 277 people perished. Other villages on the west and SW sides of the island also sustained similar damage.

1873, date unknown, Astrolabe Bay, New Guinea: An earthquake caused the collapse of many huts, especially in the mountain villages. Around sunrise, a tsunami struck that destroyed forests and threw debris onland, blocking the mouths of river estuaries. Rivers changed their courses, fissures were observed, and sinkholes and lagoons were formed.

Likely source: New Britain Trench.

1873, 19 August, Tapanuli Province, west Sumatra: Strong earthquake that lasted for 2 minutes in Padang, Sidimpuan and throughout the Tapanuli Province. In Natal, the fort, other buildings and most bridges were destroyed. Landslides were widespread.

Likely source: Sumatran Fault or downgoing plate similar to the 2009 Padang earthquake.

1875, date unknown, Luf Island, New Guinea: A tsunami struck the small islands north of New Guinea. Several of the low islands were inundated, causing 'great sickness and mortality on the Island Lub (or Hermit Island)'.

Likely source: Orphan tsunami (New Caledonia, 28 March 1875, Monterey Bay 18 March 1875) or Manus Trench.

1876, 28 May, Kajeli Bay, Buru Island: Strong shocks damaged buildings on the SW part of the island and caused a 0.3 m tsunami in the Bay of Kajeli.

Likely source: Sorong Fault system.

Reconstructing the earthquakes

Many of the events described in the Wichmann catalogue provide key observations that can be used to locate more accurately possible source regions for the events and the magnitudes of the earthquakes in these regions. The power of this approach is demonstrated in the reconstruction of the 1629 earthquake and tsunami in the southern Maluku region (Liu & Harris 2013). The earthquake was felt over a broad area and lasted for 5 min, which are characteristics of a major tectonic event. The tsunami was observed in the Banda Islands 30 min after shaking stopped. It came from the east. Acknowledging that there is the possibility that tsunamis are generated by earthquake-induced slides, the possibilities of large earthquakes should also be investigated.

If the ground shaking felt in the Banda Islands did, indeed, cause the tsunami that arrived 30 min later, then the distance of the epicentre from the Banda Islands can be reconstructed by modelling wave propagation across the Banda Sea. The bathymetry around the Banda Islands is the determining factor in how fast a tsunami can reach there. Numerical modelling of the wave propagation demonstrates that the only known fault zone of sufficient size to cause a 15 m tsunami that arrives in the Banda Islands from the east 30 min after the earthquake is the Seram Trough (Liu & Harris 2013).

In order to determine the size of the event, Liu & Harris (2013) had to estimate the geometry of the Seram Trough subduction interface that produced it. The strike and dip of the fault was estimated from the existing database of fault-plane solutions, which includes several small earthquakes along the subduction interface over the past 30 years. The size of the earthquake scales with fault-plane dimensions and the amount of slip, which can be constrained by the size of the tsunami. The length and width of the rupture zone and amounts of slip on the fault are scaled to produce a tsunami that matches what was observed.

In the case of the 1629 event, the earthquake would have to be M_w 8.2–8.8, depending on which parameters are favoured (Liu & Harris 2013). A sensitivity analysis of the reconstruction shows that changes in the dip angle of the fault influence tsunami height most, by a factor of 2 between 10° and 45°. Variations in hypocentral depth from 5 to 30 km have very little influence (10%) on tsunami height.

Reconstruction of the 1852 earthquake and tsunami, which was more widely felt than the 2011 M_w 9.0 Sendai earthquake, and the tsunami observed at many sites, points to a megathrust earthquake along the Tanimbar Trough (Fisher & Harris, in press). Many other events in the Wichmann catalogue have enough details to constrain likely source areas and magnitudes with much more accuracy than we speculated for each event.

We are in the process of translating the catalogue into Indonesian so it can be made available to those who are actually in harm's way. In November 2013, we completed and distributed the Indonesian version of earthquakes and tsunamis from 1600 to 2010 in the southern Maluku region.

Communicating the hazard

One of the most difficult, but important, lessons learned from the 2004 Banda Aceh earthquake and tsunami is the vital need for earthquake and tsunami hazard education. Although several papers were published warning of the likelihood of the recurrence of megathrust earthquakes in Sumatra, most of the people in harm's way did not know there was a threat or that they should evacuate low-lying coastal regions after prolonged ground shaking.

Since the 2004 event, a tsunami warning system has been installed for the Indian Ocean region. Four tsunamis have occurred since the warning system was installed, but the warnings are still not reaching the beaches before the waves. One of the problems with having a tsunami warning system is that people in harm's way wait until they receive word from official channels before responding to the ground shaking. Efforts to educate communities at the local level about earthquake and tsunami hazards, and to conduct evacuation drills, is proving much more effective in raising awareness at a fraction of the cost of high-tech solutions.

Key factors in persuading local communities to take the threat of earthquakes and tsunami seriously, and actively prepare for it, is to provide historical accounts of what has happened in these areas in the past. The sections of this paper on Maluku and Nusa Tengarra have already been translated into Indonesian and distributed throughout these regions by the non-profit organization 'In Harm's Way' (inharmswayhelp.org).

Conclusion

It is our hope that the English translation of the Wichmann catalogue, which we provide here, will assist in seismic and tsunami disaster prevention, reversing the mounting losses to nature in the Indonesian region. The catalogue significantly extends the temporal range of earthquake and tsunami events in Indonesia back in time by more than 400 years.

With strain rates as high as 70–110 mm a^{-1}, it is likely that some active faults have reoccurring earthquakes during the time covered by the catalogue. Although, it is more likely that the catalogue has only captured single events along active faults with recurrence intervals longer than the time since the last event. However, these data provide a way in which to predict where many active faults in the region are in their earthquake cycle. With new GPS constraints on loading rates for most of these faults, the amount of elastic strain energy that has accumulated since the last major earthquake can be determined, which provides a way to estimate expected earthquake magnitudes in various regions across Indonesia. With these estimates, it is possible to construct numerical models of tsunamis that may be generated by future earthquake events. However, these records are of limited worth if the information does not result in education and implementation of disaster mitigation strategies for those who are actually in harm's way.

This research was partially funded by NSF EAR 0337221, grants from the BYU College of Physical and Mathematical Sciences, and the BYU Office of Research and Creative Activities. We thank those who helped with the translation of the Wichmann catalogue: Aden Williamsen, Tanner Duncan, Marti Major, Carolyn Dedari, Sarah Reed, Douwe van Hinsbergen and Tobias Weisenberger. Jamie Robinson, Bradley Bishop and Stina Nyhus helped with compiling active fault data. Universitas Pembangunan Nasional (UPN) Yogyakarta sponsored our research in Indonesia. We thank Carolus Prasetyadi and Arif Rianto, and others at UPN for field support and cultural insights. Others members of the research team who helped immensely are Nova Roosmawati, Hendro Nugroho, Hanif Ibadurrahman and Stenley Loupatty. We also thank Eko Yulianto of LIPI and Juilian Fretha of the Badan Nasional Penanggulangan Bencana (BNPB) for invaluable support in the field.

References

BADER, A.G. & PUBELLIER, M. 2000. Forearc deformation and tectonic significance of the ultramafic Molucca central ridge, Talaud islands (Indonesia). *The Island Arc*, **9**, 653–663.

BAEDA, A.Y. 2011. Seismic and tsunami hazard potential in Sulawesi, Indonesia. *Journal of International Development and Cooperation*, **17**, 17–30.

Barber, A.J., Tjokrosapoetro, S. & Charlton, T.R. 1986. Mud volcanoes, shale diapirs, wrench faults and mélanges in accretionary complexes, Eastern Indonesia. *American Association of Petroleum Geologists Bulletin*, **70**, 1729–1741.

Bergman, S.C., Coffield, D.Q., Talbot, J.P. & Garrard, R.A. 1996. Tertiary tectonic and magmatic evolution of western Sulawesi and the Makassar Strait, Indonesia; evidence for a Miocene continent–continent collision. *In*: Hall, R. & Blundell, D. (eds) *Tectonic Evolution of Southeast Asia*. Geological Society, London, Special Publications, **106**, 391–429, https://doi.org/10.1144/GSL.SP.1996.106.01.25

Berry, R.F. & McDougall, I. 1986. Interpretation of $^{40}Ar/^{39}Ar$ and K/Ar dating evidence from the Aileu Formation, East Timor, Indonesia. *Chemical Geology and Isotope Geoscience*, **59**, 43–58.

Bock, Y., Prawirodirdjo, L. *et al.* 2003. Crustal motion in Indonesia from Global Positioning System measurements. *Journal of Geophysical Research*, **108**, 2367, https://doi.org/10.1029/2001JB000324

Breen, N.A., Silver, E.A. & Roof, S. 1989. The Wetar backthrust belt, eastern Indonesia: the effects of accretion against an irregularly shaped arc. *Tectonics*, **8**, 85–98.

Brune, S., Ladage, S., Babeyko, A.Y., Mueller, C., Kopp, H. & Sobolev, S. 2010. Submarine slope failures at the eastern Sunda Arc; bathymetry analysis and tsunami modeling. *Deutsche Geophysikalische Gesellschaft*, **2**, 15–18.

Budiono, K., Astjario, P., Kusnida, D. & Lubis, S. 1995. Coastal and marine geological survey after the earthquake of 12 December 1992 at Maumere Bay, Flores. *Bulletin of the Marine Geological Institute of Indonesia*, **10**, (1), 13–31.

Carter, D.J., Audley-Charles, M.G. & Barber, A.J. 1976. Stratigraphical analysis of island arc–continental margin collision in eastern Indonesia. *Journal of the Geological Society, London*, **132**, 179–198, https://doi.org/10.1144/gsjgs.132.2.0179

Cox, N.L. 2009. *Variable uplift from Quaternary folding along the northern coast of East Timor, based on U-series age determinations of coral terraces*. Unpublished MSc, Brigham Young University.

Duffy, B., Quigley, M., Harris, R.A. & Ring, U. 2013. Arc-parallel extrusion of the Timor sector of the Banda arc-continent collision. *Tectonics*, **32**, 641–660, https://doi.org/10.1002/tect.20048

Fisher, T.M. & Harris, R.A. In press. Reconstruction of the 1852 Banda Arc megathrust earthquake and tsunami. *Natural Hazards*, https://doi.org/10.1007/s11069-016-2345-6

Fritz, H., Widjo, K. *et al.* 2007. Extreme runup from the 17 July 2006 Java tsunami. *Geophysical Research Letters*, **34**, L12602, https://doi.org/10.1029/2007GL029404

Garrard, R., Supandjono, J. & Surono. 1988. The geology of the Banggai-Sula microcontinent, eastern Indonesia. *Proceedings of the Indonesian Petroleum Association 17th Annual Convention*, 23–52.

Genrich, J.F., Beck, Y., McCaffrey, R., Calais, E., Stevens, C.W. & Subarya, C. 1996. Accretion of the southern Banda arc to the Australian plate margin determined by Global Positioning System measurements. *Tectonics*, **15**, 288–295.

Ghose, R. & Oike, K. 1988. Characteristics of seismicity distribution along the Sunda Arc: some new observations. *Bulletin of Disaster Prevention Research Institute*, **38**, 29–48.

Guntoro, A. 1999. The formation of the Makassar Strait and the separation between SE Kalimantan and SW Sulawesi. *Journal of Asian Earth Sciences*, **17**, 79–98.

Hall, R. 2011. Australia–SE Asia collision: plate tectonics and crustal flow. *In*: Hall, R., Cottam, M.A. & Wilson, M.E.J. (eds) *The SE Asian Gateway: History and Tectonics of the Australia–Asia Collision*. Geological Society, London, Special Publications, **355**, 75–109, https://doi.org/10.1144/SP355.5

Hall, R., Cloke, I.R., Nur'aini, S., Puspita, S.D.&, Calvert, S.J. 2009. The North Makassar Straits; what lies beneath? *Petroleum Geoscience*, **15**, 47–158, https://doi.org/10.1144/1354-079309-829

Hamilton, W. 1979. *Tectonics of the Indonesian Region*. United States Geological Survey, Professional Papers, **1078**.

Hamzah, L., Puspito, N.T. & Imamura, F. 2000. Tsunami catalogue and zones in Indonesia. *Journal of Natural Disaster Science*, **2**, 25–43.

Harris, R.A. 1991. Temporal distribution of strain in the active Banda orogen: a reconciliation of rival hypotheses. *Journal of Asian Earth Sciences*, **6**, 373–386.

Harris, R.A. 2006. Rise and fall of the eastern Great Indonesian Arc recorded by the assembly, dispersion and accretion of the Banda Terrane, Timor. *Gondwana Research*, **10**, 207–231.

Harris, R.A. 2011. The nature of the Banda Arc–continent collision in the Timor region. *In*: D.Brown, & P.D. Ryan (eds) *Arc–Continent Collision*. Frontiers in Earth Sciences. Springer, Berlin, 163–211, https://doi.org/10.1007/978-3-540-88558-0_7

Harris, R.A. & Prasetyadi, C. 2002. Who's Next? Assessing vulnerability to geophysical hazards in densely populated region of Indonesia. *Bridges*, **2002**, 14–17.

Harris, R.A., Donaldson, K. & Prasetyadi, C. 1997. Geophysical disaster mitigation in Indonesia using GIS. *Bulletin Teknologi Mineral*, **5**, 2–10.

Harris, R.A., Vorkink, M.W., Prasetyadi, C., Roosmawati, N., Zobell, E. & Apthorpe, M. 2009. Transition from subduction to arc–continent collision: geological and neotectonic evolution of Savu, Indonesia. *Geosphere*, **5**, 152–171, https://doi.org/10.1130/GES00209.1

Hanna, W.A. 1978. *Indonesian Banda: Colonialism and its Aftermath in the Nutmeg Islands, Philadelphia*. Institute for the Study of Human Issues, Philadelphia, PA.

Heuret, A., Lallemand, S., Funiciello, F., Piromallo, C. & Faccenna, C. 2011. Physical characteristics of subduction interface type seismogenic zones revisited. *Geochemistry, Geophysics, Geosystems*, **12**, Q01004, https://doi.org/10.1029/2010GC003230

Irsyam, M., Dangkua, D. *et al.* 2008. Proposed seismic hazard maps of Sumatra and Java islands and microzonation study of Jakarta city, Indonesia. *Earth System Science*, **117**, 865–878.

JAGLA, E.A. & KOLTON, A. 2010. A mechanism for spatial and temporal earthquake clustering. *Journal of Geophysical Research*, **115**, B05312.

KARIG, D.E., BARBER, A.J., CHARLTON, T.R., KLEMPERER, S. & HUSSONG, D.M. 1987. Nature and distribution of deformation across the Banda Arc Australian collision zone at Timor. *Geological Society of America Bulletin*, **98**, 18–32.

KATILI, J.A. 1978. Past and present geotectonic position of Sulawesi, Indonesia. *Tectonophysics*, **45**, 289–322.

KATO, T., ITO, T., ABIDIN, H.Z. & AGUSTAN, 2007. Preliminary report on crustal deformation surveys and tsunami measurements caused by the July 17, 2006 South off Java Island Earthquake and Tsunami, Indonesia. *Earth, Planets and Space*, **59**, 1055–1059.

KERR, R. 2006. Stealth tsunami surprises Indonesian coastal residents. *Science*, **313**, 742–743.

KERTAPATI, E. & KOESOEMADINATA, R.P. 1983. Aftershock studies of the February 10, 1982, Sukabumi earthquake, West Java, Indonesia. *Bulletin of the International Institute of Seismology and Earthquake Engineering*, **20**, 91–101.

KOPP, H., FLUEH, E.R., KLAESCHEN, D., BIALAS, J. & REICHERT, C. 2001. Crustal structure of the central Sunda margin at the onset of oblique subduction. *Geophysical Journal International*, **147**, 449–474.

KREEMER, C., HOLT, W.E., GOES, S. & GOVERS, R. 2000. Active deformation in eastern Indonesia and the Philippines from GPS and seismicity data. *Journal of Geophysical Research*, **105**, 663–680.

LAPE, P.V. 2000. Political dynamics and religious change in the late pre-Colonial Banda islands, eastern Indonesia. *World Archaeology*, **32**, 138–155.

LINTHOUT, K., HELMERS, H., WIJBRANS, J.R. & VAN WEES, J.D.A.M. 1996. $^{40}Ar/^{39}Ar$ constraints on obduction of the Seram ultramafic complex: consequences for the evolution of the southern Banda Sea. *In*: HALL, R. & BLUNDELL, D. (eds) *Tectonic Evolution of Southeast Asia*. Geological Society, London, Special Publications, **106**, 455–464, https://doi.org/10.1144/GSL.SP.1996.106.01.28

LIU, Y.C. & HARRIS, R.A. 2013. Discovery of possible mega-thrust earthquake along the Seram Trough from numerical modeling of 1629 tsunami in the eastern Indonesian region. *Natural Hazards*, **72**, 1311–1328, https://doi.org/10.1007/s11069-013-0597-y

MAJOR, J., HARRIS, R.A. *ET AL*. 2013. Pleistocene hinterland evolution of the active Banda Arc: surface uplift and neotectonic deformation recorded by coral terraces at Kisar, Indonesia. *Journal of Asian Earth Sciences*, **73**, 149–161.

MCCAFFREY, R. 1988. Active tectonics of the eastern Sunda and Banda arcs. *Journal of Geophysical Research*, **93**, 15163–15182.

MCCAFFREY, R. 1989. Seismological constraints and speculations on Banda arc tectonics. *Netherlands Journal of Sea Research*, **24**, 141–152, https://doi.org/10.1016/0077-7579(89)90145-2

MCCAFFREY, R. 2007. The next great earthquake. *Science*, **315**, 1675–1676.

MCCAFFREY, R. & NABELEK, J. 1984. The geometry of back arc thrusting along the Eastern Sunda Arc, Indonesia – constraints from earthquake and gravity-data. *Journal of Geophysical Research*, **89**, 6171–6179.

MELTZNER, A.J., SIEH, K. *ET AL*. 2010. Coral evidence for earthquake recurrence and an A.D. 1390–1455 cluster at the south end of the 2004 Aceh–Andaman rupture. *Journal of Geophysical Research*, **115**, B10402, https://doi.org/10.1029/2010JB007499

MERRITTS, D., EBY, R., HARRIS, R., EDWARDS, R.L. & CHENG, H. 1998. Variable rates of Late Quaternary surface uplift along the Banda Arc–Australian plate collision zone, eastern Indonesia, *In*: STEWART, I.S. & VITA-FINZI, C. (eds) *Coastal Tectonics*. Geological Society, London, Special Publications, **146**, 213–224, https://doi.org/10.1144/GSL.SP.1999.146.01.12

MILTON, G. 1999. *Nathaniel's Nutmeg: How One Man's Courage Changed the Course of History*. Hodder & Stoughton, London.

MUSSON, R.M.W. & JIMENEZ, M.J. 2008. *Procedures for Macroseismic Estimation of Earthquake Parameters. Deliverable D3*. Network of Research Infrastructure for European Seismology (NERIES), Sixth Framework Programme, EC Project number 026130.

NATAWIDJAJA, D.H., SIEH, K., WARD, S.N., CHENG, H., EDWARDS, R.L., GALETZKA, J. & SUWARGADI, B.W. 2004. Paleogeodetic records of seismic and aseismic subduction from central Sumatran microatolls, Indonesia. *Journal of Geophysical Research*, **109**, B04306, https://doi.org/10.1029/2003JB002398

NATAWIDJAJA, D.H., SIEH, K. *ET AL*. 2006. Source parameters of the great Sumatran mega-thrust earthquakes of 1797 and 1833 inferred from coral microatolls. *Journal of Geophysical Research*, **111**, B06403, https://doi.org/10.1029/2005JB004025

NEWCOMB, K.R. & MCCANN, W.R. 1987. Seismic history and seismotectonics of the Sunda Arc. *Journal of Geophysical Research*, **92B**, 421–439.

NUGROHO, H., HARRIS, R.A., AMIN, W.L. & BILAL, M. 2009. Active plate boundary reorganization in the Banda arc–continent collision: insights from new GPS measurements. *Tectonophysics*, **479**, 52–65, https://doi.org/10.1016/j.tecto.2009.01.026

OKAL, E.A. & REYMOND, D. 2003. The mechanism of great Banda Sea earthquake of 1 February 1938: applying the method of preliminary determination of focal mechanism to a historical event. *Earth and Planetary Science Letters*, **216**, 1–15.

OPPENHEIMER, C. 2011. *Eruptions that Shook the World*. Oxford University Press, New York.

PELINOVSKY, E., YULIADI, D., PRASETYA, G. & HIDAYAT, R. 1997. The 1996 Sulawesi Tsuanmi. *Natural Hazards*, **16**, 29–38.

POWNALL, J., HALL, R., ARMSTRONG, R.A. & FORSTER, M.A. 2014. Earth's youngest known ultrahigh-temperature granulites discovered on Seram, eastern Indonesia. *Geology*, **42**, 279–282.

PRASETYA, G.S., DE LANGE, W.P. & HEALY, T.R. 2001. The Makassar Strait tsunamigenic region, Indonesia. *Natural Hazards*, **24**, 295–307.

RANGIN, C., LE PICHON, X. *ET AL*. 1999. Plate convergence measured by GPS across the Sundaland/Philippine Sea Plate deformed boundary: the Philippines and eastern Indonesia. *Geophysical Journal International*, **139**, 296–316.

RASTOGI, B.K. & JAISWAL, R.K. 2006. A catalogue of tsunamis in the Indian Ocean. *Science of Tsunami Hazards*, **25**, 128–143.

ROOSMAWATI, N. & HARRIS, R.A. 2009. Surface uplift history of the incipient Banda arc–continent collision: geology and synorogenic foraminifera of Rote and Savu Islands, Indonesia. *Tectonophysics*, **479**, 95–110, https://doi.org/10.1016/j.tecto.2009.04.009

ROSIDI, H.M.D., TJOKOSAPOETRO, S., GAFOER, S. & SUWITODIRDJO, K. 1979. *Geologic Map of the Kupang–Atambua Quadrangles, Timor, 1: 250,000*. Geological Research Development Centre, Bandung, Indonesia.

RUFF, L. & KANAMORI, H. 1980. Seismicity and the subduction process. *Physics of the Earth and Planetary Interiors*, **23**, 240–252.

SIEH, K. & NATAWIDJAJA, D. 2000. Neotectonics of the Sumatran fault, Indonesia. *Journal of Geophysical Research*, **105**, 28,295–28,326, https://doi.org/10.1029/2000JB900120

SIEH, K., NATAWIDJAJA, D.H., MELTZNER, A.J. & SHEN, C.-C. 2008. Earthquake supercycles inferred from sea-level changes recorded in the corals of west Sumatra, *Science*, **322**, 1674–1678.

SILVER, E.A. & MOORE, J.C. 1978. The Molucca Sea collision zone, Indonesia. *Journal of Geophysical Research*, **83**, 1681–1691.

SILVER, E.A., MCCAFFREY, R., JOYODIWIRYO, Y. & STEVENS, S. 1983*a*. Ophiolite emplacement and collision between the Sula Platform and the Sulawesi island arc, Indonesia. *Journal of Geophysical Research*, **88**, 9419–9435.

SILVER, E.A., REED, R., MCCAFFREY, R. & JOYODIWIRYO, Y. 1983*b*. Back arc thrusting in the eastern Sunda arc, Indonesia: a consequence of arc continent collision. *Journal of Geophysical Research*, **88**, 7429–7448, https://doi.org/10.1029/JB088iB09p07429

SIMANDJUNTAK, T.O. 1993. Neogene tectonics and orogenesis of Indonesia. *In*: TEH, G.H. (ed.) *Proceedings of the Symposium on Tectonic Framework and Energy Resources of the Western Margin of the Pacific Basin*. Geological Society of Malaysia Bulletin, Geological Society of Malaysia, Kuala Lumpar, **33**, 43–64.

SMITH, R.B. & SILVER, E.A. 1991. Geology of a Miocene collision complex, Buton, eastern Indonesia. *Geological Society of America Bulletin*, **103**, 660–678, https://doi.org/10.1130/0016-7606(1991)103<0660:goamcc>2.3.co;2

SNYDER, D.B., PRASETYO, H., BLUNDELL, D.J., PIGRAM, C.J., BARBER, A.J., RICHARDSON, A. & TJOKROSAPOETRO, S. 1996. A dual doubly vergent orogen in the Banda arc continent collision zone as observed on deep seismic reflection profiles. *Tectonics*, **15**, 34–53.

SOCQUET, A., SIMONS, W. ET AL. 2006. Microblock rotations and fault coupling in SE Asia triple junction (Sulawesi, Indonesia) from GPS and earthquake slip vector data. *Journal of Geophysical Research*, **111**, B08409, https://doi.org/10.1029/2005JB003963

SOLOVIEV, S.L. & GO, Ch.N. 1974. *A Catalogue of Tsunamis on the Western Shore of the Pacific Ocean (173-1968)*. Nauka Publishing House, Moscow (Canada Institute for Scientific and Technical Information. 1984. Canadian Translation of Fish and Aquatic Sciences, 5077).

STEIN, S. & OKAL, E.A. 2007. Ultra-long period seismic study of the December 2004 Indian Ocean earthquake and implications for regional tectonics and the subduction process. *Bulletin of the Seismological Society of America*, **97**, S279–S295.

STEIN, S., GELLER, R.J. & LIU, M. 2012. Why earthquake hazard maps often fail and what to do about it. *Tectonophysics*, **562–563**, 1–25.

TJIA, H.D. 1977. Tectonic depressions along the transcurrent Sumatra fault zone. *Geology of Indonesia*, **4**, 13–27.

TJIA, H.D. 1983. Earthquake stress directions in the Indonesian Archipelago. *In*: HILDE, T.W.C. & UYEDA, S., (eds) *Geodynamics of the Western Pacific–Indonesian Region*. American Geophysical Union, Geodynamics Series, **11**, 413–422.

TSUJI, T., YAMAMOTO, K. ET AL. 2009. Earthquake fault of the 26 May 2006 Yogyakarta earthquake observed by SAR interferometry. *Earth, Planets and Space*, **61**, 29–32.

VAN BEMMELEN, R.W. 1949. *The geology of Indonesia; vol. IA, General geology of Indonesia and adjacent archipelagos*. Martinus Nijnhoff, Government Printing Office sole Agent, The Hague.

VAN LEEUWEN, T.M. 1981. The geology of southwest Sulawesi with special reference to the Biru area. *In*: BARBER, A. & WIRYOSUJONO, S. (eds) *The Geology and Tectonics of Eastern Indonesia*. Geological Research and Development Centre, Special Publications, **2**, 277–304.

WATKINSON, I., HALL, R. & FERDIAN, F. 2011. Tectonic re-interpretation of the Banggai-Sula–Molucca Sea margin, Indonesia. *In*: HALL, R., COTTAM, M.A. & WILSON, M.E.J. (eds) *The SE Asian Gateway: History and Tectonics of the Australia–Asia Collision*. Geological Society, London, Special Publications, **355**, 197–218 https://doi.org/10.1144/SP355.10

WELC, J.L. & LAY, T. 1987. The source rupture process of the Great Banda Sea earthquake of November 4, 1963. *Physics of the Earth and Planetary Interiors*, **45**, 242–254.

WICHMANN, C.E.A. 1918. *Die Erdbeben des indischen Archipels bis zum Jahre 1857*. Muller, Amsterdam.

WICHMANN, C.E.A. 1922. *Die Erdbeben des Indischen Archipels von 1858 bis 1877*. Koninklijke Akademie van Wetenschappen, Amsterdam.

WOOD, G.D. 2014. *Tambora*. University Press, Princeton, NJ.

Assessing tsunami hazard using heterogeneous slip models in the Mentawai Islands, Indonesia

JONATHAN D. GRIFFIN[1,2]*, IGNATIUS R. PRANANTYO[1,3], WIDJO KONGKO[4], AFIF HAUNAN[5], RAHAYU ROBIANA[5], VICTORIA MILLER[2], GARETH DAVIES[2], NICK HORSPOOL[2], IMUN MAEMUNAH[5], WISNU B. WIDJAJA[6], DANNY H. NATAWIDJAJA[7] & HAMZAH LATIEF[8]

[1]*Australia–Indonesia Facility for Disaster Reduction, Menara Thamrin Suite 1505, Jalan M.H. Thamrin Kav. 3, Jakarta 10250, Indonesia*

[2]*Geoscience Australia, Cnr Jerrabomberra Avenue and Hindmarsh Drive, Symonston, ACT 2609, Australia*

[3]*Present address: Research School of Earth Sciences, Australian National University, Canberra, ACT 0200, Australia*

[4]*Coastal Dynamics Research Centre, Agency for the Assessment and Application of Technology, Jalan Grafika 2, Sekip, Yogyakarta 55281, Indonesia*

[5]*Geological Agency of Indonesia, Jalan Diponegoro No. 57, Bandung 40122, Indonesia*

[6]*Indonesian National Disaster Management Agency, Jalan Ir. H. Juanda No. 36, Jakarta 10120, Indonesia*

[7]*Indonesian Institute of Science, Kompleks LIPI, Jalan Sangkuriang, Bandung 40135, Indonesia*

[8]*Bandung Institute of Technology, Jalan Ganesha 10, Bandung 40132, Indonesia*

**Correspondence: jonathan.griffin@ga.gov.au*

Abstract: Tsunami hazard maps are generated for the coastline of the Mentawai Islands, West Sumatra, Indonesia, to support evacuation and disaster response planning. A random heterogeneous slip generator is used to forward model a suite of earthquake rupture scenarios on the Mentawai Segment of the Sunda Subduction Zone. Up to 1000 rupture models that fit constraints provided by coral and geodetic records of coseismic vertical deformation from major earthquakes in 1797, 1833 and 2007 are used to model inundation and to define a maximum inundation zone that envelopes all of these scenarios. Comparison with single-scenario hazard assessments developed by experts and agreed through scientific consensus shows that there is value in modelling a suite of scenarios in order to obtain a more robust and conservative estimate of potential inundated areas. Although both the model presented here and the single-scenario models are based on assumptions about the characteristics of future events using knowledge of past events, by sampling a range of plausible outcomes we gain a more robust estimate of which areas may be inundated during a tsunami within the bounds of the assumptions applied.

Purpose of the hazard assessment

To plan for tsunami evacuation at the community level, information is needed about which areas of the coast may be inundated during a tsunami event. Based on this information, communities can identify safe evacuation areas outside the inundation zone and the most efficient routes to reach these safe areas. For a government considering life safety as the first priority in managing a disaster, it is reasonable to plan evacuation areas based on a deterministic assessment of 'maximum credible' or 'worst-case' scenarios, to the extent that scientific knowledge, conservatism and the assumptions that we make can define these scenarios. In contrast, disaster management strategies targeting other losses (e.g. economic losses) accept a different level of risk and are designed for a certain level of risk tolerance based on probabilistic hazard assessments.

From: Cummins, P. R. & Meilano, I. (eds) 2017. *Geohazards in Indonesia: Earth Science for Disaster Risk Reduction*. Geological Society, London, Special Publications, **441**, 47–70.
First published online June 2, 2016, https://doi.org/10.1144/SP441.3

The tsunami hazard assessment described in this paper was undertaken under direction from the Indonesian National Disaster Management Agency (Badan Nasional Penanggulangan Bencana (BNPB)) to support evacuation and response planning. A key requirement articulated by BNPB was that it should determine as close to a worst-case scenario as possible. This requirement was based on lessons from the 2011 Tohoku earthquake and tsunami, recognized to be a much larger event than had been planned for by authorities in Japan (Stein *et al.* 2012), leading to some evacuation shelters failing to save lives (Suppasri *et al.* 2013). In particular, the tsunami hazard assessment described here aims to address two key disaster management objectives:

(1) The development of tsunami hazard maps to support evacuation planning at the local level. Detailed tsunami hazard maps were provided to the Disaster Management Agency for the Mentawai Islands District to be used for planning evacuation routes and shelter sites.

(2) The development of impact scenarios for contingency planning and response simulation. Inundation results were combined with exposure data and vulnerability functions to generate estimates of potential impacts using the InaSAFE impact assessment software (inasafe.org). The impact scenario was used to underpin the Mentawai Megathrust international disaster response simulations held in 2013–14 involving member nations of the Association of SE Asian Nations (ASEAN) and the East Asia Summit. In addition, these impact assessments detailed revision of West Sumatra's provincial disaster contingency plan for tsunami.

The fundamental requirement in producing tsunami inundation information considering these objectives was to delineate those areas that have a potential to be inundated by tsunami and those that do not. Information regarding the probability of different magnitude tsunami events was not developed, nor were detailed, accurate estimates of flow characteristics (as might inform engineering design).

Earthquake and tsunami potential of the Mentawai Segment of the Sunda Subduction Zone

The potential for the Mentawai Segment of the Sunda Subduction Zone (Fig. 1) to generate a large earthquake and subsequent tsunami threat to the coastline of western Sumatra is well established by extensive palaeogeodetic, geodetic and numerical modelling studies (Borrero *et al.* 2006; Natawidjaja *et al.* 2006; McCloskey *et al.* 2008; Sieh *et al.* 2008; Schlurmann *et al.* 2010). Palaeogeodetic records of uplift, and subsequent death, of coral microatolls record coseismic deformation of the Mentawai Islands above the megathrust. Upwards growth of coral microatolls records interseimic subsidence as a result of the locking of the seismogenic zone (Natawidjaja *et al.* 2006). Analysis of the timing and magnitude of uplift events has demonstrated that major megathrust earthquakes occur on this segment of the Sunda Subduction Zone approximately every 200 years, on average (Sieh *et al.* 2008), including a pair of M_w 8.5–8.9 events occurring in 1797 and 1833 that both generated tsunamis that inundated the city of Padang, West Sumatra (Natawidjaja *et al.* 2006). On 12 September 2007, a pair of earthquakes (M_w 8.4 and 7.9) ruptured part, but not all, of the patch thought to have ruptured in 1833 (Konca *et al.* 2008), with the conclusion being that there is a high likelihood of another major earthquake (M_w *c.* 8.8) occurring on the same patch as the 1797 event, extending to the unruptured portion of the 1833 patch, within the next few decades (Sieh 2006; Konca *et al.* 2008; Sieh *et al.* 2008). The subsequent 2009 M_w 7.9 intraslab Padang earthquake, and the 2010 M_w 7.8 near-trench Mentawai earthquake and tsunami both caused hundreds of fatalities, yet neither has been the expected great megathrust rupture that will release the slip deficit on the Mentawai Segment (McCloskey *et al.* 2010; Hill *et al.* 2012).

A great rupture occurring on the Mentawai Segment would complete a series of great megathrust earthquakes along the Sunda Subduction Zone from the Andaman Islands to offshore Bengkulu since the 2004 Sumatra–Andaman earthquake (Fig. 1). Very little is known of the rupture history of the next segment to the east, offshore of Lampung and the Sunda Strait. However, it seems prudent to assume a similar hazard here in the absence of strong evidence to the contrary. In fact, the lack of any protection from offshore islands means that this section of the Sumatra's west coast can expect even larger tsunami than the mainland coastline of West Sumatra Province for offshore earthquakes of the same magnitude (Horspool *et al.* 2014).

The greatest tsunami risk posed by the Mentawai Segment of the Sunda Subduction Zone is to the densely populated city of Padang, the capital of West Sumatra Province. The city occupies a wide coastal plain with several hundred thousand people living just a few metres above sea level. Furthermore, high population densities and limited routes landwards, due to the number of river channels bounding the city, mean that at present evacuation of only a small portion of the population to safe areas is possible within the 20–30 min that it is expected to take a tsunami to arrive following an earthquake (Schlurmann *et al.* 2010; Di Mauro

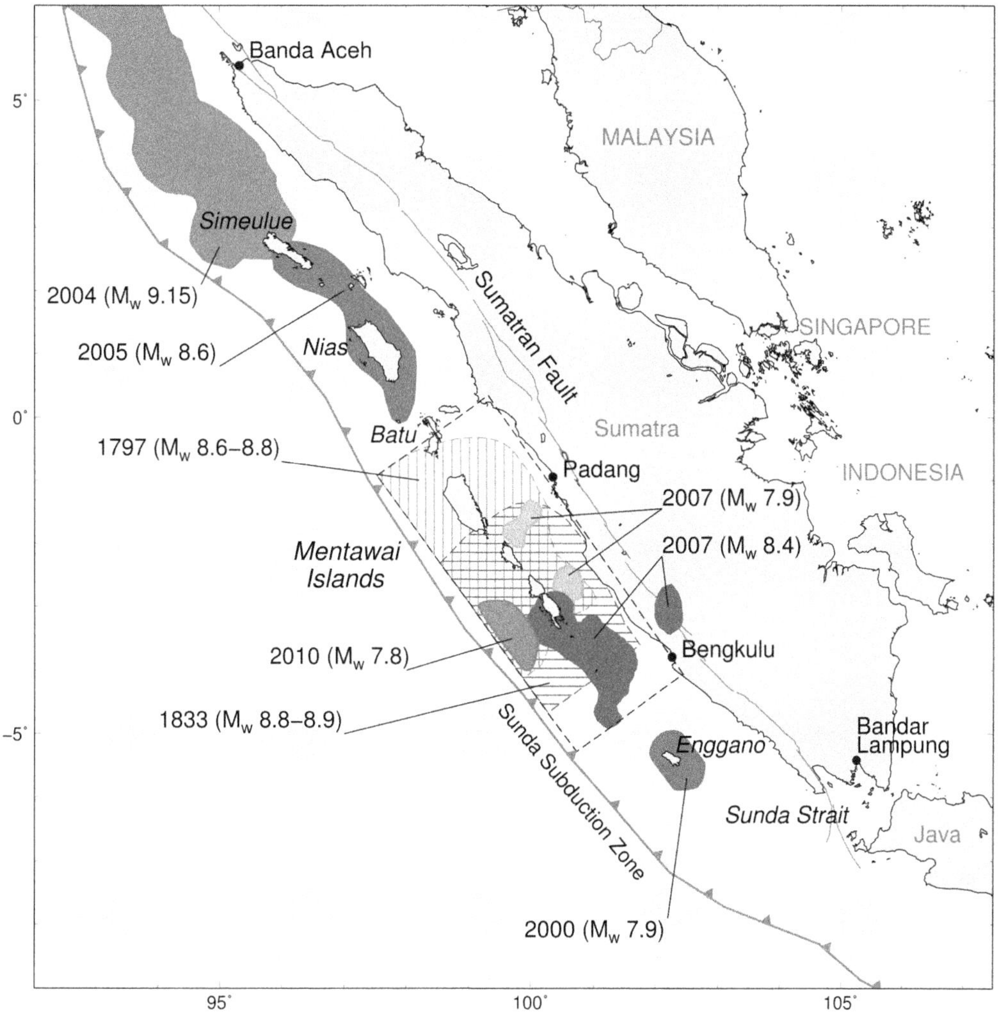

Fig. 1. Regional setting showing the location of selected recent and historical ruptures of the Sunda Subduction Zone interface. Rupture models for recent events are redrawn from Briggs *et al.* (2006), Chlieh *et al.* (2007), Konca *et al.* (2008) and Hill *et al.* (2012). Rupture areas for the 1797 and 1833 events are taken from models presented by Philibosian *et al.* (2014) that assume that slip extends to the trench. The rectangle with a dashed black outline shows the approximate extent of the Mentawai Segment and the extent of the fault model used in this study.

et al. 2013). The Government of Indonesia is attempting to address this problem through development of vertical evacuation shelters and preparedness simulations as part of its Master Plan for Tsunami Risk Reduction (BNPB 2012).

Tsunami hazard assessment approach

Although tsunami inundation has been modelled using very high-resolution and highly accurate elevation data (Anko *et al.* 2008) for the city of Padang (Schlurmann *et al.* 2010) to support tsunami preparedness activities (Taubenböck *et al.* 2009), detailed tsunami inundation information for other regions of West Sumatra had not been produced until this work.

In 2012, the Indonesian Geospatial Information Agency (Badan Informasi Geospasial (BIG)) acquired an airborne interferometric synthetic aperture radar (IFSAR) elevation dataset with 5 m resolution and 3 m vertical accuracy for the majority of the island of Sumatra. A comparison of this data type with higher-resolution Light Detection And Ranging (LiDAR) and High-Resolution Stereo Camera (HRSC) data has determined that airborne

IFSAR is sufficiently accurate for use in modelling the extent of tsunami inundation for the purposes of defining evacuation zones, even if not sufficient for more detailed engineering purposes, using a model mesh resolution of 25 m (Griffin *et al.* 2015). The availability of IFSAR elevation data for the coastline of West Sumatra Province, including the Mentawai Islands, provides an opportunity to develop tsunami inundation maps that extend beyond the city of Padang.

Previous modelling of tsunami inundation for Padang has used a limited number of earthquake source scenarios (Schlurmann *et al.* 2010) based on best-fits to coral and GPS deformation data (Chlieh *et al.* 2008), with additional slip added to best-fit earthquake rupture models to conservatively account for uncertainty. However, the coral and GPS data provide limited constraints on the source away from the Mentawai Islands, particularly on the trenchwards side (Hill *et al.* 2012). Furthermore, heterogeneity of earthquake-slip distributions can lead to a significantly different tsunami height for different areas, particularly in the near field (Geist 2002; McCloskey *et al.* 2007; Løvholt *et al.* 2012). This is particularly so for the Mentawai Islands, sitting directly above the main seismogenic zone of the subduction interface. McCloskey *et al.* (2007) demonstrated that there is, in fact, a large range of possible slip distributions that could produce the deformation observed in the coral data, and showed how this translates into local variability of the coastal tsunami height, although these authors do not consider inundation.

To address the uncertainty associated with source heterogeneity, the hazard maps developed here are constructed from multiple earthquake scenarios, with the tsunami inundation area defined by an envelope covering all areas that are inundated by at least one scenario. We make the assumption that the amount of surface deformation that will occur in a future earthquake will be similar to the cumulative surface deformation of the great earthquakes of 1797 and 1833 (Philibosian *et al.* 2014), minus coseismic deformation from recent earthquakes in September 2007 (Konca *et al.* 2008). Random distributions of slip are generated using the Gallovič & Brokešová (2004) slip generator and coseismic deformation calculated to find source models that fit the combined coseismic vertical deformation data. Other random slip generators have been proposed (e.g. Geist 2002; McCloskey *et al.* 2008; Løvholt *et al.* 2012; Lavallée *et al.* 2011). No assessment is undertaken here with regard to which method best reproduces the characteristics of megathrust earthquakes, and further research is needed to resolve this question.

For Sipora Island, a subset of 1000 slip distributions from 5000 models that fit the deformation data and have the greatest vertical deformation is used to drive tsunami inundation models. Based on the analysis of these results, it was determined that a smaller subset of 200 slip distributions with greatest vertical deformation is sufficient for capturing maximum tsunami depths at more than 99.5% of locations, and therefore this subset of 200 slip distributions is used to model the tsunami hazard for Siberut and the Pagai Islands.

The approach presented here captures the potential for large slip to occur near the trench in regions of low shear strength that can contribute significantly to the height of tsunami owing to the proximity to the surface of the crust and the depth of water at the trench. We do not account for the longer rupture duration for so-called 'tsunami earthquakes' (Kanamori 1972; Satake 1994) nor the effect of horizontal displacement (Tanioka & Satake 1996) or the dynamics of surface rupture (Rudnicki & Wu 1995). Concurrently, we also consider events that cause large amounts of deformation east of the Mentawai Islands that may pose a greater threat to the east coast of the Mentawai Islands and the west coast of West Sumatra. In addition, we also model tsunami from the 'Padang Consensus Model' (Schlurmann *et al.* 2010), which is a M_w 8.9 event derived from the analysis of Chlieh *et al.* (2008), and a modified version of this model with increased slip near the trench.

This study presents the first detailed assessment of an onshore tsunami hazard for the Mentawai Islands, and the first tsunami hazard assessment that utilizes a suite of heterogeneous earthquake rupture scenarios to define an inundation area. Modelling tsunami inundation from multiple source models developed by multiple authors increases confidence that the resulting hazard maps have captured a robust inundation zone, even if our assumptions and knowledge limit us from finding the absolute 'worst case'.

The coral data

Philibosian *et al.* (2014) reanalysed coral data from the Mentawai Islands, recording palaeoseismic uplift during the 1797 and 1833 events, and extending the analysis of Natawidjaja *et al.* (2006). The cumulative surface deformation of the great earthquakes of 1797 and 1833 (Philibosian *et al.* (2014), minus coseismic deformation from recent earthquakes on the same patch in September 2007, (Konca *et al.* 2008) is shown in Table 1 and Figure 2. Uncertainties in coral uplift measurements are added together for the 1797 and 1833 events. GPS and coral uplifts and uncertainties from the 2007 event (Konca *et al.* 2008) are interpolated to the coral sites of Philibosian *et al.* (2014) using the kriging method of Delhomme (1978) for spatial

Table 1. *Net remaining coseismic deformation and uncertainties from the 1797, 1833 and 2007 earthquakes (in cm)*

Latitude (°N)	Longitude (°E)	1797 coseismic uplift (cm)	1798 σ (cm)	1833 coseismic uplift (cm)	1833 σ (cm)	2007 Coseismic uplift (cm)	2007 σ (cm)	Net remaining uplift (cm)	Combined σ (cm)
−0.539	98.464	20	5	5	5	−0.4	19.6	25	30
−1.226	99.034	96	13			−0.7	19.2	97	32
−1.982	99.6	163	20			21.2	3.2	142	23
−2.132	99.536	149	25			20.7	4.7	128	30
−2.256	99.604	263	43			17.8	18.7	245	62
−2.286	99.785	169	14			12.6	25.2	156	39
−2.29	99.794	175	14			12.2	25.5	163	40
−2.37	99.741	71	5	182	55	12.5	28.0	241	88
−2.564	99.976			148	35	3.0	13.1	145	48
−2.55	100.044			191	9	0.7	9.9	190	19
−2.603	100.11	15	5	152	7	−0.7	6.3	168	18
−2.752	99.995	80	10	251	76	7.2	4.8	324	91
−2.831	100.008	90	5	254	84	8.9	11.3	335	100
−2.826	100.283	38	5	117	30	−0.4	6.5	155	42
−3.067	100.246	50	5	209	90	66.3	5.9	193	101
−2.981	100.423			134	17	34.3	5.8	100	23
−3.128	100.312	28	4	193	17	79.4	8.0	142	29
−3.039	100.463	38	10	59	30	41.7	12.7	55	53
−3.211	100.332	5	5	212	16	89.5	6.6	128	28
−3.163	100.505			103	43	49.5	4.7	54	48
−3.207	100.451	15	5	124	17	68.8	9.3	70	31
−3.192	100.486	124*	21			58.9	8.3	65	29
−3.216	100.487			110	14	62.8	10.0	47	24
−3.285	100.446	3	3	161	54	79.3	10.9	85	68
−3.268	100.572			88	28	55.5	16.0	33	44
−3.486	100.638	0		155	40	67.0	6.4	88	46

*Total uplift from both 1797 and 1833 earthquakes.
Data for the 1797 and 1833 events from Philibosian *et al.* (2014), and the 2007 event interpolated from Konca *et al.* (2008).

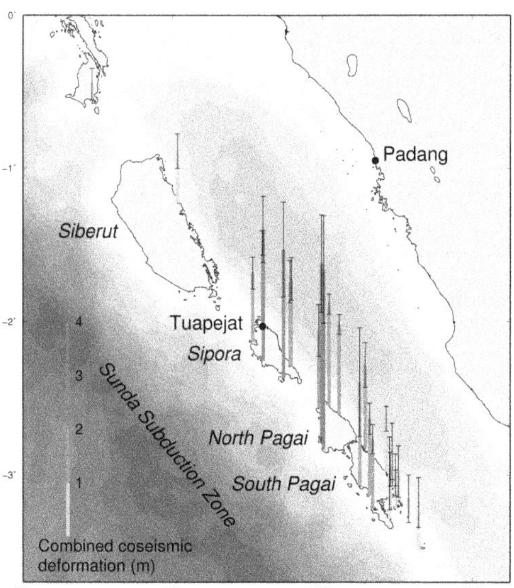

Fig. 2. Combined coseismic deformation for the 1797, 1833 and 2007 events with associated errors.

interpolation of data with known, varying, uncertainties. This method is implemented using the *gstat* package for the R programming language (Pebesma & Wesseling 1998) fitting an exponential variogram model.

The assumption that the amount of surface deformation for future earthquakes will be similar to that from previous events is based on coral records of several past earthquakes (Sieh *et al.* 2008), which suggest approximately cyclic rupture of the Mentawai Segment of the Sunda Subduction Zone in discrete periods of activity that record similar total vertical coseismic displacement (during one or more earthquakes clustered in time). Persistent behaviour of rupture barriers across several seismic cycles observed further north on the Sumatran Megathrust in Simeulue also supports this assumption (Meltzner *et al.* 2012). Deformation recorded by GPS stations on the Mentawai Islands for the near-trench M_w 7.7 event of 25 October 2010 is small (<5 cm: Hill *et al.* 2012), much less than the uncertainties associated with coseismic deformation from other events and of opposite sign (i.e. subsidence).

Furthermore, this event did not rupture the main portion of the megathrust beneath the Mentawai Islands and, therefore, we ignore vertical deformation from this event.

Methodology

Revision of the Padang Consensus Model

The original Padang Consensus Model (hereafter called the 'original consensus model') was based on the analysis of Chlieh *et al.* (2008) and the product of consensus workshops held in 2009 and 2010, as outlined in Schlurmann *et al.* (2010). A key aspect of the original consensus model was that, following analysis of coral deformation and GPS data to determine a best-fit source model, the estimated slip was increased by an additional 50% as a 'Safety Factor' to account for inherent complexity and unpredictability (Schlurmann *et al.* 2010). The resultant original consensus model has a moment magnitude of 8.9.

The greatest amount of slip and subsequent vertical deformation in the original consensus model is beneath Siberut, with upwards vertical deformation occurring on both the west and east sides of the island. The coastline of West Sumatra subsides by over 1 m, and resulting inundation modelling shows the potential for catastrophic inundation in Padang (Schlurmann *et al.* 2010). In the original consensus model, several metres of slip occur near the trench along much of the extent of the fault. In contrast, the 2011 Tohoku event has shown the potential for much larger (30–70 m) near-trench slip to occur (e.g. Satake *et al.* 2013). Furthermore, Hill *et al.* (2012) showed that GPS data from the Mentawai Islands does not constrain well the amount of slip occurring on the shallowest part of the megathrust near the trench. The preferred slip model of Hill *et al.* (2012) for the 2010 M_w 7.8 Mentawai Islands earthquake and tsunami has more slip occurring near the trench west of the Pagai Islands than the original consensus model, despite the much larger magnitude of this model (M_w 8.9). Therefore, the original consensus model was revised in order to generate a new model (hereafter called the 'revised consensus model') with increased near-trench slip and, therefore, increased potential for very large tsunami to be generated west of the Mentawai Islands.

For the revised consensus model, near-trench slip (at depths <20 km) in the original consensus model is multiplied by 3, resulting in slip of 15–20 m near the trench along much of the megathrust strike and a maximum slip value of 28 m. This multiplication is equivalent to decreasing the crustal shear strength in the shallow region of the megathrust from a crustal average value of 3×10^{10} GPa to a more realistic value of 1×10^{10} GPa (Bilek & Lay 1999), while preserving the seismic moment of the original consensus model. Other aspects of the original consensus source model remain unchanged. This new model was developed in response to concerns about 'tsunami earthquakes' raised during a science consensus workshop hosted by BNPB in Jakarta during March 2013. A comparison between surface vertical deformation for the original and revised consensus models is shown in Figure 3, which shows the much greater deformation near the trench.

Generation of heterogeneous slip models

The steps involved in the generation of our random slip model are described in detail below. In summary, they are:

(1) Generate a set of subfaults on the dipping megathrust interface.
(2) Calculate the surface deformation for 1 m of slip on each subfault using the Wang *et al.* (2003) software and the one-dimensional (1D) crustal structure of Collings *et al.* (2012).
(3) Randomly generate millions of spatially heterogeneous slip distributions (events) using the Gallovič & Brokešová (2004) model and Somerville (2003) slip heterogeneity scaling relationships, then map slip to the subfault model generated in step (1). Each slip distribution is for an earthquake of magnitude 8.7–9.1 randomly sampled from a uniform distribution.
(4) For each slip distribution, scale the surface deformation for each subfault by total slip on that subfault, then sum the deformation from all subfaults to generate a surface deformation model for that event.
(5) For each event, test whether the surface deformation fits the coral deformation data within 2σ uncertainties (Table 1).
(6) Events that fit the coral data are sorted by maximum deformation and tsunami inundation modelled for the 1000 (for Sipora) or 200 (for other regions) events with the largest vertical deformation (out of 5000 models that fit the coral data). Tsunami inundation is modelled by adding surface deformation to the initial sea surface and elevation model.
(7) At every on-shore point in the area of interest, calculate the maximum tsunami flow depth from all possible source models (including the consensus and revised consensus models) to generate a maximum tsunami inundation area.

Subfault model. We sample slip on a rectangular grid of subfaults with depth-varying dip. Fault

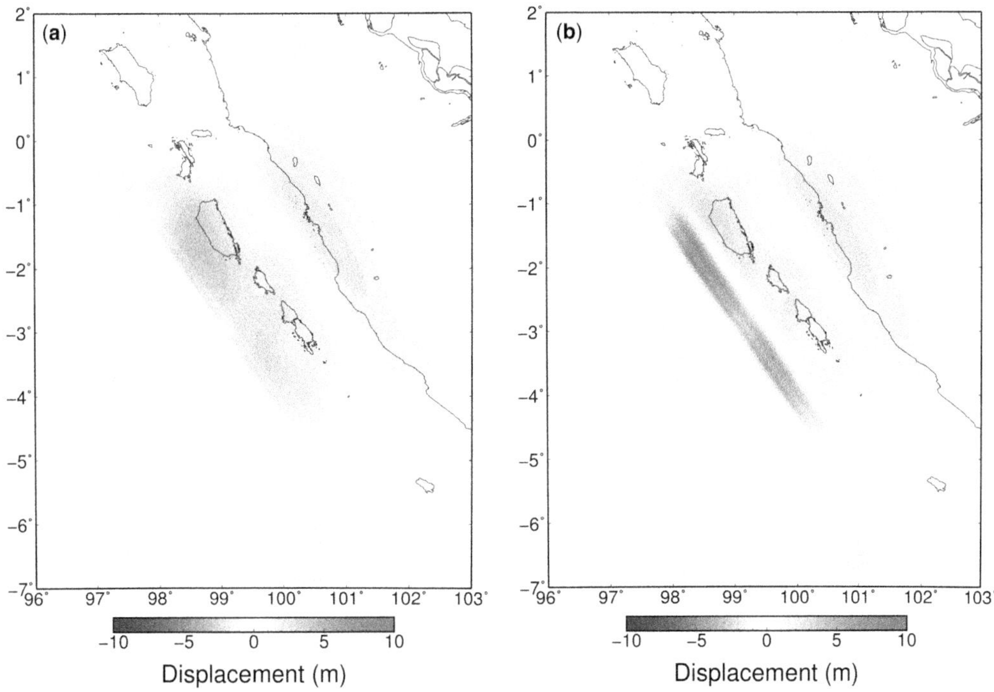

Fig. 3. Comparison of modelled surface deformation for (**a**) the original Padang Consensus Model and (**b**) the revised source model with increased near-trench slip.

dimensions are taken from the Padang Consensus Model (Chlieh *et al.* 2008; Schlurmann *et al.* 2010). Dips for the upper portion of the megathrust are determined from seismic reflection profiles (Singh *et al.* 2011) and, for the deeper portion, from the Slab1.0 subduction zone geometry model of Hayes *et al.* (2012). The depth-varying dip model (relative to sea level) is given in Table 2.

Subfaults have dimensions of 19 km in length by 7.8 km in width. This allows us to create a 32 × 28 grid of subfaults with the same fault dimensions as the Padang Consensus Model. The Gallovič & Brokešová (2004) random slip generator (described later in this section) generates slip on grids that must be a power of 2: however, the outer subfaults on each edge of the grid are constrained to go to zero slip. Therefore, we generate slip over a 32 × 32 grid of subfaults with dimensions of 19 × 7.8 km and then clip the uppermost four rows of subfaults. The remaining subfaults are then applied to our 32 × 28 subfault model of the Mentawai Megathrust, allowing greater slip near the trench, and with slip on each subfault rescaled to preserve the total moment of the unclipped subfault model.

Table 2. *Dip and depth range for the fault-plane geometry*

Minimum depth (km)	Maximum depth (km)	Dip (°)
7	9	6
9	14	7
14	20	11
20	40	20
40	60	25
60	80	39

Depths are relative to sea level.

Surface deformation calculations. Surface deformation is calculated using a layered Earth model and the EDGRN/EDCMP software of Wang *et al.* (2003). The layered Earth model is the 1D model for the Mentawai Islands region of Collings *et al.* (2012). The decision to use a layered Earth model is based on Hill *et al.* (2012), who noted that this was important for simulating the 2010 Mentawai Islands earthquake and tsunami accurately. Static calculations of surface deformation do not account for the effects of slower rupture velocities within low shear strength layers (Kanamori 1972; Satake 1994) and the effects of surface rupture (Rudnicki & Wu 1995) are ignored.

Random slip generation. We generated random slip distributions using the hybrid k^{-2} rupture model of Gallovič & Brokešová (2004). This model is based on the Herrero & Bernard (1994) model of k^{-2} decay of the spatial Fourier spectrum for wave numbers $k > k_c = 1/L$, where k_c is the corner wave number and L is the length of the fault. The 2D slip distribution, $D(k_x, k_y)$, is described in terms of its spatial fourier spectrum as:

$$D(k_x, k_y) = \frac{\Delta \bar{u} LW}{\sqrt{1 + \left[(k_x L/K_x)^2 + (k_y L/K_y)^2 \right]^2}} e^{-i\phi(k_x, k_y)},$$
$$k_x > k_{cx},\ k_y > k_{cy}$$

where $\Delta \bar{u}$ is the mean slip, W is the fault width, k_x and k_y are the wave numbers in the x (along-strike) and y (down-dip) directions, and ϕ is a random phase term. The parameters $K_x = k_{cx}L$ and $K_y = k_{cy}W$, where k_{cx} and k_{cy} are corner wave numbers in the x and y directions, control the roughness of the random part of the slip distribution. Larger values of K give rougher slip distributions (Gallovič & Brokešová 2004). Corner wave numbers are sampled from the scaling relationships of Somerville (2003):

$$\log k_{cx} = 1.28 - 0.5\,\mathrm{M_w}, \quad \sigma \approx 0.3$$
$$\log k_{cy} = 1.60 - 0.5\,\mathrm{M_w}, \quad \sigma \approx 0.3$$

We conservatively estimate uncertainty ($\sigma = 0.3$), as these scaling relationships were developed using data from only seven subduction zone events and before a number of significant events occurred, including the 2004 Indian Ocean tsunami and the 2011 Tohoku tsunami. In applying the Somerville (2003) relationships, we randomly sample uncertainty from a normal distribution between $\pm 2\sigma$. We assume no correlation between uncertainties in the x and y directions.

For each random slip distribution, we forward model vertical surface deformation at each of the sites for which coral data are available. Models that fit all coral data points within 2σ are retained, while others are rejected. Two example slip and deformation models are shown in Figure 4. The constraints from the Mentawai Islands mean that there is stronger control on slip distributions near the islands and less control further away (Hill *et al.* 2012). Therefore, we sample ruptures with high slip near the trench, as well as those with high slip east of the Mentawai Islands. Events with high slip to the west of the Mentawai Islands generate a significant hazard on the west coast of the Mentawai Islands: events with high slip on the eastern side, however, will result in a smaller tsunami overall (owing to deeper slip and shallower water) but may generate locally higher tsunami for the eastern side of the Mentawai Islands and/or the west coast of West Sumatra.

Tsunami modelling. Tsunami inundation modelling was conducted using the ANUGA hydrodynamic software (Nielsen *et al.* 2005; Jakeman *et al.* 2010). ANUGA solves the 2D non-linear shallow-water wave (NLSWW) equations in Cartesian coordinates using a finite volume method. The NLSWW equations are solved in conservative form, such that discontinuities (hydraulic jumps) are accommodated in the solution. This means that, although wave breaking is not explicitly modelled by the 2D NLSWW equations, propagating bores are simulated as the wave front steepens during shoaling. Li & Raichlen (2002) have shown, for a uniform plane beach, that modelling bore propagation is sufficient for capturing the general run-up properties (including maximum run-up) of post-breaking waves even if details of the internal flow structure are missing. Furthermore, Baldock *et al.* (2007) performed a series of numerical experiments with ANUGA that showed reductions in wave amplification once theoretical breaking criteria were exceeded that are consistent with theoretical expectations. ANUGA has been validated against tsunami field observations from the 2004 Indian Ocean tsunami at Patong Beach, Thailand (Jakeman *et al.* 2010).

Tsunami inundation models are extremely sensitive to elevation data. For bathymetry data, the global GEBCO dataset was used in deep water and a 90 m resolution commercial grid, developed by TCarta Marine based on chart data, was used in water shallower than 1000 m. In a few areas near the Pagai Islands, these data were supplemented with additional higher-resolution navy chart data (DISHIDROS) in shallow-water areas. The number of bathymetry observations available in the near-shore area to generate the Tcarta Marine dataset is quite limited in the Mentawai Islands. This is supplemented by interpretation of the coastline and shallow areas from Landsat imagery: however, the resolution of bathymetry data is a significant and unquantified source of uncertainty.

Onshore, we used an airborne IFSAR digital terrain model (courtesy of BIG) with 5 m horizontal resolution and 3 m vertical accuracy to model onshore inundation. Griffin *et al.* (2015) showed that similar results for inundation extent could be obtained with this dataset when compared with higher-resolution and higher-accuracy LiDAR (generally the most accurate elevation dataset commercially available). All bathymetry and elevation data were projected into UTM Zone 47 South for input into ANUGA's Cartesian coordinate system.

The effect of surface roughness was parameterized using a Manning's coefficient of 0.025 in the

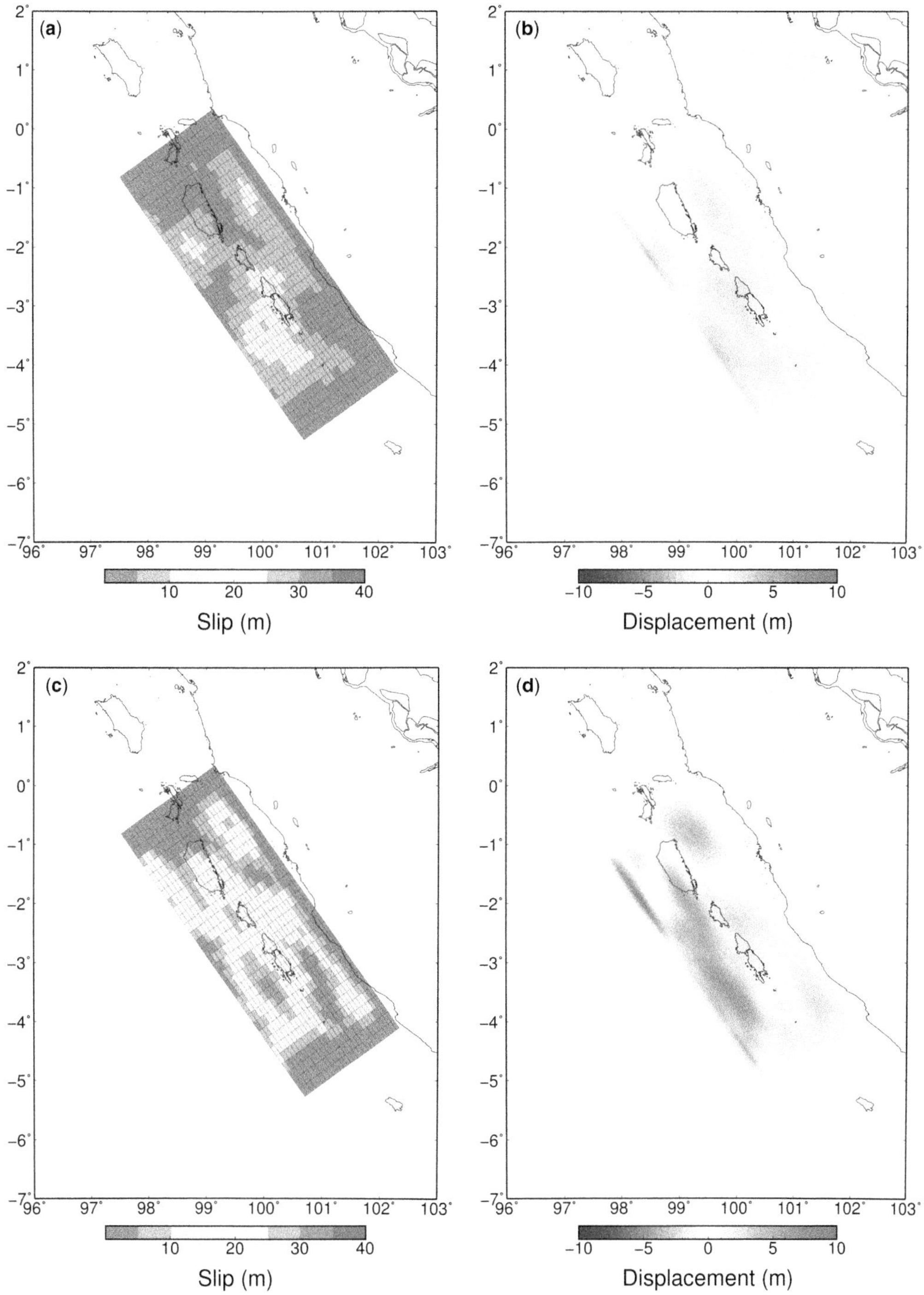

Fig. 4. Example slip and vertical deformation for two models that fit the coral data: (**a**) slip distribution and (**b**) surface deformation for M_w 8.83; and (**c**) slip distribution and (**d**) surface deformation for M_w 9.02.

ocean, following Imamura (2009). A constant Manning's coefficient of 0.036 was used on land based on the value for grassland from Kaiser *et al.* (2011). This value is taken as a conservative minimum onshore value, and is expected to underestimate roughness in more heavily vegetated and urban areas.

ANUGA uses an unstructured triangular mesh to construct the model domain. For the highest-resolution coastal areas, each triangular mesh element has a maximum allowable area of 625 m^2 (equivalent to a 25 m rectangular grid). In order to model the entirety of the coastline of the Mentawai Islands within the constraints of available computing power, we ran the model eight times for each source model with the highest mesh resolution located in a different region. The model extent was the same in all models.

Deformation calculated using the method outlined above was directly added to both the initial water height and the topography to simulate the tsunami. The hydrodynamic model was then run for 9000 s to ensure reflections from the Sumatran mainland were captured, and the maximum tsunami height tracked. The internal time step was controlled by the Courant–Friedrichs–Lewey (CFL) criterion.

Results

Heterogeneous slip models

After generating approximately 15 million random slip distributions for M_w 8.7–9.1, we found 5000 source models that generate coseismic deformation that fits all of the coral deformation data points within 2σ. The magnitude distribution of the 5000 models (Fig. 5) has a mode of M_w 8.78 and a maximum of M_w 9.05. We compare the resulting slip distributions with 5000 source models for the same uniformly distributed magnitude range that are not constrained by the coral data. Figure 6 shows convergence of maximum slip values on each of the 896 subfaults in our subfault model for the constrained and unconstrained slip distributions. Although neither shows complete convergence, the constrained models converge more rapidly than

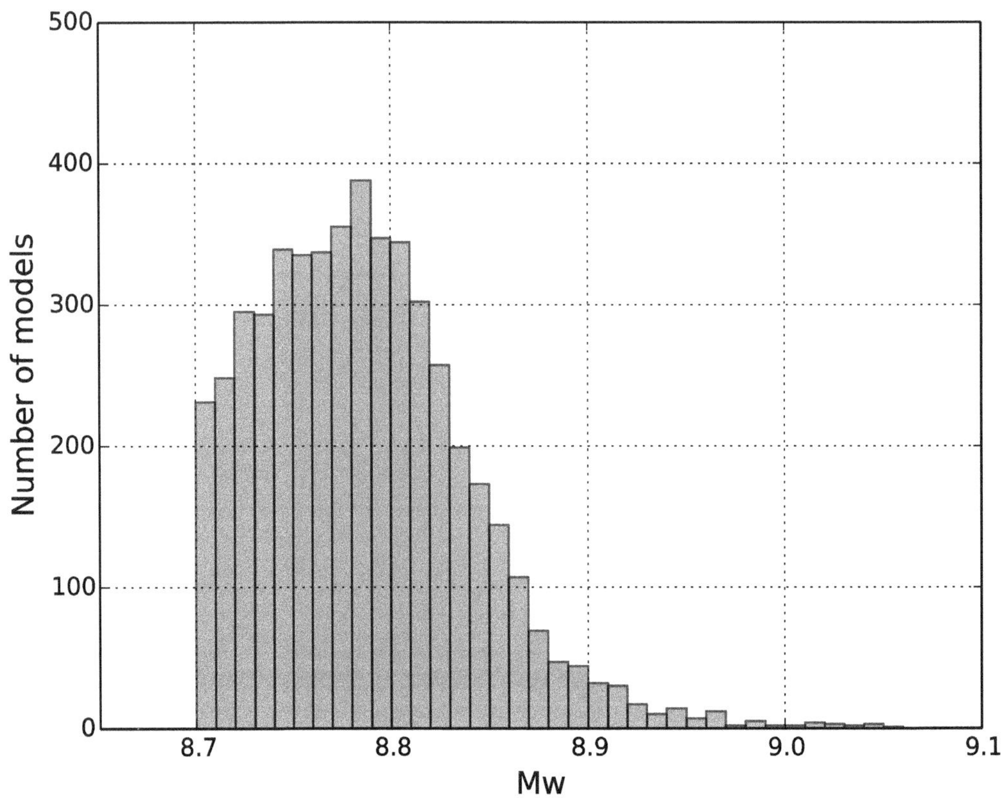

Fig. 5. Distribution of earthquake magnitude for the 5000 source models that fit the coral data within 2σ. The mode is 8.78 and the maximum magnitude is 9.05. Magnitudes were originally sampled from a uniform distribution in the range M_w 8.7–9.1.

those without constraints. Furthermore, maximum slip for the constrained model is about half that of the unconstrained model. Similarly, maximum vertical deformation in the suite of models constrained by the coral data is 8.8 m compared with 16.8 m in the unconstrained models. Figure 7 compares the distributions of K_{cx} and K_{cy} for the constrained and unconstrained models. For the constrained models, the distribution is narrower and the peak value is shifted to the right, showing that some values of K_{cx} (0.4–0.8) and K_{cy} (0.3–0.6) are more likely to generate slip models consistent with the coral data.

Plotting statistics of the source models at the level of each subfault (Figs 8 & 9) demonstrates how the slip distribution is controlled by both the coral data and the methodology implemented in the Gallovič & Brokešová (2004) code. Mean slip is higher towards the centre of the fault in the fitted models (Fig. 8a) and appears to be controlled by the coral data. The maximum slip at each subfault across the 5000 models that fit the data is constrained beneath and to the east of the Mentawai Islands by the coral data. Larger slip values can occur further away from the coral constraints, both to the west and east: however, slip is reduced again near the trenchwards edge of the fault model for the fitted models. This is less pronounced in the unconstrained models, suggesting that it is caused by an interaction of the Gallovič & Brokešová (2004) slip generator with the coral data, perhaps by localizing high slip near the location of the islands. Maximum values (Figs 8c & 9c) vary between neighbouring subfaults away from the Mentawai Islands, and particularly near the trench: this suggests incomplete sampling of models that fit the data rather than any strong constraint from the coral data. The standard deviation plot (Fig. 8b) shows that directly beneath and east of the islands, the slip is relatively more constrained by the coral data, with a standard deviation of less than 2.5 m slip. Further away, however, the standard deviation is larger (>3 m), before being constrained again near the edges of the fault model, compared with a smoother distribution of standard deviation for the unconstrained models (Fig. 9b).

The distribution of slip in the models that fit the data demonstrates how vertical deformation data can be used to constrain slip beneath and downdip of the data, but provides limited constraint updip. Considering vertical deformation due to a single, buried thrust fault, Løvholt *et al.* (2012) showed that, for dip angles of less than 30°, the greatest upwards vertical deformation occurs above the updip edge of the fault, while subsidence of approximately half the magnitude of vertical uplift occurs above the downdip edge of the fault. Furthermore, the wavelength of the vertical deformation for deeper events is longer, extending the zone of uplift trenchwards from the updip edge of the fault. These effects can be shown if we consider the vertical deformation for a uniform 1 m of slip on each of five subfaults beneath a given data point (Fig. 10). The subfaults that generate the greatest vertical deformation at the data point for uniform slip are those immediately beneath and downdip of the data point, and therefore the amount of slip on these subfaults will be more tightly constrained by the data point. Slip on subfaults updip of the data point generates subsidence of a lower magnitude than the uplift from subfaults a similar distance downdip, allowing for a greater variation in slip in the absence of other constraints.

Tsunami inundation models

A comparison of inundation results for the two 'consensus' models shows that the extent of inundation is almost identical for Padang and other areas along the west coast of West Sumatra (Fig. 11a). On the western coast of the Mentawai Islands, the revised source model, with greater near-trench slip, gives more inundation compared with the original model (Fig. 11b). This suggests that uncertainty around the amount of slip near the trench does not translate into significant uncertainty in inundation for Padang, and that the hazard here is predominantly driven by slip occurring deeper on the megathrust that generates vertical deformation to the east of the Mentawai Islands. There are two factors that may explain this. First, the Mentawai Islands provide a physical barrier for tsunami propagating from near the trench to the west coast of Sumatra. Secondly, as Padang sits on a flat coastal plain without clear topographical barriers to tsunami inundation extent, it is likely that the similarity in inundation extent between the two models is because inundation distance is controlled by longer-period tsunami waves generated by slip on the deeper portion of the megathrust, rather than by shorter-period but higher-amplitude tsunami generated near the trench (see Fig. 3). Conversely, for the steeper west coast of the Mentawai Islands, short-period large-amplitude tsunami generated in the near field drive large local run-ups in the revised consensus model. This interpretation is consistent with the findings of Satake *et al.* (2013), who found that large tsunami inundation distances on the Sendai and Ishinomaki plains in the 2011 Tohoku tsunami were due to long-period tsunami generated by slip deep on the plate interface, whereas short-period tsunami generated by slip near the trench caused the large run-up heights observed along the Sanriku coast.

The 5000 models that fit the coral data were sorted in order of maximum vertical deformation, and the 1000 models with the largest vertical

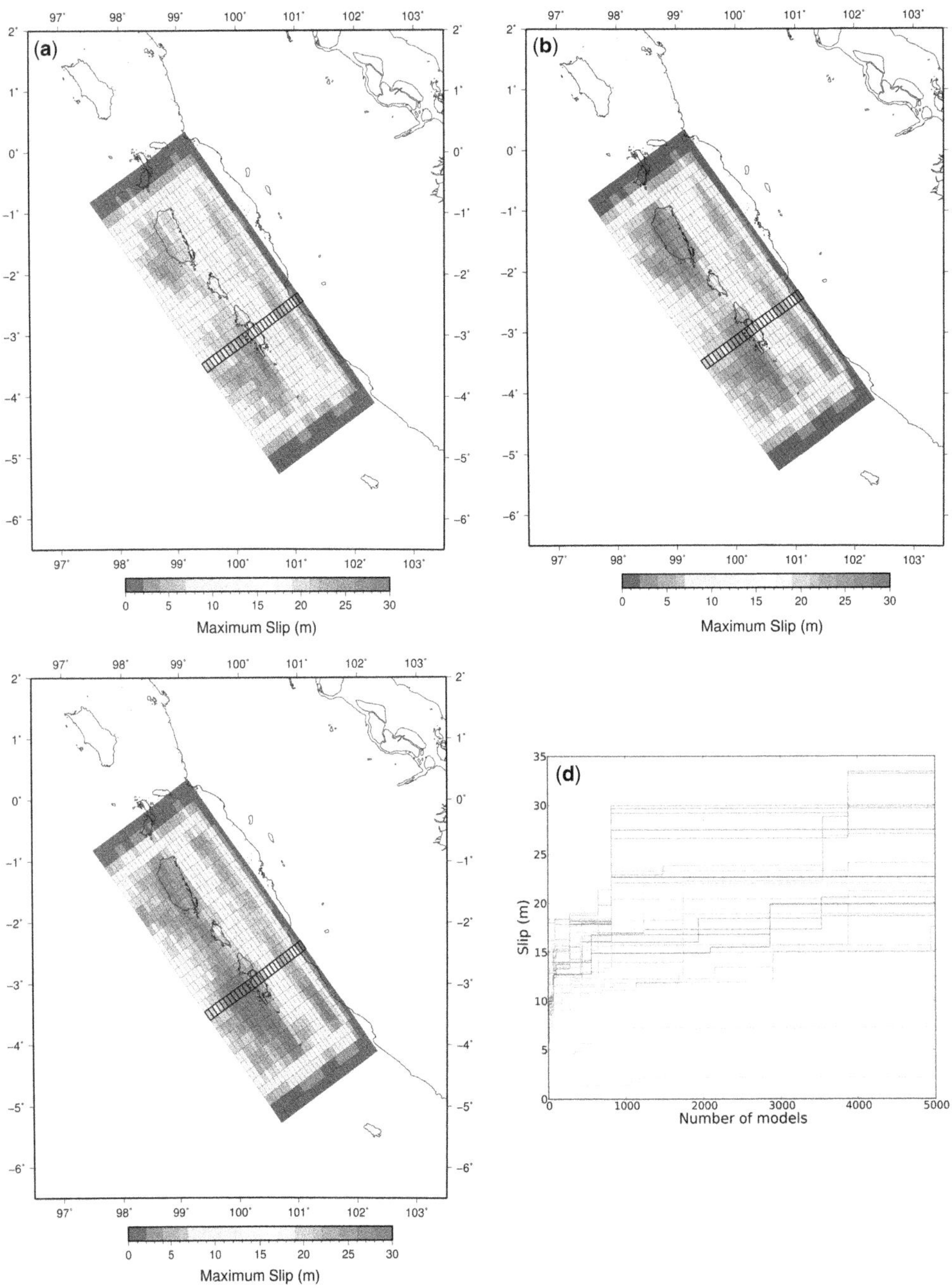

Fig. 6. Convergence of maximum slip as more random subfaults are sampled. Maximum slip values for (**a**) 1000, (**b**) 2500 and (**c**) 5000 slip models constrained by the coral data. For subfaults outlined in black, a running maximum slip value is plotted in (**d**), where each line represents the maximum slip value on one subfault as the number of slip models is increased.

deformation used to model tsunami generation, propagation and inundation for Sipora. By determining which model generates the greatest tsunami depth at each point within the area of interest, we found that the first 100 slip models generate maximum tsunami depth at more than 99.0% of locations,

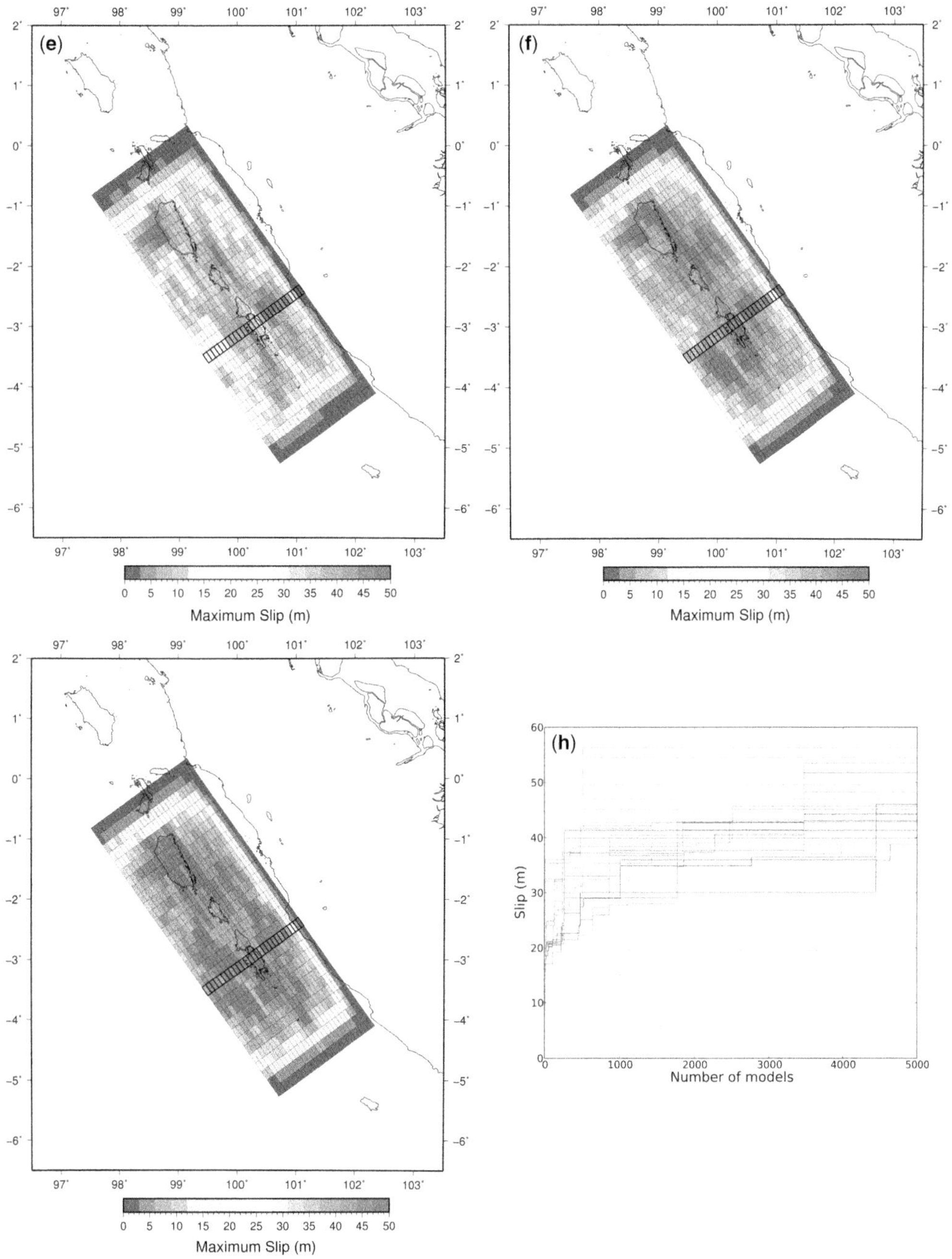

Fig. 6. (**e**)–(**h**) as for (a–d), but for the unconstrained models.

and the first 200 slip distributions at more than 99.5% of locations. Based on this analysis, we only ran tsunami inundation models for the 200 slip distributions with the greatest vertical deformation for Siberut and the Pagai Islands.

Figure 12a shows the maximum tsunami height (i.e. the maximum height above sea level reached by the tsunami flow) from the suite of heterogeneous slip models for the Mentawai Islands. Tsunami heights exceed all those observed by Hill *et al.*

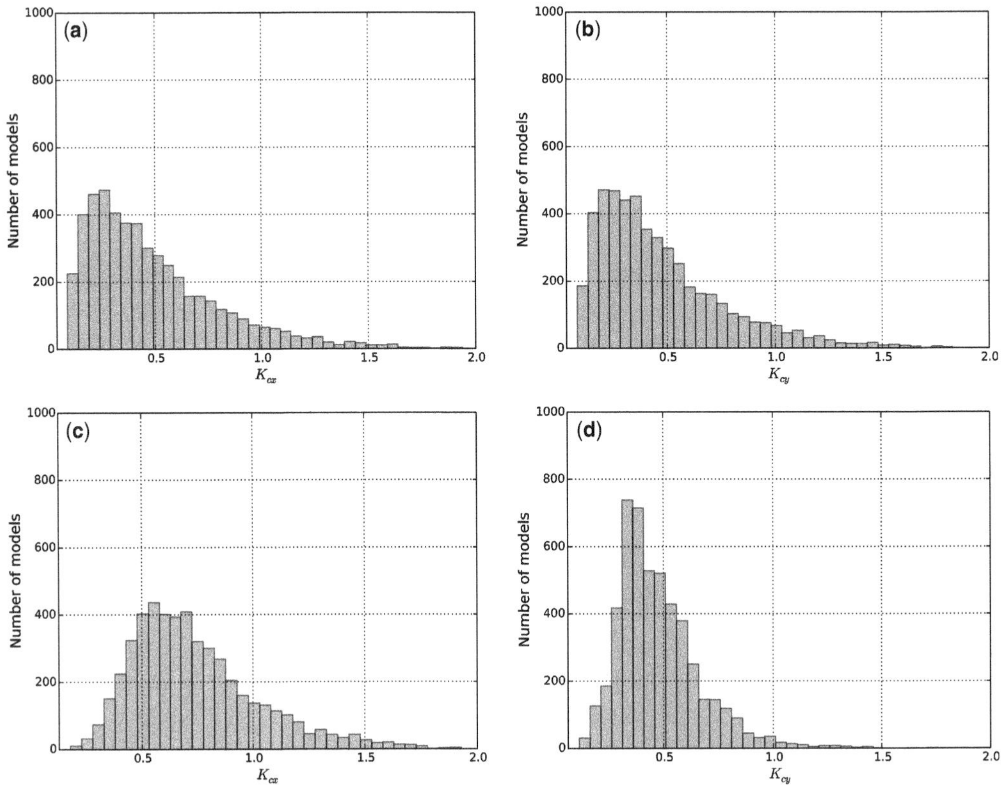

Fig. 7. Distribution of (**a**) K_{cx} and (**b**) K_{cy} for unconstrained models; (**c**) & (**d**) as for (a) & (b), but for the models constrained with the coral data. Higher K_{cx} and K_{cy} values will give rougher slip surfaces.

(2012) in the field survey of the 2010 Mentawai Islands tsunami (Fig. 12b) and, in general, modelled inundation envelopes the observation points. There are, however, a few locations on South Pagai and its offshore islands where inundation from the 2010 tsunami is observed that are not inundated by our suite of models. The maximum horizontal distance by which our models are exceeded is 190 m on Sibigau, a small island off the west coast of South Pagai. There are a number of possible explanations for this discrepancy, including uncertainties in onshore elevation data and parameterization of friction, and features of the 2010 tsunami source that are not captured in our models (e.g. the generation of coseismic subsidence at the Mentawai Islands).

Figure 13 compares maximum inundation from the original and revised consensus models with the maximum inundation from 1000 heterogeneous slip models for part of western Sipora. It can be seen that inundation extents are similar between the revised consensus model and the maximum footprint of the heterogeneous model on the west coast of Sipora, varying along the coast in terms of which model inundates the furthest inland. The suite of heterogeneous slip models tends to inundate further inland than either consensus model on the east coast of Sipora. The area inundated by the original consensus model is much less than that of the other models, with no inundation occurring on the east coast of Sipora. Figure 14 shows the percentage of models from the suite of 1000 heterogeneous slip models that cause inundation at each point. For western Sipora, the original consensus model inundation limit plots near the region where 50% of the heterogeneous slip models cause inundation, while the revised consensus model closely aligns with the limit of inundation.

Discussion

The coral uplift data as a model constraint

After forward modelling over 15 million slip realizations, we find only 5000 that fit the coral data at 2σ, implying a strong constraint on the slip distribution. Furthermore, maximum slip and vertical deformation values are about half that of a suite

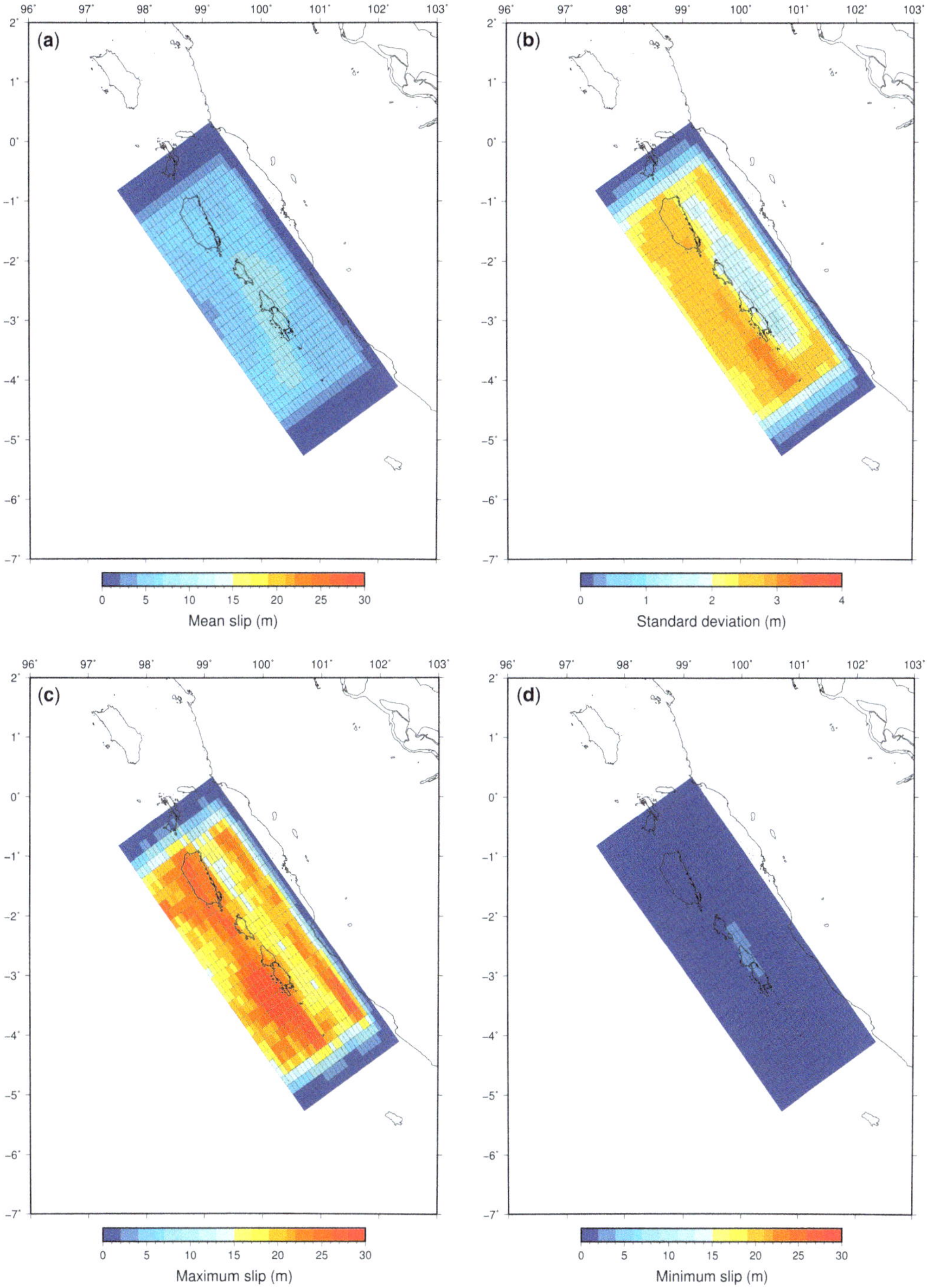

Fig. 8. Statistics of the magnitude of slip on each subfault for the 5000 models that fit the coral data showing: (**a**) mean slip; (**b**) standard deviation; (**c**) maximum slip; and (**d**) minimum slip.

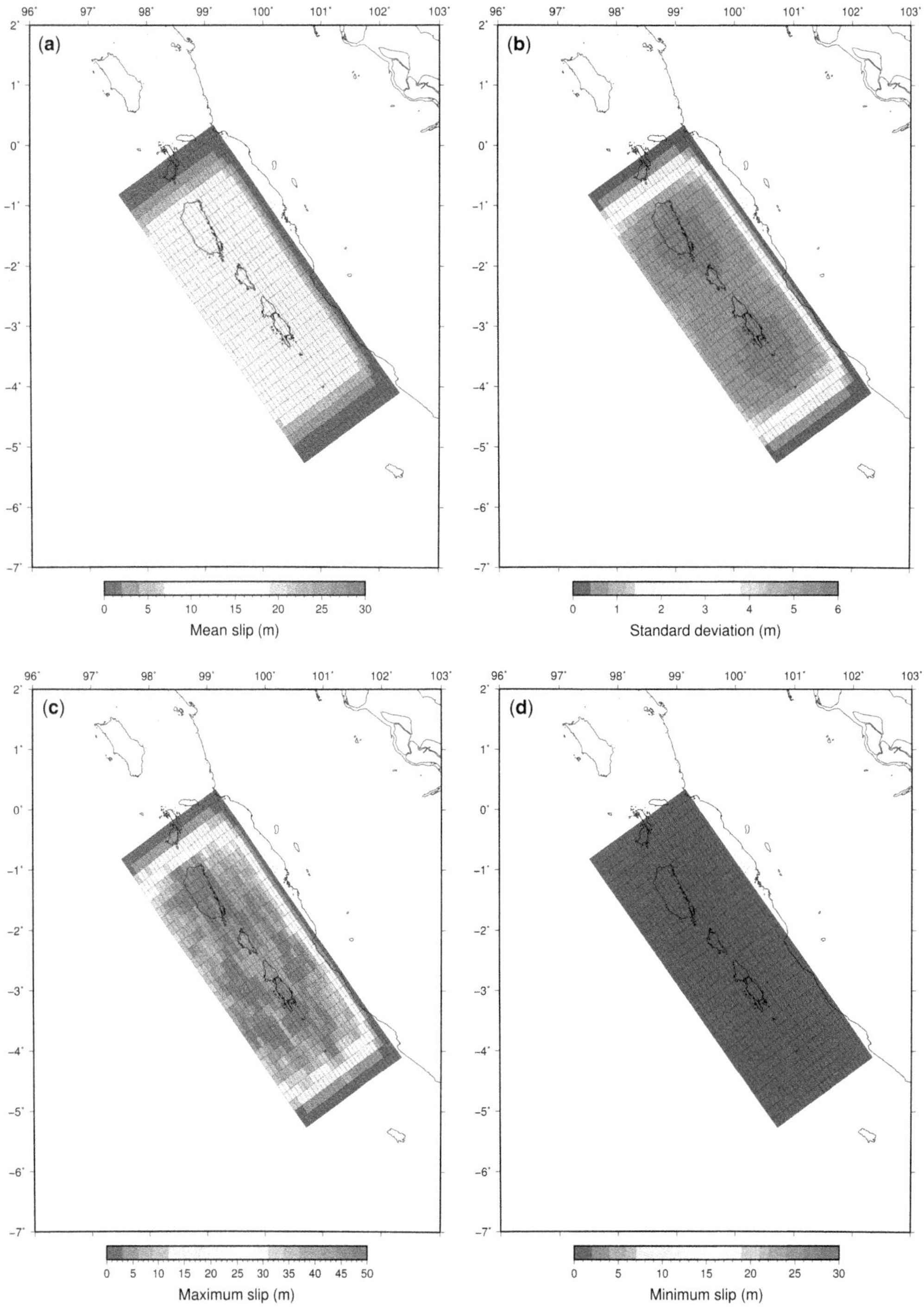

Fig. 9. Statistics of the magnitude of slip on each subfault for 5000 models that are not constrained by the coral data showing: (**a**) mean slip; (**b**) standard deviation; (**c**) maximum slip; and (**d**) minimum slip. Note the different scale compared with Figure 8 for plots of standard deviation (b) and maximum slip (c).

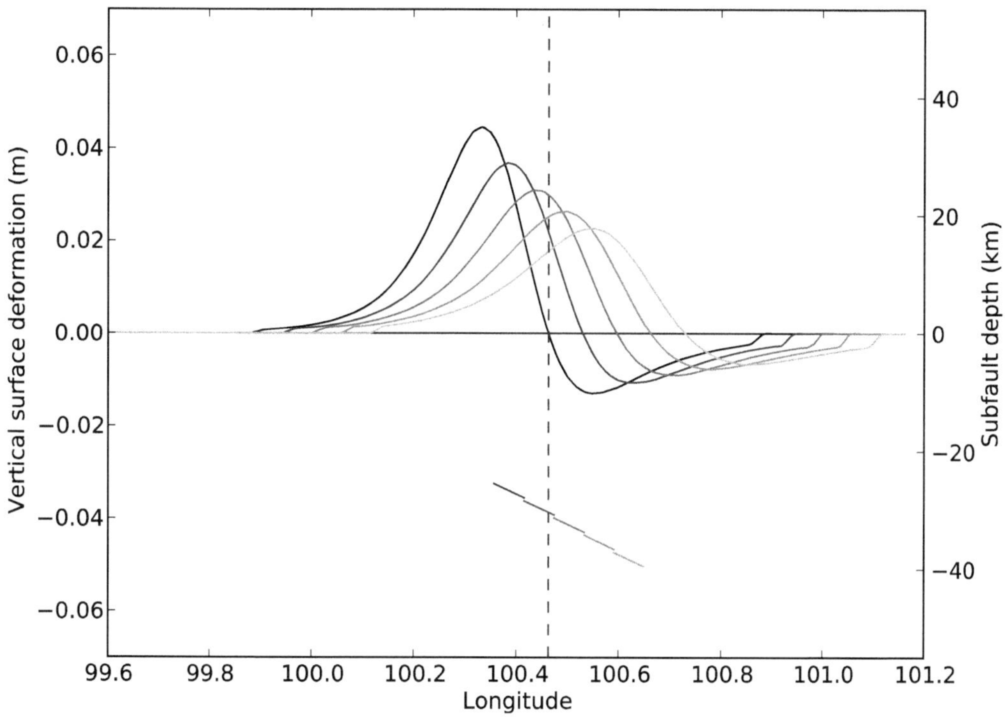

Fig. 10. Illustration of the greater control of the coral deformation data on slip occurring downdip (i.e. to the right) of the data point. The figure shows vertical surface deformation for 1 m of slip for five adjacent 19 × 8.5 km subfaults dipping at 20°. Considering a coseismic deformation data point indicated by the location of the dashed line, it can be seen that the subfault immediately downdip of the data point has the largest surface vertical deformation at the data point, followed by those to either side. Subfaults further updip give negligible or negative surface deformation of smaller magnitude compared with the vertical deformation from subfaults that are an equal distance downdip of the data point.

of 5000 models generated without the coral constraint on vertical deformation (Fig. 6). However, it must be emphasized that the use of the coral data is based on assumptions about future events based on previous events, and it is possible that larger magnitude events that do not fit the coral data could occur.

We do not use GPS observations of present interseismic subsidence rates to extrapolate total interseismic subsidence and potential future uplifts, as was done in the generation of the original Padang Consensus Model (Chlieh *et al.* 2008; Schlurmann *et al.* 2010). Although consistent patterns of uplift are demonstrated for several seismic supercycles by Sieh *et al.* (2008), interseismic subsidence can vary during and between megathrust cycles (Natawidjaja *et al.* 2007; Meltzner *et al.* 2012), and therefore extrapolation of present subsidence rates is uncertain. By restricting our analysis to models that fit observed coseismic uplifts (coral measurements for the 1797, 1833 and 2007 events, and GPS measurements for the 2007 event) and associated uncertainties, we avoid uncertain extrapolation of present interseismic rates at the expense of fewer data points to constrain our model.

Reliance on the coral data means there are more constraints in the south (South Pagai, North Pagai and the Sipora Islands) compared to the north, with only one data point for uplift from the 1797 event in Siberut and one further north in the Batu Islands. This results in greater variance in slip beneath Siberut compared with Sipora and the Pagai Islands in our random slip models (Fig. 8). Furthermore, the data only span a narrow portion of the megathrust above the main seismogenic zone (i.e. zone B of Lay *et al.* 2012). Geodetic data recording vertical deformation from the Mentawai Islands do not provide strong constraints at the trench (Hill *et al.* 2012), and this is confirmed in our model by the greater variance of total slip on subfaults trenchwards of the islands compared with those beneath and directly downdip of the islands (Fig. 8b): variance also increases further away from the islands towards the most downdip portion of the fault.

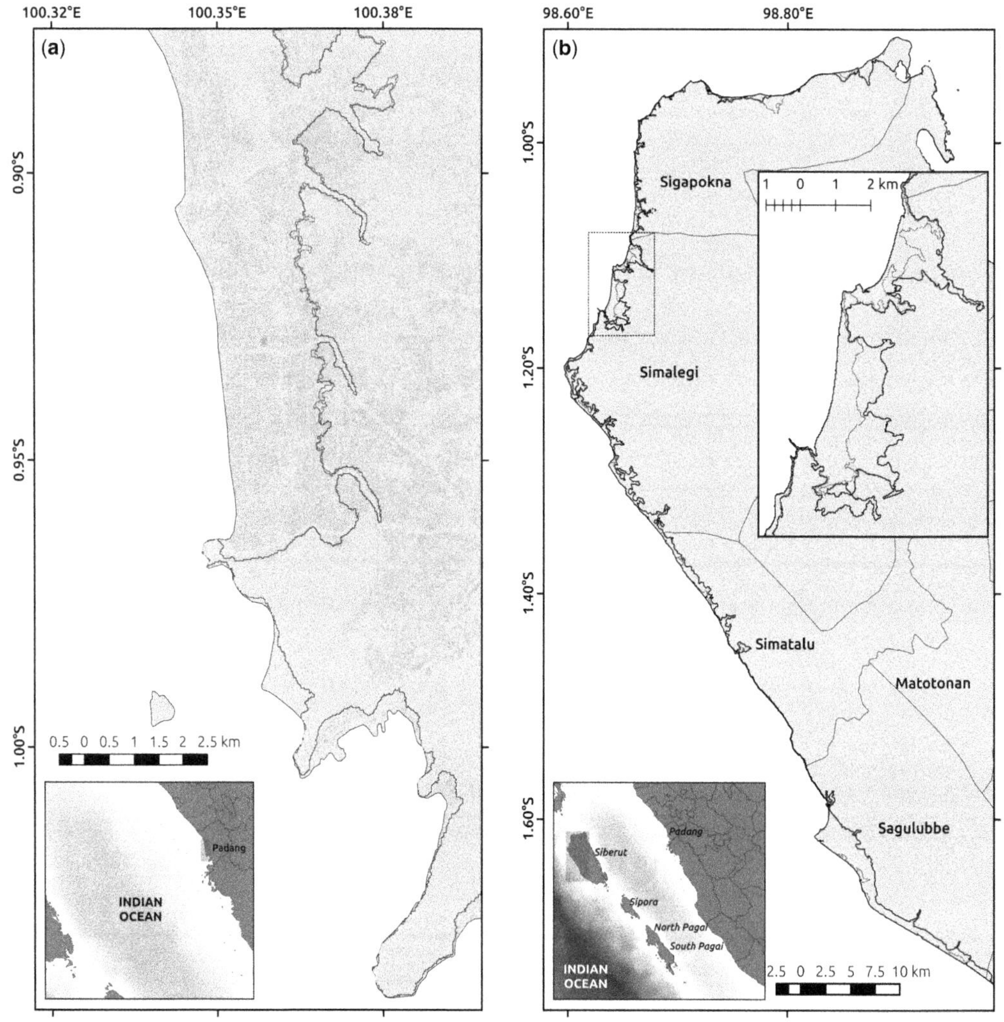

Fig. 11. Comparison of inundation results for the original consensus model (dashed line) and the revised consensus model (solid line) for (**a**) Padang and (**b**) NW Siberut. Note that inundation modelling does not consider the effect of buildings and other roughness elements, and is undertaken purely to compare the two source models. Buildings shown on images are taken from OpenStreetMap.

Constraining slip near the trench is further complicated as near-trench slip will generate subsidence at most locations on the Mentawai Islands (Hill *et al.* 2012; Philibosian *et al.* 2012): yet, all the coral data points record coseismic uplift. Therefore, while our model constrains slip on the main seismogenic portion of the megathrust based on previous rupture cycles, weak constraints trenchwards of the islands allow for more complex patterns of interseismic and coseismic slip near the trench (e.g. as suggested for Tohoku, Japan by Perfettini & Avouac 2014) that are not yet well resolved by the palaeogeodetic record (Philibosian *et al.* 2012).

The approach presented here does not capture near-trench tsunami earthquakes such as the 2010 M_w 7.8 Mentawai Islands event that do not rupture the main portion of the megathrust beneath the Mentawai Islands. Coseismic deformation from the 2010 event measured at the islands was small and mostly negative (Hill *et al.* 2012), whereas events used in our hazard assessment uplift the islands by 0.5–3 m (Table 1).

The 2007 Bengkulu event was smaller than the 1833 rupture (Konca *et al.* 2008): therefore, we cannot preclude that the remainder of the present seismic cycle will be expressed in a number of

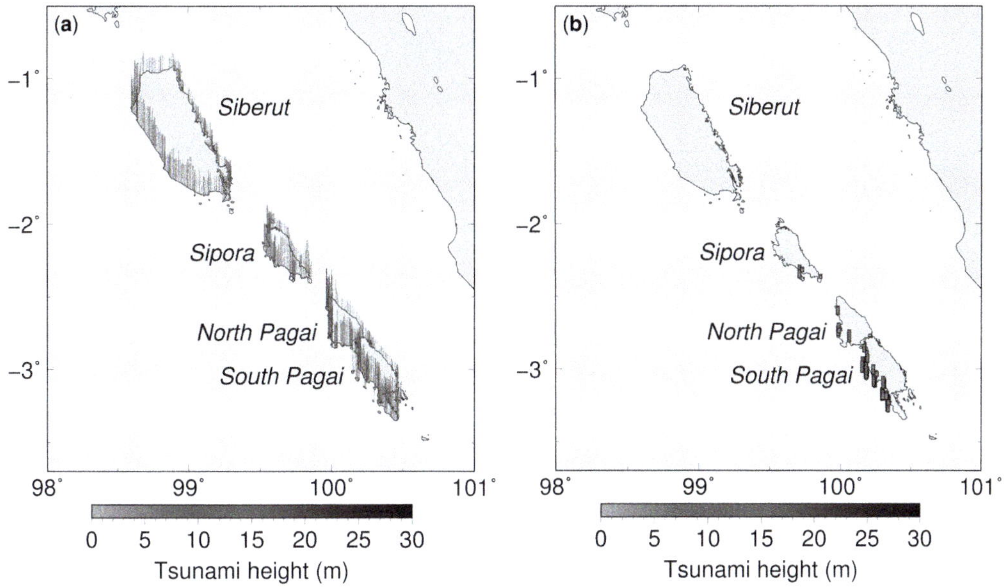

Fig. 12. Maximum tsunami heights from (**a**) the suite of heterogeneous slip models and (**b**) the field survey of tsunami inundation by Hill *et al.* (2012) for the 2010 Mentawai Islands tsunami.

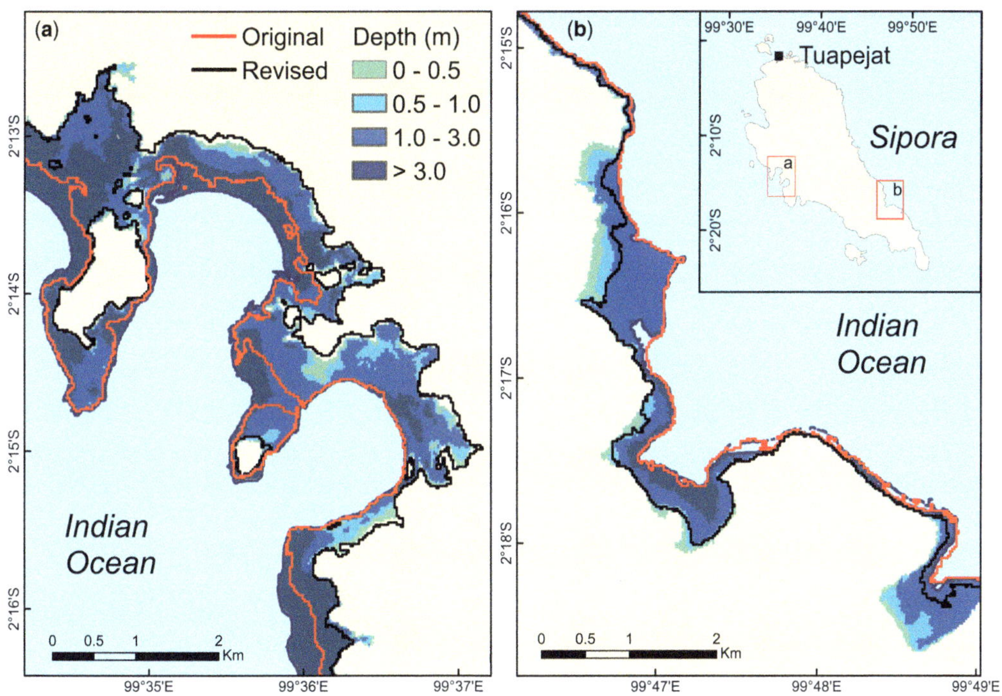

Fig. 13. Comparison of inundation for areas in (**a**) western Sipora and (**b**) eastern Sipora. Coloured regions show the maximum inundation depth from 1000 heterogeneous slip models. The red line is the inundation limit from the original consensus model and the black line is the inundation limit from the revised consensus model.

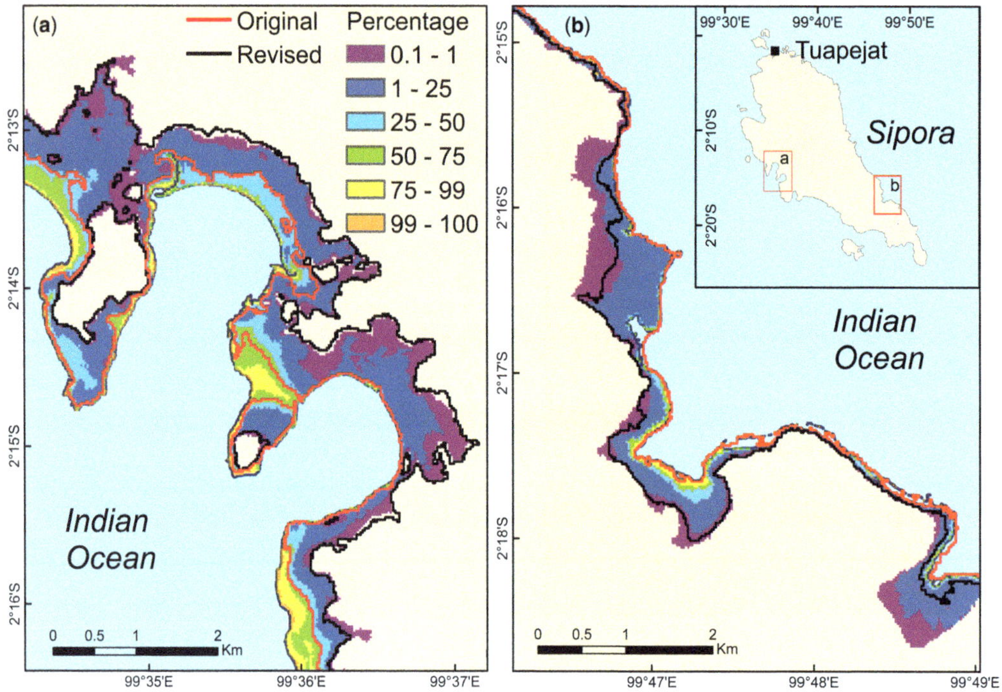

Fig. 14. Percentage of the suite of 1000 heterogeneous slip models that cause inundation at each point for areas on (**a**) western Sipora and (**b**) eastern Sipora. The red line is the inundation limit from the original consensus model and the black line is the inundation limit from the revised consensus model.

smaller events. However, considering the aims of this hazard assessment, it is reasonable to base our assessment on the assumption that the cycle will be completed in one future approximately M_w 8.8 rupture, and that limited constraints trenchwards of the Mentawai Islands mean that larger amounts of slip could have accumulated here compared with the main seismogenic zone below the Mentawai Islands.

Heterogeneous slip model

In generating random heterogeneous slip distributions, we attempt to sample possible ruptures that may occur, within the constraints of the coral deformation data. Analysis of the resulting slip distributions (Fig. 6) shows that we have not achieved convergence of the maximum slip distribution, in part due to wide sampling (2σ) of the Somerville (2003) scaling relationships for slip roughness. These relationships are poorly constrained and do not take into account recent tsunamigenic earthquakes for which high-resolution source models are available, including the 2011 Tohoku earthquake. We address this limitation by sampling the scaling relationship at 2σ, meaning that the coral data provide the main constraint on the plausibility of the slip distribution. This is shown in the difference in the resulting K_{cx} and K_{cy} distributions for the unconstrained (Fig. 7a, b) and constrained models (Fig. 7c, d).

Decay of the slip distribution at the edges of the fault means that maximum slip values at the trench are smaller than those further towards the centre of the fault. Although we adjust for this by clipping four rows of subfaults on the trenchwards edge of the slip distribution and rescaling slip to our subfault model, Figure 8 shows that the highest slip values do not extend all the way to the trench. We model maximum slip greater than 15 m along the trench, exceeding slip in models of the 2010 event (Hill *et al.* 2012), but do not generate extreme Tohoku-style events with high slip (up to 70 m) near the trench (Satake *et al.* 2013), even in the unconstrained model (Fig. 10). Therefore, application of the Gallovič & Brokešová (2004) random slip generator to surface rupturing earthquakes may not fully capture the potential for high slip to occur at the surface.

It is not yet clear how the distribution of models converges, both in terms of slip distributions and maximum inundation. Furthermore, for application of heterogeneous source models in areas without the same constraints that exist for the Mentawai Islands through the availability of the coral data, further research is needed to determine how many slip realizations are needed to robustly assess the inundation threat within certain levels of confidence along with a refinement of roughness scaling laws. Clearly, very large slips and vertical deformations are possible within the framework presented here if there is an absence of palaeoseismic constraints, such as the coral data (cf. Figs 8 & 9). A better understanding of how inundation from heterogeneous slip distributions converges will guide more robust future tsunami hazard assessment that considers source uncertainty.

Other sources of tsunami hazard

The hazard assessment presented here does not consider sources of tsunami hazard other than a full rupture of the Mentawai Segment of the Sunda Subduction Zone. We assume that this feature will be most important for a conservative assessment of local tsunami hazard, and that barriers to rupture imposed at each end of the megathrust will remain. It is possible that a lengthier rupture of the Sunda Subduction Zone could exceed these barriers: however, for the Mentawai Islands, the near-field slip on the Mentawai Segement of the Sumatran Megathrust is still expected to dominate the local hazard. Other sources of tsunami hazard in the region must also be noted. Outer-rise earthquakes are another source of near-field tsunami hazard, although further away than megathrust events. Similarly, rupture of other segments of the Sunda Subduction Zone may generate tsunami that reach the Mentawai Islands, but these are expected to be less significant in terms of hazard than the Mentawai Segment. The Mentawai Backthrust (Singh *et al.* 2010; Wiseman *et al.* 2011) has the potential to generate tsunami on the eastern side of the Mentawai Islands that could be locally significant for the east coast of the Mentawai Islands and the west coast of West Sumatra. Furthermore, Singh (*et al.* 2010) observed evidence of submarine mass failures to the east of Siberut and, therefore, local tsunami run-ups may be increased if a mass failure event is triggered by an earthquake on either the megathrust or the backthrust.

Which approach best answers disaster managers' questions?

In developing these hazard maps, clear requirements were articulated by the Indonesian National Disaster Management Agency (BNPB) regarding a worst-case scenario, specifically influenced by BNPB's perception of the failures of tsunami hazard assessments prior to the 2011 Tohoku tsunami. A key challenge in meeting these requirements is interpreting and then defining what is considered to be a realistic worst case, as any chosen scenario incorporates a subjective assessment of likelihood. For example, we do not consider the low probability, but extreme impact, of tsunami generated by meteor impact. In developing our hazard assessment, we assume that the coseismic deformation generated by future events will be similar to that of previous events and then use this assumption as a bound for a range of models. Development of a single scenario is a simpler approach computationally, as well as conceptually simpler to communicate to a non-scientific audience. However, by simulating many heterogeneous slip models within the bounds of our assumptions, we consider a greater range of possible scenarios and identify scenarios that generate more inundation than the original and revised consensus models in some areas. In the case presented here, the original consensus model generates much less inundation for the Mentawai Islands compared with the suite of random models, whereas the inundation footprint of the revised consensus model is similar, particularly on the west coast of the islands.

The original consensus model was developed using modern coupling rates to estimate the accumulated slip deficit and, hence, a future rupture model (Chlieh *et al.* 2008). Implicit in this model are the assumptions that interseismic deformation rates are constant and that previously accumulated slip deficit was released in the 1797–1833 earthquake sequence. In our random heterogeneous slip models, we use the coral palaeogeodetic data to constrain vertical coseismic deformation with the assumption that the pattern of coseismic deformation during the present earthquake supercycle will be similar to that of previous supercycles. The assumptions in both the Chlieh *et al.* (2008) model and ours are reasonable given what we know about previous ruptures of the Sunda Subduction Zone, but, nonetheless, they are not certain.

The original consensus model gave devastating inundation for Padang, but much less on the Mentawai Islands, compared with the revised consensus model with increased near-trench slip. Therefore, the revised model comes much closer to providing a maximum inundation zone for all areas, meeting a key requirement articulated by disaster managers. This demonstrates that deformation between the Mentawai Islands and the coast of West Sumatra is most significant as a driving hazard in Padang. It is this region where our slip models are most strongly controlled by the coral data (Fig. 8) and therefore, although we have not run tsunami inundation models using our suite of random heterogeneous models for Padang, the original consensus model may well prove to be a reasonable model here, even if it underestimates possible inundation on the Mentawai Islands.

Any tsunami scenario, or suite of scenarios, used for disaster management is based on assumptions about the characteristics of future events based on our present theoretical understanding of the underlying mechanisms and our observational record of past events. The method presented here of randomly sampling heterogeneous slip models allowed us to sample a wide range of possible outcomes within the bounds provided by our assumptions. In particular, considering a full suite of source models allows us to more robustly assess tsunami hazard in all areas of interest, due to the sensitivity of local tsunami inundation to source heterogeneity (Geist 2002). The method can be used to assess single-scenario models and, in this case, we find that the revised consensus model is a reasonable approximation to the maximum inundation zone, whereas the original consensus model underpredicts potential inundation. Therefore, even if a single scenario is presented in order to simplify communication of the hazard product, randomly generated heterogeneous source models can be used to assess its performance. Further work that reduces the uncertainty in the parameterization of the random slip generator will improve confidence in future hazard assessments and allow for better quantification of uncertainty in tsunami inundation footprints, including through application within a probabilistic tsunami hazard assessment framework.

Conclusions

We have developed a tsunami hazard assessment for the Mentawai Islands, West Sumatra, Indonesia based on a suite of randomly generated heterogeneous slip models. Slip models are constrained to have vertical deformation similar to that obtained from coral records of coseismic deformation from the Mentawai Islands for major earthquakes in 1797 and 1833, minus that observed during partial rupture of the Mentawai Segment of the Sunda Subduction Zone in 2007. The resulting maximum inundation zone is much greater than that using the original Padang consensus source model, but similar in many areas to a revision of the model that considers greater near-trench slip. While all tsunami hazard assessments are based on assumptions about the characteristics of future events, the approach presented here allows us to consider the effect of uncertainty and heterogeneity in the spatial slip distribution and therefore more robustly, although not yet completely, samples a range of possible outcomes consistent with our assumptions and models of future tsunami events. The resulting inundation footprint therefore provides a conservative indication of the potential wetted area to inform life safety measures, such as in evacuation planning.

This work was supported by the Indonesian National Disaster Management Agency (BNPB) and the Australian Department of Foreign Affairs and Trade (DFAT) through the Australia–Indonesia Facility for Disaster Reduction (AIFDR). The Indonesian Geospatial Information Agency (BIG) kindly provided airborne IFSAR elevation data for the inundation modelling. Belle Philibosian (California Institute of Technology) kindly shared coral uplift data ahead of the publication of her results and provided advice on its interpretation. František Gallovič shared code for generating random slip models. The authors wish to acknowledge productive conversations with Irina Rafliana and Nugroho Hananto (Indonesian Institute of Sciences), Emma Hill, Aron Meltzner and Jamie McCaughey (Earth observatory Singapore), Suhardjono (Agency for Meteorology, Climatology and Geophysics, Indonesia), Hong Kie Thio and Paul Somerville (URS Corporation), Trevor Dhu (AFIDR), and Phil Cummins (Geoscience Australia) during development of the tsunami models. Lesley Wyborn's (Geoscience Australia) support for our use of the National Computational Infrastructure (NCI) at the Australian National University was critical in achieving this work. Some data processing and figure preparation was done using the Generic Mapping Tools (Wessel & Smith 1998). The manuscript was improved by reviews from David Burbidge, David Robinson, Andrew Jones, Aditya Gusman and an anonymous reviewer. This paper is published with the permission of the Chief Executive Officer of Geoscience Australia.

References

Anko, B., Heiko, H., Karsten, S., Michael, S. & Jurgen, W. 2008. MFC – a modular line camera for 3D world modulling. *In*: Sommer, G. & Klette, R. (eds) *RobVis 2008, LNCS 4931*. Springer, Berlin, 319–326.

Baldock, T.E., Barnes, M.P. *et al.* 2007. Modelling tsunami inundation on coastlines with characteristic form. *In*: *Proceedings of the 16th Australasian Fluid Mechanics Conference*, Crown Plaza, Gold Coast, Australia, 2–7 December 2007, School of Engineering, The University of Queensland, Brisbane, Australia, 939–942.

Bilek, S.L. & Lay, T. 1999. Rigidity variations with depth along interplate megathrust faults in subduction zones. *Nature*, **400**, 443–446.

BNPB. 2012. *Masterplan for Tsunami Risk Reduction [Masterplan Pengurangan Risiko Bencana Tsunami]* [in Indonesian]. Badan Nasional Penanggulangan Bencana (Indonesian National Disaster Management Agency), Jakata, http://bnpb.go.id/uploads/migration/pubs/578.pdf

Borrero, J.C., Sieh, K., Chlieh, M. & Synolakis, C.E. 2006. Tsunami inundation modeling for western Sumatra. *Proceedings of the National Academy of Sciences of the United States of America*, **103**, 19673–19677.

BRIGGS, R.W., SIEH, K. *ET AL.* 2006. Deformation and slip along the Sunda megathrust in the great 2005 Nias-Simeulue earthquake. *Science*, **311**, 1897–1901.

CHLIEH, M., AVOUAC, J.P. *ET AL.* 2007. Coseismic slip and afterslip of the great Mw 9.15 Sumatra–Andaman earthquake of 2004. *Bulletin of the Seismological Society of America*, **97**, S152–S173.

CHLIEH, M., AVOUAC, J.P., SIEH, K., NATAWIDJAJA, D.H. & GALETZKA, J. 2008. Heterogeneous coupling of the Sumatran megathrust constrained by geodetic and paleogeodetic measurements. *Journal of Geophysical Research: Solid Earth*, **113**, B05305.

COLLINGS, R., LANGE, D. *ET AL.* 2012. Structure and seismogenic properties of the Mentawai segment of the Sumatra subduction zone revealed by local earthquake traveltime tomography. *Journal of Geophysical Research: Solid Earth*, **117**, B01305.

DELHOMME, J.P. 1978. Kriging in the hydrosciences. *Advances in Water Resources*, **1**, 251–266.

DI MAURO, M., MEGAWATI, K., CEDILLOS, V. & TUCKER, B. 2013. Tsunami risk reduction for densely populated Southeast Asian cities: analysis of vehicular and pedestrian evacuation for the city of Padang, Indonesia, and assessment of interventions. *Natural Hazards*, **68**, 373–404.

GALLOVIČ, F. & BROKEŠOVÁ, J. 2004. On strong ground motion synthesis with k^{-2} slip distributions. *Journal of Seismology*, **8**, 211–224.

GEIST, E.L. 2002. Complex earthquake rupture and local tsunamis. *Journal of Geophysical Research: Solid Earth*, **107**, ESE 2-1–ESE 2-16, https://doi.org/10.1029/2000JB000139

GRIFFIN, J., LATIEF, H. *ET AL.* 2015. An evaluation of onshore digital elevation models for modeling tsunami inundation zones. *Frontiers in Earth Science*, **3**.

HAYES, G.P., WALD, D.J. & JOHNSON, R.L. 2012. Slab1.0: a three-dimensional model of global subduction zone geometries. *Journal of Geophysical Research: Solid Earth*, **117**, B01302.

HERRERO, A. & BERNARD, P. 1994. A kinematic self-similar rupture process for earthquakes. *Bulletin of the Seismological Society of America*, **84**, 1216–1228.

HILL, E.M., BORRERO, J.C. *ET AL.* 2012. The 2010 M_w 7.8 Mentawai earthquake: very shallow source of a rare tsunami earthquake determined from tsunami field survey and near-field GPS data. *Journal of Geophysical Research: Solid Earth*, **117**, B06402.

HORSPOOL, N., PRANANTYO, I.R. *ET AL.* 2014. A probabilistic Tsunami Hazard assessment for Indonesia. *Natural Hazards and Earth Systems Science*, **14**, 3105–3122.

IMAMURA, F. 2009. Tsunami modelling: calculating inundation and hazard maps. *In*: BERNARD, E. & ROBINSON, A. (eds) *Tsunamis. The Sea, Ideas and Observations on Progress in the Studies of the Sea, Volume 15*. Harvard University Press, Cambridge, MA, 321–332.

JAKEMAN, J., NIELSEN, O., VAN PUTTEN, K., MLECZKO, R., BURBIDGE, D. & HORSPOOL, N. 2010. Towards spatially distributed quantitative assessment of tsunami inundation models. *Ocean Dynamics*, **60**, 1115–1138, https://doi.org/10.1007/s10236-010-0312-4

KAISER, G., SCHEELE, L., KORTENHAUS, A., LØVHOLT, F., RÖMER, H. & LESCHKA, S. 2011. The influence of land cover roughness on the results of high resolution tsunami inundation modeling. *Natural Hazards and Earth Systems Science*, **11**, 2521–2540.

KANAMORI, H. 1972. Mechanism of tsunami earthquakes. *Physics of the Earth and Planetary Interiors*, **6**, 346–359.

KONCA, A.O., AVOUAC, J.P. *ET AL.* 2008. Partial rupture of a locked patch of the Sumatra megathrust during the 2007 earthquake sequence. *Nature*, **456**, 631–635.

LAVALLÉE, D., MIYAKE, H. & KOKETSU, K. 2011. Stochastic model of a subduction-zone earthquake: sources and ground motions for the 2003 Tokachi-oki, Japan, earthquake. *Bulletin of the Seismological Society of America*, **101**, 1807–1821.

LAY, T., KANAMORI, H. *ET AL.* 2012. Depth-varying rupture properties of subduction zone megathrust faults. *Journal of Geophysical Research: Solid Earth*, **117**, B04311.

LI, Y. & RAICHLEN, F. 2002. Non-breaking and breaking solitary wave run-up. *Journal of Fluid Mechanics*, **456**, 295–318.

LØVHOLT, F., PEDERSEN, G., BAZIN, S., KÜHN, D., BREDESEN, R.E. & HARBITZ, C. 2012. Stochastic analysis of tsunami runup due to heterogeneous coseismic slip and dispersion. *Journal of Geophysical Research: Oceans*, **117**, C03047.

MCCLOSKEY, J., ANTONIOLI, A. *ET AL.* 2007. Near-field propagation of tsunamis from megathrust earthquakes. *Geophysical Research Letters*, **34**, L14346.

MCCLOSKEY, J., ANTONIOLI, A. *ET AL.* 2008. Tsunami threat in the Indian Ocean from a future megathrust earthquake West of Sumatra. *Earth and Planetary Science Letters*, **265**, 61–81.

MCCLOSKEY, J., LANGE, D., TILMANN, F., NALBANT, S.S., BELL, A.F., NATAWIDJAJA, D.H. & RIETBROCK, A. 2010. The September 2009 Padang earthquake. *Nature Geoscience*, **3**, 70–71.

MELTZNER, A.J., SIEH, K. *ET AL.* 2012. Persistent termini of 2004- and 2005-like ruptures of the Sunda megathrust. *Journal of Geophysical Research: Solid Earth*, **117**, B04405.

NATAWIDJAJA, D.H., SIEH, K. *ET AL.* 2006. Source parameters of the great Sumatran megathrust earthquakes of 1797 and 1833 inferred from coral microatolls. *Journal of Geophysical Research: Solid Earth*, **111**, 1–37.

NATAWIDJAJA, D.H., SIEH, K., GALETZKA, J., SUWARGADI, B.W., CHENG, H., EDWARDS, R.L. & CHLIEH, M. 2007. Interseismic deformation above the Sunda Megathrust recorded in coral microatolls of the Mentawai islands, West Sumatra. *Journal of Geophysical Research: Solid Earth*, **112**, B02404.

NIELSEN, O., ROBERTS, S., GRAY, D., MCPHERSON, A. & HITCHMAN, A. 2005. Hydrodynamic modelling of coastal inundation. *In*: *MODSIM 2005 International Congress on Modelling and Simulation*. Modelling and Simulation Society of Australia and New Zealand (MSSANZ), Canberra, 518–523, http://www.mssanz.org.au/modsim05/papers/nielsen.pdf

PEBESMA, E.J. & WESSELING, C.G. 1998. Gstat, a program for geostatistical modelling, prediction and simulation. *Computers and Geosciences*, **24**, 17–31.

PERFETTINI, H. & AVOUAC, J.P. 2014. The seismic cycle in the area of the 2011 M_w 9. 0 Tohoku-Oki Earthquake. *Journal of Geophysical Research: Solid Earth*, **119**, 4469–4515, https://doi.org/10.1002/2013JB010697

PHILIBOSIAN, B., SIEH, K. *ET AL.* 2012. An ancient shallow slip event on the Mentawai segment of the Sunda megathrust, Sumatra. *Journal of Geophysical Research: Solid Earth*, **117**, B05401.

PHILIBOSIAN, B., SIEH, K. *ET AL.* 2014. Rupture and variable coupling behavior of the Mentawai segment of the Sunda Megathrust during the supercycle culmination of 1797–1833. *Journal of Geophysical Research: Solid Earth*, **119**, 7258–7287, https://doi.org/10.1002/2014JB011200

RUDNICKI, J.W. & WU, M. 1995. Mechanics of dip-slip faulting in an elastic half-space. *Journal of Geophysical Research: Solid Earth*, **100**, 22 173–22 186.

SATAKE, K. 1994. Mechanism of the 1992 Nicaragua tsunami earthquake. *Geophysical Research Letters*, **21**, 2519–2522.

SATAKE, K., FUJII, Y., HARADA, T. & NAMEGAYA, Y. 2013. Time and space distribution of coseismic slip of the 2011 Tohoku earthquake as inferred from tsunami waveform data. *Bulletin of the Seismological Society of America*, **103**, 1473–1492.

SCHLURMANN, T., KONGKO, W., GOSEBERG, N., NATAWIDJAJA, D.H. & SIEH, K. 2010. Near-Field Tsunami Hazard Map Padang, West Sumatra: utilizing high resolution geospatial data and reasonable source scenarios. *In*: MCKEE SMITH, J. & LYNETT, P. (eds) *Proceedings of 32nd Conference on Coastal Engineering*, Shanghai, China, 2010. Curran Associates, Red Hook, NY, 3686–3702.

SIEH, K. 2006. Sumatran megathrust earthquakes: from science to saving lives. *Philosophical Transactions: Mathematical, Physical and Engineering Sciences*, **364**, 1947–1963.

SIEH, K., NATAWIDJAJA, D.H. *ET AL.* 2008. Earthquake supercycles inferred from sea-level changes recorded in the corals of West Sumatra. *Science*, **322**, 1674–1678.

SINGH, S.C., HANANTO, N.D., CHAUHAN, A.P., PERMANA, H., DENOLLE, M., HENDRIYANA, A. & NATAWIDJAJA, D. 2010. Evidence of active backthrusting at the NE Margin of Mentawai Islands, SW Sumatra. *Geophysical Journal International*, **180**, 703–714.

SINGH, S.C., HANANTO, N., MUKTI, M., PERMANA, H., DJAJADIHARDJA, Y. & HARJONO, H. 2011. Seismic images of the megathrust rupture during the 25th October 2010 Pagai earthquake, SW Sumatra: frontal rupture and large tsunami. *Geophysical Research Letters: Solid Earth*, **38**, L16313.

SOMERVILLE, P. 2003. *Scaling Relations of Asperity Size, Number and Stress Drop in Crustal and Subduction Earthquakes*. Report prepared for the Ohsaki Research Institute.

STEIN, S., GELLER, R.J. & LIU, M. 2012. Why earthquake hazard maps often fail and what to do about it. *Tectonophysics*, **562–563**, 1–25.

SUPPASRI, A., SHUTO, N., IMAMURA, F., KOSHIMURA, S., MAS, E. & YALCINER, A.C. 2013. Lessons learned from the 2011 Great East Japan tsunami: performance of tsunami countermeasures, coastal buildings, and tsunami evacuation in Japan. *Pure and Applied Geophysics*, **170**, 993–1018.

TANIOKA, Y. & SATAKE, K. 1996. Tsunami generation by horizontal displacement of ocean bottom. *Geophysical Research Letters*, **23**, 861–864.

TAUBENBÖCK, H., GOSEBERG, N. *ET AL.* 2009. 'Last-Mile' preparation for a potential disaster – Interdisciplinary approach towards tsunami early warning and an evacuation information system for the coastal city of Padang, Indonesia. *Natural Hazards and Earth Systems Sciences*, **9**, 1509–1528, https://doi.org/10.5194/ nhess-9-1509-2009

WANG, R., MARTÍN, F.L. & ROTH, F. 2003. Computation of deformation induced by earthquakes in a multilayered elastic crust – FORTRAN programs EDGRN/EDCMP. *Computers and Geosciences*, **29**, 195–207.

WESSEL, P. & SMITH, W.H.F. 1998. New, improved version of generic mapping tools released. *Eos, Transactions of the American Geophysical Union*, **79**, 579, https://doi.org/10.1029/98EO00426

WISEMAN, K., BANERJEE, P., SIEH, K., BÜRGMANN, R. & NATAWIDJAJA, D.H. 2011. Another potential source of destructive earthquakes and tsunami offshore of Sumatra. *Geophysical Research Letters: Solid Earth*, **38**, L03111.

Fault systems of the eastern Indonesian triple junction: evaluation of Quaternary activity and implications for seismic hazards

IAN M. WATKINSON* & ROBERT HALL

SE Asia Research Group, Department of Earth Sciences, Royal Holloway University of London, Egham, Surrey TW20 0EX, UK

**Correspondence: ian.watkinson@rhul.ac.uk*

Abstract: Eastern Indonesia is the site of intense deformation related to convergence between Australia, Eurasia, the Pacific and the Philippine Sea Plate. Our analysis of the tectonic geomorphology, drainage patterns, exhumed faults and historical seismicity in this region has highlighted faults that have been active during the Quaternary (Pleistocene to present day), even if instrumental records suggest that some are presently inactive. Of the 27 largely onshore fault systems studied, 11 showed evidence of a maximal tectonic rate and a further five showed evidence of rapid tectonic activity. Three faults indicating a slow to minimal tectonic rate nonetheless showed indications of Quaternary activity and may simply have long interseismic periods. Although most studied fault systems are highly segmented, many are linked by narrow (<3 km) step-overs to form one or more long, quasi-continuous segment capable of producing $M > 7.5$ earthquakes. Sinistral shear across the soft-linked Yapen and Tarera–Aiduna faults and their continuation into the transpressive Seram fold–thrust belt represents perhaps the most active belt of deformation and hence the greatest seismic hazard in the region. However, the Palu–Koro Fault, which is long, straight and capable of generating super-shear ruptures, is considered to represent the greatest seismic risk of all the faults evaluated in this region in view of important strike-slip strands that appear to traverse the thick Quaternary basin-fill below Palu city.

During the last decade several devastating earthquakes occurred on faults around the world that were either poorly understood or not recognized at all. The M_w 6.6 Bam earthquake (Iran) of 26 December 2003 ruptured a section of the Bam Fault that had a poor surface expression and had not caused a destructive earthquake for 2000 years (Eshghi & Zare 2003; Fu *et al.* 2004). The M_w 8.0 Wenchuan earthquake (China) of 12 May 2008 resulted from complex rupture of part of the Lonmen Shan tectonic belt (Burchfiel *et al.* 2008), an area that was previously considered not to be at risk from large earthquakes (Chen & Hsu 2013). The M_w 7.1 Haiti earthquake of 12 January 2010 occurred on the well-known Enriquillo Fault, part of the fault system marking the northern boundary of the Caribbean plate, but which had previously been mapped as having low seismic hazard based on recent seismicity (Stein *et al.* 2012). The Canterbury earthquake sequence (New Zealand) ruptured the Greendale Fault, which was previously unrecognized because it was buried beneath alluvial sediments (Quigley *et al.* 2012). The Canterbury sequence culminated in the M_w 6.3 Christchurch earthquake of 22 February 2011. These events emphasize the need for the accurate identification of faults that have been active during the Quaternary and have the potential for modern tectonic activity.

Eastern Indonesia is a region of complex and rapid neotectonics. Convergence between Australia, Eurasia, the Pacific and the Philippine Sea plates (e.g. Hamilton 1979; DeMets *et al.* 1994; Hall 1996, 2012; Bock *et al.* 2003; Charlton 2010) results in both contraction and extension from subduction hinge rollback, lithospheric delamination and slab break-off (e.g. Harris 1992; Spakman & Hall 2010; Hall 2012).

Great uncertainty surrounds the position, tectonic role and modern activity of eastern Indonesia's many Quaternary faults (e.g. Hamilton 1979; Okal 1999; Bailly *et al.* 2009; Charlton 2010). New fault systems continue to be identified using both modern geophysical/remote sensing and conventional field techniques (e.g. Stevens *et al.* 2002; Spencer 2010, 2011; Watkinson *et al.* 2011; Pownall *et al.* 2013) and it is likely that many others remain unknown, with important implications for seismic hazard analysis.

Despite intense seismicity in eastern Indonesia, there have been few catastrophic earthquake disasters in the last 100 years compared with other rapidly deforming areas such as China, Iran, Japan and Pakistan (e.g. Holzer & Savage 2013; National Geophysical Data Center/World Data Service). Significant events include: the 25 June 1976 M_w 7.1 Papua earthquake, which killed 3000–6000

From: CUMMINS, P. R. & MEILANO, I. (eds) 2017. *Geohazards in Indonesia: Earth Science for Disaster Risk Reduction*. Geological Society, London, Special Publications, **441**, 71–120.
First published online December 19, 2016, updated April 19, 2017, https://doi.org/10.1144/SP441.8

people; the 12 December 1992 M_w 7.8 Flores earthquake, which killed 2500 people and destroyed 31 800 houses; the 17 February 1996 M_w 8.2 Biak earthquake, which caused a 7 m tsunami and killed at least 100 people (Okal 1999); the 16 November 2008 M_w 7.4 Minahasa earthquake, which killed six people and displaced 10 000; and the 16 June 2010 M_w 7.0 Yapen earthquake, which killed 17 people and destroyed 2556 houses (National Geophysical Data Center/World Data Service; USGS Earthquake Hazards Program). With increasing urban development and the replacement of traditional wooden dwellings with concrete constructions, it is likely that damaging earthquakes will become more frequent in the future (e.g. Wyss 2005).

This paper catalogues Quaternary fault systems onshore eastern Indonesia from Sulawesi to Papua, providing evidence for Quaternary tectonic activity and a reconnaissance evaluation of the seismic hazard of the faults (Fig. 1).

Methods

Definitions and extent of study

This study was concerned with evaluating Quaternary (Pleistocene and Holocene, 2.59–0 Ma) fault activity. Quaternary activity lies within the realm of neotectonics, the study of broadly post-Miocene, 'young' and still-active tectonic events, the effects of which are compatible with modern seismotectonics (Pavlides 1989). Neotectonics is distinct from palaeoseismology – the study of deformation related to specific past earthquakes (e.g. Michetti *et al.* 2005). Thus faults that show evidence of Quaternary activity may or may not also show evidence of palaeoseismicity, depending on whether they have recently ruptured the surface, the rates of sedimentation and erosion, and whether they are truly 'active' in the sense that they have failed during the Holocene. Equally, Quaternary faults may or may not be present in records of instrumental or historical seismicity, depending on whether they have recently become inactive, have a long interseismic period, or have yielded historical earthquakes in locations where there was no written documentation. Quaternary fault activity is therefore distinct from, but influential in, the field of active tectonics, which includes future fault activity that may affect human society (Wallace 1986).

Quaternary fault activity was evaluated in this study by the following criteria: (1) instrumental/historical seismicity and geodetic observations; (2) deformation of Quaternary sediments, often indicated by topographic lineaments that could be linked to an underlying fault; (3) the systematic offset of modern streams across a topographic lineament; (4) evidence of structurally controlled drainage network modification where signs of an earlier arrangement were preserved; (5) geomorphic indices recording the relative youthfulness of fault-controlled mountain fronts; and (6) evidence of landslips localized to faults.

The study extent was a 2200 × 800 km swath of the Indonesian archipelago centred on the triple junction between Australia, Eurasia, the Pacific and the Philippine Sea plates. It includes much of eastern Indonesia from Sulawesi eastwards, except

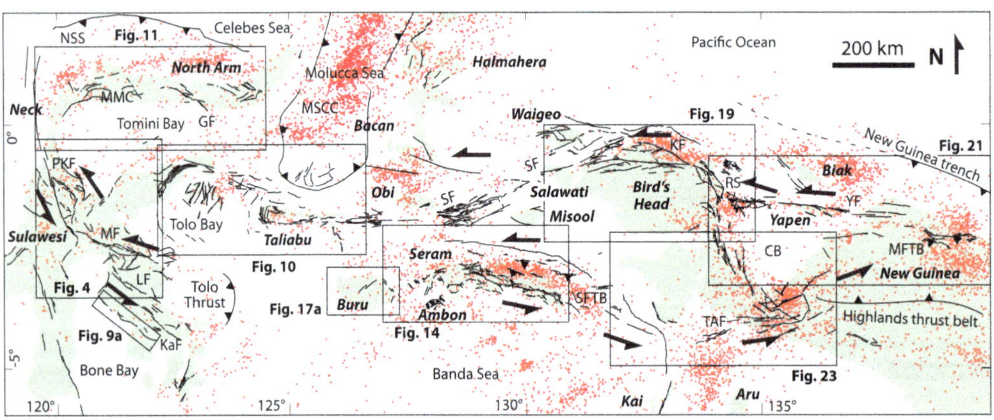

Fig. 1. Map of eastern Indonesia showing upper crustal structures with geomorphic evidence of Quaternary tectonic activity and seismicity (1973–2014, focal depths <35 km). CB, Cenderawasih Bay; GF, Gorontalo Fault; KaF, Kolaka Fault; KF, Koor Fault; LF, Lawanopo Fault; MF, Matano Fault; MFTB, Mamberamo fold–thrust belt; MMC, Molino Metamorphic Complex; MSCC, Molucca Sea Collision Complex; NSS, North Sulawesi Subduction; PKF, Palu–Koro Fault; RS, Ransiki Fault; SF, Sorong Fault; SFTB, Seram fold–thrust belt; TAF, Tarera–Aiduna Fault; YF, Yapen Fault. Locations of other figures as indicated.

the islands of the southern Banda Arc. Because of the focus on geomorphic expression, the study mainly dealt with onshore faults, except where multibeam bathymetry was available.

Datasets

Our interpretations of Quaternary fault activity are based on a variety of remote sensing data, field observations by both authors and their students over several years (e.g. Roques 1999; Watkinson 2011; Pownall *et al.* 2013; Hennig *et al.* 2014) and published geodetic/geophysical data. Digital elevation models (DEMs) based on Shuttle Radar Topography Mission (SRTM) 3 arc second/90 m resolution and ASTER 30 m resolution data were processed using ERMapper software. These data were also used to extract topographic contours and drainage networks using ArcGIS. Landsat TM and ETM+ scenes composed of the 30 m resolution bands 432, 451, 531 and 742 (red–green–blue combinations) were used and, where appropriate, sharpened with ETM+ band 8 panchromatic 15 m resolution data. Where available, high-resolution visible spectrum imagery from Google Earth and Bing Maps (compiled from a variety of sources) and the ESRI World Imagery compilation, which includes 2.5 m SPOT and <1 m DigitalGlobe imagery, was also interpreted. ESRI World Imagery is compiled from Esri, DigitalGlobe, GeoEye, Earthstar Geographics, CNES/Airbus DS, USDA, USGS, AEX, Getmapping, Aerogrid, IGN, IGP, swisstopo, and the GIS User Community.

Multibeam bathymetric data (kindly provided by TGS) from parts of the offshore Sorong Fault Zone and Cenderawasih Bay were interpreted in the same way as the DEMs. The multibeam data were acquired using a Kongsberg Simrad EM120 Multibeam Echo Sounder using 191 beams at equidistant spacing. Positioning control used a C-Nav Starfire DGPS. During processing, positioning, tidal and calibration corrections were applied, random noise and artefacts were removed, and a terrain model using a 25 m bin size was gridded and exported to ESRI format. The multibeam data were further processed in ERMapper to remove voids.

All data were integrated in ArcGIS together with previously published georeferenced maps. The CMT focal mechanisms were from the International Seismological Centre catalogue, plotted using Mirone software. We considered only earthquakes with a focal depth $\leq$35 km to avoid contamination from deeper structures that have little surface expression.

Geomorphic indices

Geomorphic indices are a valuable tool to rapidly evaluate the relative tectonic rate of surface faults on a reconnaissance scale (Keller 1986). We utilized mountain front sinuosity (S_{mf}) and the valley floor width to valley height index (V_f) following the method of Bull & McFadden (1977) and Bull (1978). The key parameters are summarized in Table 1. An excellent description of the method and its uncertainties is given in Bull (2007). Although conventionally applied to normal faults, geomorphic indices can been used in any setting where there is vertical motion, including regions of transpression and transtension. However, they are of little value in regions of pure strike-slip and were not applied to pure strike-slip segments in this study.

Mountain front sinuosity is the ratio $S_{mf} = L_{mf}/L_s$, where L_{mf} is the straight line length of the mountain front and L_s is the true, or sinuous, length along the mountain front following topographic contours at the contact between alluvial fans and the solid geology of the range front (Table 1). This method assumes that a fault-bounded range front will become more sinuous over time in the absence of tectonic activity (e.g. Bull & McFadden 1977; Rockwell *et al.* 1984). The method is well established for Quaternary fault evaluation in regions of extension (e.g. Ramírez-Herrera 1998), contraction and strike-slip (e.g. Dehbozorgi *et al.* 2010), transtension (e.g. Silva *et al.* 2003; Yıldırım 2014), combined extension and contraction (Wells *et al.* 1988) and differential uplift (e.g. Sohoni *et al.* 1999). Critical uncertainties include the interpreter's definition of the sinuous mountain front, which is partly dependent on the quality of the input satellite data, and the recognition of discrete mountain front segments. Climate also has an impact on S_{mf} independent of the tectonic rate. In a humid environment like eastern Indonesia it is expected that erosion and hence S_{mf} will be higher than in an arid region for a given tectonic rate.

The valley floor width to valley height index, V_f, measures the ratio between the valley floor width and the valley depth: $V_f = 2V_{fw}/(E_{ld} - E_{sc}) - (E_{rd} - E_{sc})$, where V_{fw} is the valley floor width, E_{ld} and E_{rd} are the topographic elevations of the left and right valley watersheds and E_{sc} is the elevation of the valley floor (Table 1). The method assumes that recently excavated river channels (i.e. those into which a river has incised as a result of recent uplift) are V-shaped and become more U-shaped over time (e.g. Bull & McFadden 1977; Rockwell *et al.* 1984). Like S_{mf}, V_f has been applied in a wide range of tectonic settings (e.g. Wells *et al.* 1988; Ramírez-Herrera 1998; Yıldırım 2014). V_f is sensitive to a number of variables apart from tectonic rate, so we standardized as much as possible by: measuring V_f in all cases 1 km upstream from the mountain front; measuring the valley width as the width of the river channel visible on the highest resolution satellite imagery available or the width

Table 1. *Summary of geomorphic indices used in mountain front analysis, modified after Wells* et al. *(1988)*

Parameter	Definition	Derivation*	Measurement†	Purpose	Potential difficulties
S_{mf}	Sinuosity of topographic mountain fronts	L_{mf}/L_s	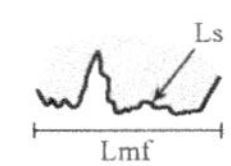	Defines the degree of topographic modification of mountain front from the position of possible controlling structures	Defining actual topographic junction Defining discrete mountain front segments
V_f	Valley floor width to valley height index	$2V_{fw}/[(E_{ld} - E_{sc}) - (E_{rd} - E_{sc})]$	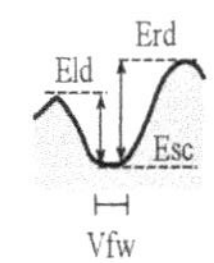	Defines the ratio of the valley floor width to the mean height of two adjacent divides, measured at given locations along a stream channel within the range block	Resolution of satellite imagery in defining V_{fw} and divide elevations Effects of changes in lithology Need to minimize variations in stream size (length and area)

*L_{mf}, straight line length of mountain front; L_s, sinuous length along mountain front; V_{fw}, valley floor width; E_{ld} and E_{rd}, topographic elevations of left and right valley watersheds; E_{sc}, elevation of valley floor.
†Schematic map view for S_{mf}, schematic cross-section view for V_f.
Both indices after Bull & McFadden (1977) and Bull (1978).

of the valley to the point where the floor rose 10 m above the minimum elevation in individual transects; measuring only streams that reached the mountain front without joining a higher order stream; and measuring only streams oriented $\geq 70^\circ$ from the mountain front. Noise in the V_f signal was reduced by averaging between three and ten separate V_f measurements along each fault segment.

High-quality topographic maps are not available for eastern Indonesia, so both S_{mf} and V_f were measured in ArcGIS software using a combination of 30 m ASTER GDEM satellite data and the ESRI World Imagery compilation. This allowed the finest possible resolution of L_s and V_{fw}, which are crucial, but potentially subjective, parameters. High-quality satellite imagery may be better for such measurements than conventional maps (Bull 2007).

Schemes for the classification of relative tectonic activity based on a combination of geomorphic indices have been proposed (e.g. Bull & McFadden 1977; Bull 1978, 2007). Here we applied a modified scheme from McCalpin (2009). This uses S_{mf} and V_f to classify relative tectonic activity as follows: $S_{mf} < 1.1$, mean $V_f < 0.15$, maximal activity; S_{mf} 1.1–1.3, mean V_f 0.15, rapid activity; S_{mf} 1.6–2.3, mean V_f 1.5, slow activity; $S_{mf} \geq 2.5$, V_f 1.7–2.5, minimal activity; and S_{mf} 2.6–4.0, mean V_f 7.4, inactive. This classification allows a comparison between faults with different relative tectonic rates and corresponding geomorphic expression. Because the indices record undated Quaternary fault activity expressed by geomorphology, the classes also correspond to a Quaternary tectonic rate and not necessarily to a modern tectonic rate comparable with geodetic measurements. It should also be remembered that the schemes were developed using faults in arid areas of the western USA where tectonic landforms are preserved for longer than in humid areas (e.g. Bull 1978), meaning faults in the tropics will generally be classified as tectonically 'slower' than equivalent faults at higher latitudes.

We analysed both S_{mf} and V_f for a total of 111 segments from 24 fault systems across the study area (Fig. 2a–r, Table 2) and found a good correlation between S_{mf} and V_f (Fig. 3), supporting the reliability of each method. A previous study of geomorphic indices along a segment of the Palu–Koro Fault (Vecchiotti 2008) obtained similar results to those presented here. However, we used these indices only as a simple quantitative means to support other evidence for Quaternary fault activity and did not classify faults on the basis of these data alone.

Sulawesi

Sulawesi lies at the triple junction between the Australian, Eurasian and Philippine Sea plates (e.g. Hamilton 1979; Silver *et al.* 1983*a*, *b*; Hall 1996). North of Sulawesi, the Celebes Sea is being subducted beneath Sulawesi (e.g. Hamilton 1979; Silver *et al.* 1983*a*). Convergence across the subduction margin increases from 20 ± 4 mm a^{-1} in the east to 54 ± 10 mm a^{-1} in the west, associated with a clockwise rotation of about 4° Ma^{-1} about a pole close to Manado (Walpersdorf *et al.* 1998; Rangin *et al.* 1999; Stevens *et al.* 1999; Beaudouin *et al.* 2003). Immediately east of Sulawesi's north 'arm', convergence between the Philippine Sea plate and Sundaland is partly accommodated by the Molucca Sea double subduction and the overlying Sangihe and Halmahera thrusts (e.g. Rangin *et al.* 1999; Hall 2002; Beaudouin *et al.* 2003).

Despite its setting within a collisional orogen, Sulawesi is subject to widespread and young extension. Tomini Bay encloses a deep, enigmatic basin containing up to 10 km of late Cenozoic sediments (Jablonski *et al.* 2007; Pholbud *et al.* 2012). Medium- to high-K Pliocene to modern volcanism in the Togian Islands within the bay results from Pliocene to Recent extension (Cottam *et al.* 2011) and onshore metamorphic core complexes are in the process of being exhumed by processes related to crustal thinning (Kavalieris *et al.* 1992; van Leeuwen *et al.* 2007; Spencer 2011).

Active strike-slip faults (e.g. Bellier *et al.* 2001), with left-lateral slip rates of up to 39 mm a^{-1} (Socquet *et al.* 2006), characterize much of Sulawesi's onshore Quaternary deformation. Often considered to result from NW-directed collision between the Sula platform and Sulawesi (e.g. Silver *et al.* 1983*b*; Simandjuntak 1986), modern reconstructions emphasize the process of subduction hinge rollback related to the substantial amounts of oceanic crust that have been, and continue to be, subducted around Sulawesi (e.g. Spakman & Hall 2010; Hall 2012). The occurrence of Late Miocene to apparently modern north–south-directed continental extension (e.g. Spencer 2011) in a broad region adjacent to the south-directed Celebes Sea subduction means that a rollback mechanism must be considered.

Palu–Koro Fault

The Palu–Koro Fault (Fig. 4) is the most prominent active fault of Sulawesi and is of particular importance because it is straddled by Palu city (population 340 000). The Palu–Koro Fault appears to pass from the SW corner of the Celebes Sea to a diffuse termination onshore at the northern end of Bone Bay, a distance of 500 km, of which 220 km is onshore.

The fault's tectonic role is disputed: sinistral shear along a joint Palu–Koro–Matano Fault system has been thought to accommodate clockwise rotation and the northwards movement of a rigid

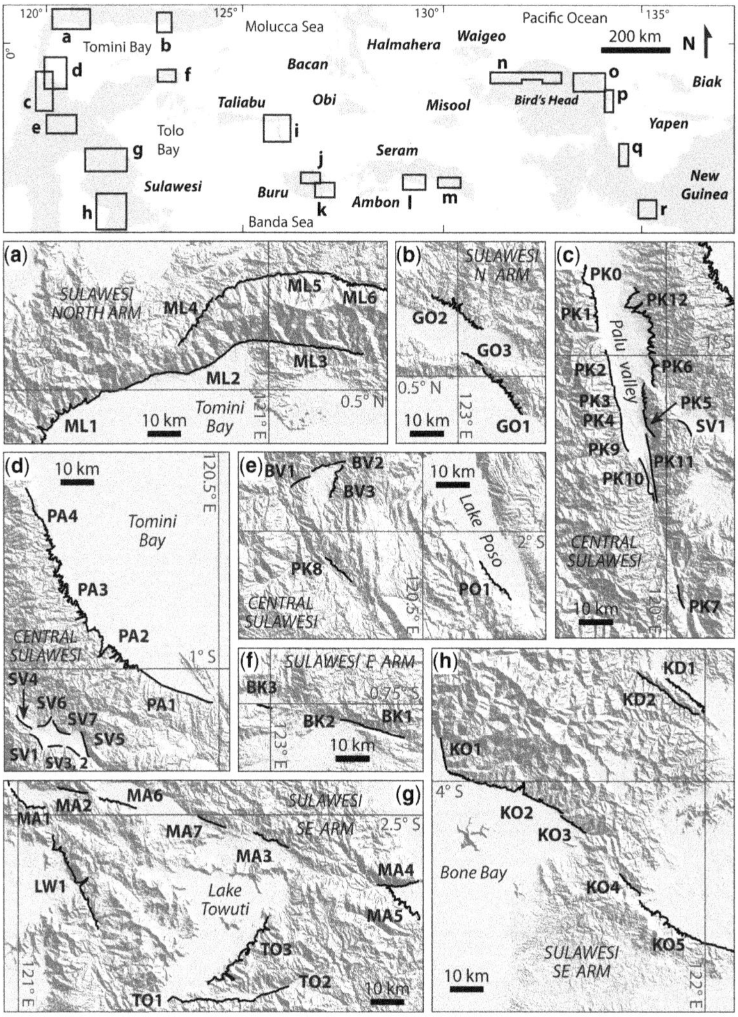

Fig. 2. Maps showing fault segments analysed for geomorphic indices. Index map at top. Bold lines are the sinuous mountain front trace (L_s) used in mountain front sinuosity calculations. Base map is a 90 m SRTM digital elevation model. All maps (a–r) drawn to the same scale. Fault segment codes correspond to the codes used in Table 2.

eastern Sulawesi block driven by collision of the Banggai–Sula block in the east (e.g. Hamilton 1979; Silver *et al.* 1983*b*; Beaudouin *et al.* 2003). However, it is significant that the Palu–Koro Fault and the North Sulawesi Trench form the western and northern limits, respectively, of a region of late Cenozoic extreme continental extension that includes deep sedimentary basins (e.g. Jablonski *et al.* 2007; Pholbud *et al.* 2012), exhumation of the mid- to lower crust in settings similar to metamorphic core complexes (e.g. van Leeuwen *et al.* 2007; Watkinson 2011), exhumed low-angle normal faults (Spencer 2011) and decompression-related mantle melts (Cottam *et al.* 2011). These features can be associated with the overriding plate above a retreating subduction hinge, particularly in the early stages of continent–continent collision (Royden 1993). The orientation and kinematics of the Palu–Koro Fault are compatible with an interpretation that it is passively bounding a region of lithospheric extension driven by northwards rollback in the Celebes Sea, although it is unclear whether there is a hard linkage between the fault and the trench.

It is not disputed that the fault is an active zone of high strain. Geodetic measurements suggest a 39 mm a^{-1} sinistral slip rate together with 11–14 mm a^{-1} of extension (Socquet *et al.* 2006),

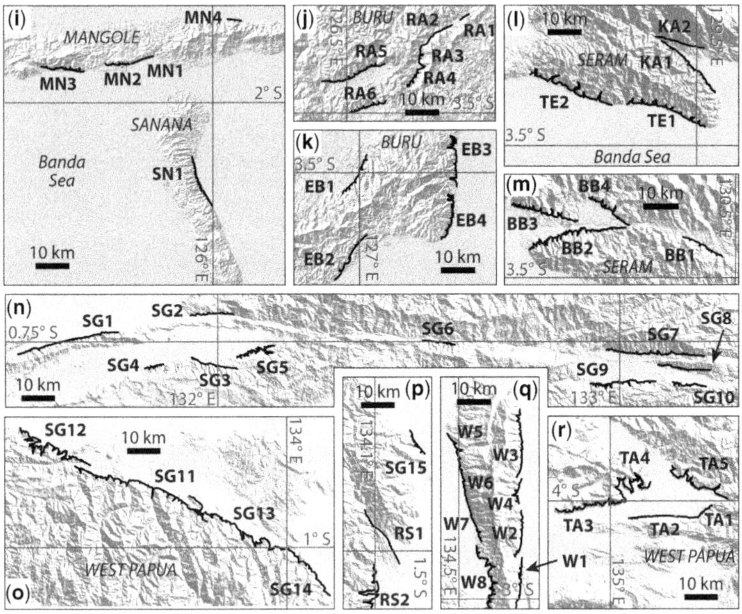

Fig. 2. *Continued.*

consistent with a 35 ± 8 mm a^{-1} strike-slip rate determined from displaced alluvial fans 11 000 ± 2300 years old (Bellier *et al.* 2001).

There is palaeoseismic evidence for three M_w 6.8–8.0 earthquakes during the last 2000 years, suggesting a recurrence interval of about 700 years (Beaudouin 1998; Bellier *et al.* 1998). However, even allowing for 10 m slip for each M_w 6.8–8.0 event, the resultant 30 m total displacement in 2000 years is less than the 54–86 m predicted from Holocene slip rates (Bellier *et al.* 2001). Although it has been proposed by these earlier researchers that the deficit is accommodated by aseismic creep, it is equally possible that large, undetected earthquakes occurred on unobserved fault strands and that the total recurrence interval for all the Palu–Koro Fault strands is much less than 700 years. Socquet *et al.* (2006) proposed four parallel strands across a zone *c.* 50 km wide, locked at depths between 0 and 5 km.

Records of historical seismicity in Sulawesi are poor. Damaging earthquakes occurred along the Palu–Koro Fault in 1905, 1907, 1909, 1927, 1934, 1968 (*c.* M_s 6.7), 1985 and 1993 (*c.* M_s 5.7) (Katili 1970; Hamilton 1979; Beaudouin 1998), but little detail is known. Large earthquakes close to the fault zone occurred in 1996 (M_w 7.7) and 1998 (M_w 6.6 and 6.0); the former caused a 2–4 m high tsunami in the Toli Toli region (Pelinovsky *et al.* 1997). However, these earthquakes originated offshore, did not clearly lie on the active Palu–Koro Fault and none had a focal mechanism indicating left-lateral slip along the Palu–Koro trend.

The Palu–Koro Fault has the clearest geomorphological expression of any eastern Indonesian fault. It occupies a steep-sided, narrow valley along much of its path through central Sulawesi, before branching into the Palu valley, which is up to 15 km wide (Fig. 5a). Two prominent scarps bound the valley and form the base of mountains that rise to >2.3 km elevation. The western scarp is highly linear, particularly the remarkable central segments *c.* 15 km south of Palu city. Mountain front sinuosity values are consistently low at 1.08–1.09, indicating maximal tectonic activity, increasing to 1.28–1.56 at the northern and southern ends of the valley, indicating rapid to moderate tectonic activity (Fig. 5a). The valley floor curvature is generally correspondingly tight, with an average V_f of 0.24 along the western scarp.

Features such as prominent triangular facets, hanging valleys and steep-sided, deeply incised streams are also focused along the central western basin-bounding segment (Fig. 5b). These landforms support dominantly rapid normal faulting along the basin margin faults. Wine glass canyons, in particular, indicate that the tectonic subsidence/uplift rate is faster than erosion. Lateral offset of the alluvial fans and rivers across the mountain front have been observed, notably in the northern and southern segments of the fault system (e.g. Hamilton 1979; Bellier *et al.* 2006).

Table 2. *Summary of measurements of mountain front sinuosity and average valley width to height ratio for analysed fault segments*

Fault	Segment	L_{mf}*	$L_s^{\dagger}$	$S_{mf}^{\ddagger}$	Ave. $V_f^{\S}$	Figure 2[‖]	Fault	Segment	L_{mf}*	$L_s^{\dagger}$	$S_{mf}^{\ddagger}$	Ave. $V_f^{\S}$	Figure 2[‖]
Malino boundary	ML1	45.16	27.17	1.66	1.01	a	Kolaka	KO1	8.79	8.38	1.05	1.25	h
Malino boundary	ML2	52.00	49.60	1.05	0.33	a	Kolaka	KO2	52.12	33.80	1.54	1.12	h
Malino boundary	ML3	28.79	27.53	1.05	0.64	a	Kolaka	KO3	10.07	8.24	1.22	1.68	h
Malino boundary	ML4	22.99	16.08	1.43	0.29	a	Kolaka	KO4	10.25	7.91	1.30	0.23	h
Malino boundary	ML5	38.22	33.86	1.13	0.22	a	Kolaka	KO5	49.44	30.21	1.64	1.19	h
Malino boundary	ML6	27.41	17.27	1.59	0.29	a		**Average**			**1.35**	**1.09**	
	Average			**1.32**	**0.46**		Mangole	MN1	7.07	6.33	1.12	0.49	i
Gorontalo	GO1	33.24	16.37	2.03	0.88	b	Mangole	MN2	9.64	6.60	1.46	0.49	i
Gorontalo	GO2	39.29	16.64	2.36	1.27	b	Mangole	MN3	18.96	12.08	1.57	0.55	i
Gorontalo	GO3	12.03	6.58	1.83	1.69	b	Mangole	MN4	4.56	3.86	1.18	N/A	i
	Average			**2.07**	**1.28**		Sanana	SN1	2.45	1.93	1.27	0.44	i
Palu–Koro	PK0	16.94	10.99	1.54	0.31	c		**Average**			**1.32**	**0.49**	
Palu–Koro	PK1	8.94	6.60	1.35	0.22	c	Rana	RA1	2.86	2.83	1.01	0.35	j
Palu–Koro	PK2	10.48	9.64	1.09	0.29	c	Rana	RA2	3.68	3.12	1.18	0.23	j
Palu–Koro	PK3	7.24	6.69	1.08	0.21	c	Rana	RA3	8.19	7.73	1.06	0.18	j
Palu–Koro	PK4	4.33	3.99	1.09	0.19	c	Rana	RA4	20.19	10.31	1.96	1.53	j
Palu–Koro	PK5	9.43	7.90	1.19	0.99	c	Rana	RA5	24.35	18.31	1.33	0.28	j
Palu–Koro	PK6	11.02	6.91	1.59	0.35	c	Rana	RA6	15.49	10.40	1.49	0.68	j
Palu–Koro	PK7	7.15	6.44	1.11	0.56	c	East Buru	EB1	18.09	12.56	1.44	0.47	k
Palu–Koro	PK8	10.78	9.61	1.12	0.89	e	East Buru	EB2	18.61	15.08	1.23	0.50	k
Palu–Koro	PK9	9.72	6.22	1.56	0.20	c	East Buru	EB3	26.93	12.53	2.15	1.13	k
Palu–Koro	PK10	16.34	12.80	1.28	1.10	c	East Buru	EB4	25.16	12.62	1.99	0.75	k
Palu–Koro	PK11	27.15	19.02	1.43	0.32	c		**Average**			**1.48**	**0.61**	
Palu–Koro	PK12	64.23	27.88	2.30	0.80	c	Southern Seram	TE1	42.32	23.05	1.84	1.88	l
	Average			**1.36**	**0.47**		Southern Seram	TE2	49.68	23.89	2.08	1.36	l
Parigi boundary	PA1	69.67	21.44	3.25	0.90	d		**Average**			**1.96**	**1.62**	
Parigi boundary	PA2	72.43	26.17	2.77	0.78	d	Kawa	KA1	27.55	20.75	1.33	0.28	l
Parigi boundary	PA3	62.33	19.92	3.13	1.45	d	Kawa	KA2	15.16	13.76	1.10	0.26	l
Parigi boundary	PA4	17.86	13.48	1.32	0.50	d		**Average**			**1.21**	**0.27**	
	Average			**2.62**	**0.91**		Bobol	BB1	14.59	11.61	1.26	1.57	m
Sapu valley	SV1	6.17	5.67	1.09	0.40	d	Bobol	BB2	51.83	26.37	1.97	2.60	m
Sapu valley	SV2	3.64	3.38	1.08	N/A	d	Bobol	BB3	35.56	17.84	1.99	1.44	m
Sapu valley	SV3	4.99	3.90	1.28	N/A	d	Bobol	BB4	23.05	12.73	1.81	1.04	m
Sapu valley	SV4	6.30	5.60	1.13	N/A	d		**Average**			**1.76**	**1.66**	
Sapu valley	SV5	6.89	4.74	1.45	N/A	d	Sorong	SG1	33.89	28.77	1.18	1.15	n
Sapu valley	SV6	4.67	3.78	1.24	N/A	d	Sorong	SG2	13.90	11.94	1.16	1.54	n
Sapu valley	SV7	5.27	3.84	1.37	N/A	d	Sorong	SG3	15.11	13.16	1.15	0.48	n
	Average			**1.23**	**0.40**		Sorong	SG4	8.54	5.35	1.60	0.59	n

Bada valley	BV1	5.97	5.73	1.04	0.57	e	Sorong	SG5	29.39	10.53	2.79	4.90	n
Bada valley	BV2	10.09	9.47	1.07	0.36	e	Sorong	SG6	10.19	8.94	1.14	0.27	n
Bada valley	BV3	13.27	9.30	1.43	0.73	e	Sorong	SG7	36.50	27.47	1.33	0.45	n
	Average			**1.18**	**0.55**		Sorong	SG8	17.73	14.83	1.20	0.44	n
Poso area	PO1	18.15	13.74	1.32	0.29	e	Sorong	SG9	31.90	18.34	1.74	1.33	n
	Average			**1.32**	**0.29**		Sorong	SG10	15.44	9.57	1.61	0.40	n
Balantak	BK1	13.02	11.80	1.10	0.25	f	Sorong	SG11	65.53	40.33	1.62	0.42	o
Balantak	BK2	6.85	6.59	1.04	0.47	f	Sorong	SG12	62.92	22.04	2.85	5.10	o
Balantak	BK3	5.16	4.24	1.22	N/A	f	Sorong	SG13	54.48	30.12	1.81	8.68	o
	Average			**1.12**	**0.36**		Sorong	SG14	22.17	17.39	1.27	0.32	o
Matano	MA1	15.06	11.78	1.28	1.12	g	Sorong	SG15	11.44	7.61	1.50	0.49	p
Matano	MA2	9.21	8.66	1.06	0.25	g		**Average**			**1.60**	**1.77**	
Matano	MA3	12.54	10.57	1.19	0.45	g	Ransiki	RS1	18.83	17.71	1.06	N/A	p
Matano	MA4	12.56	11.63	1.08	0.23	g	Ransiki	RS2	31.31	11.87	2.64	N/A	p
Matano	MA5	23.51	12.35	1.90	0.79	g		**Average**			**1.85**	**N/A**	
Matano	MA6	12.34	10.52	1.17	0.72	g	Wandaman boundary	WM1	17.22	13.76	1.25	N/A	q
Matano	MA7	8.46	8.31	1.02	0.84	g	Wandaman boundary	WM2	15.55	12.08	1.29	N/A	q
	Average			**1.24**	**0.63**		Wandaman boundary	WM3	23.73	18.57	1.28	N/A	q
Kendari	KD1	24.61	14.03	1.75	0.52	h	Wandaman boundary	WM4	11.58	6.74	1.72	N/A	q
Kendari	KD2	24.62	20.42	1.21	0.58	h	Wandaman boundary	WM5	19.75	13.83	1.43	N/A	q
	Average			**1.48**	**0.55**		Wandaman boundary	WM6	9.84	9.33	1.05	N/A	q
Lawanopo	LW1	43.13	28.29	1.52	0.83	g	Wandaman boundary	WM7	9.99	7.80	1.28	N/A	q
	Average			**1.52**	**0.83**		Wandaman boundary	WM8	45.67	19.61	2.33	N/A	q
Towuti bounding	TO1	27.98	24.39	1.15	0.41	g		**Average**			**1.45**	N/A	
Towuti bounding	TO2	9.58	9.30	1.03	0.56	g	Tarera–Aiduna	TA1	6.11	5.05	1.21	N/A	r
Towuti bounding	TO3	51.16	25.10	2.04	1.22	g	Tarera–Aiduna	TA2	20.74	19.19	1.08	N/A	r
	Average			**1.41**	**0.73**		Tarera–Aiduna	TA3	32.26	19.75	1.63	N/A	r
							Tarera–Aiduna	TA4	38.83	8.48	4.58	N/A	r
							Tarera–Aiduna	TA5	35.31	18.46	1.91	N/A	r
								Average			**2.08**	**N/A**	

*Straight line length of mountain front.
†Sinuous length of mountain front.
‡Mountain front sinuosity ($S_{mf} = L_{mf}/L_s$).
§Average valley floor width to valley depth ratio ($V_f = 2V_{fw}/(E_{ld} - E_{sc}) - (E_{rd} - E_{sc})$, where V_{fw} is the valley floor width, E_{ld} and E_{rd} are the topographic elevations of the left and right valley watersheds and E_{sc} is the elevation of the valley floor).
||Location of sinuosity segment on Figure 2.

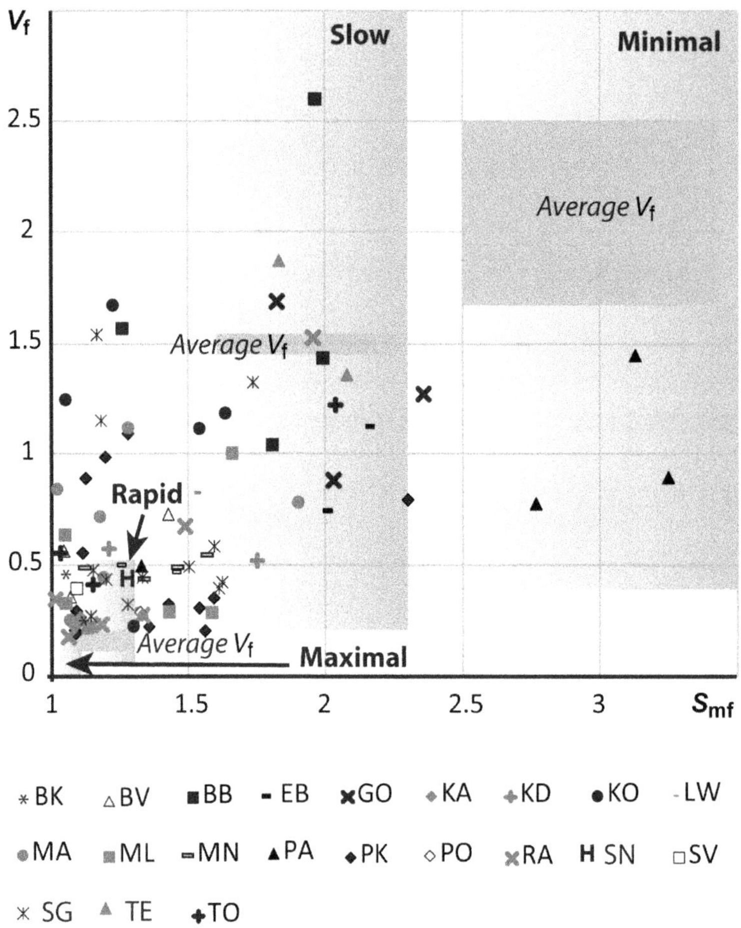

Fig. 3. Graph of mountain front sinuosity (S_{mf}) v. valley floor width to valley height index (V_f) for studied faults. Grey boxes indicate tectonic activity rates, after McCalpin (2009), with average V_f marked by the darker grey bar. BB, Bobol Fault; BK, Balantak Fault; BV, Bada valley faults; EB, East Buru faults; GO, Gorontalo Fault; KA, Kawa Fault; KD, Kendari faults; KO, Kolaka Fault; LW, Lawanopo Fault; MA, Matano Fault; ML, Malino boundary faults; MN, Mangole faults; PA, Parigi faults; PK, Palu–Koro Fault; PO, Poso faults; RA, Rana Fault; SG, Sorong Fault; SN, Sanana faults; SV, Sapu valley faults; TE, Southern Seram faults; TO, Towuti faults.

A 5° releasing bend/step-over is required to link the southern segments of the Palu–Koro Fault, where it emerges from its narrow valley at Pakuli, with the northern segments NW of Palu city. In analogue models and other non-linear strike-slip faults, such releasing geometries are often associated with well-defined oblique-normal sidewall faults and a cross-basin fault system with a more subtle surface expression that accommodates most of the strike-slip strain (e.g. Mann *et al.* 1995; Mann 2007; Wu *et al.* 2009) (Fig. 6a inset).

Analysis of Palu River channels since 2003 from satellite imagery and the pattern of older filled oxbow lakes on the valley floor indicates that long reaches of the river rarely deviate from a linear path directly along-strike from the strike-slip fault where it enters the Palu valley in the south (Fig. 6a, b). Many meanders have a square aspect with linear longitudinal segments parallel to the projected fault (Fig. 6c). In the south of the valley a linear braided reach is similarly parallel to the projected fault; individual braid channels are anomalously linear (Fig. 6d). Strands of the Palu–Koro Fault cutting an alluvial fan and offsetting its incised drainage directly along-strike to the south confirm that the river is structurally controlled. It is more reasonable to project this southern fault strand directly north across the basin than it is to consider strike-slip strain transferring immediately to the western sidewall fault between Pakuli and Bolongga, particularly as

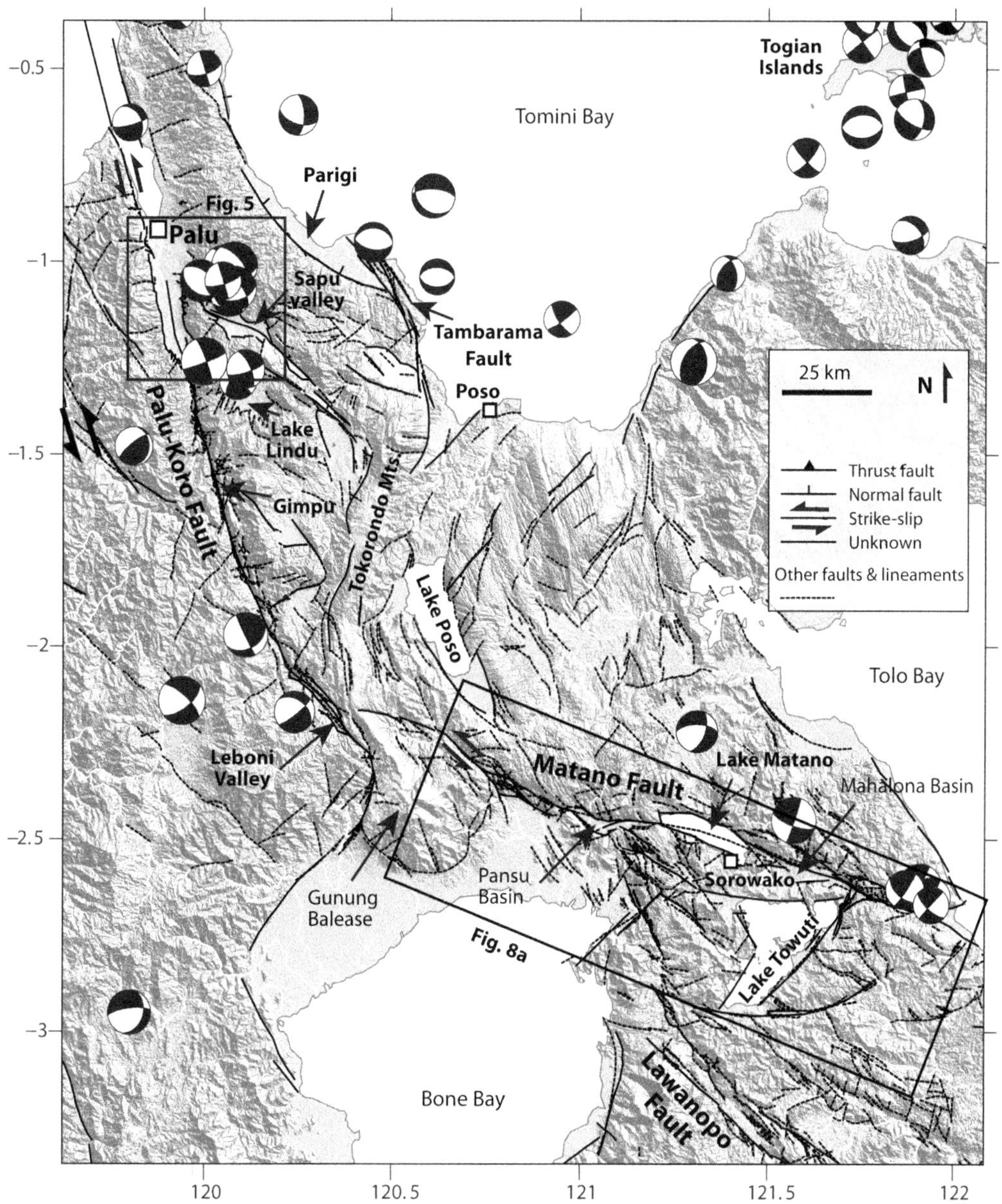

Fig. 4. Central Sulawesi overview digital elevation model (SRTM), CMT catalogue earthquakes <35 km depth and structures that show geomorphic evidence of Quaternary tectonic activity. Rivers marked in white. Illumination from NE. Location shown in Figure 1.

geomorphic indices in that region indicate a relatively low tectonic rate (Fig. 5a).

Thus we propose that much of the Palu–Koro Fault strike-slip strain through the Palu valley is not accommodated on the prominent sidewall faults, but on a cross-basin fault system that is obscured by fluvial deposits during interseismic periods (as it is now) (Figs 5a & 6a). The sidewall faults are largely an extensional partition, explaining the lateral slip deficit across them, noted by Bellier *et al.* (2001). Confinement of the Palu River meander belts within the strike-slip cross-basin fault system may be due to the development of a subtle graben, or to changes in permeability, cementation or compaction in the

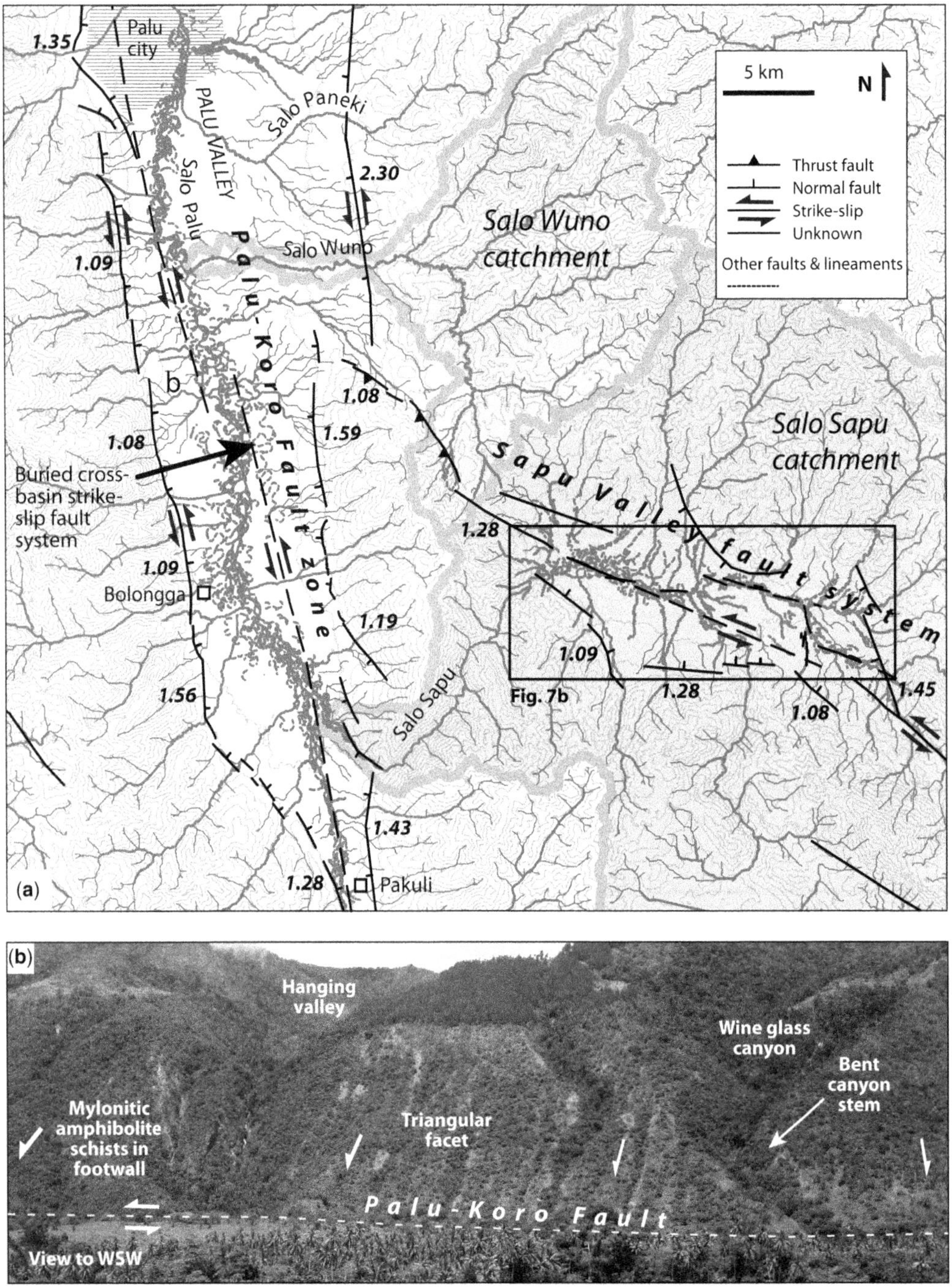

Fig. 5. (**a**) The Palu and Sapu valleys showing structures that with geomorphic evidence of Quaternary tectonic activity, plus topography and drainage. Mountain front sinuosity values in bold italic text. For location, see Figure 4. Major drainage basins for Salo Sapu and Salo Wuno are marked, separated by uplift at the western end of the Sapu valley fault system. (**b**) View of the Palu–Koro Fault scarp from the Palu valley, showing geomorphic evidence of Quaternary tectonic activity.

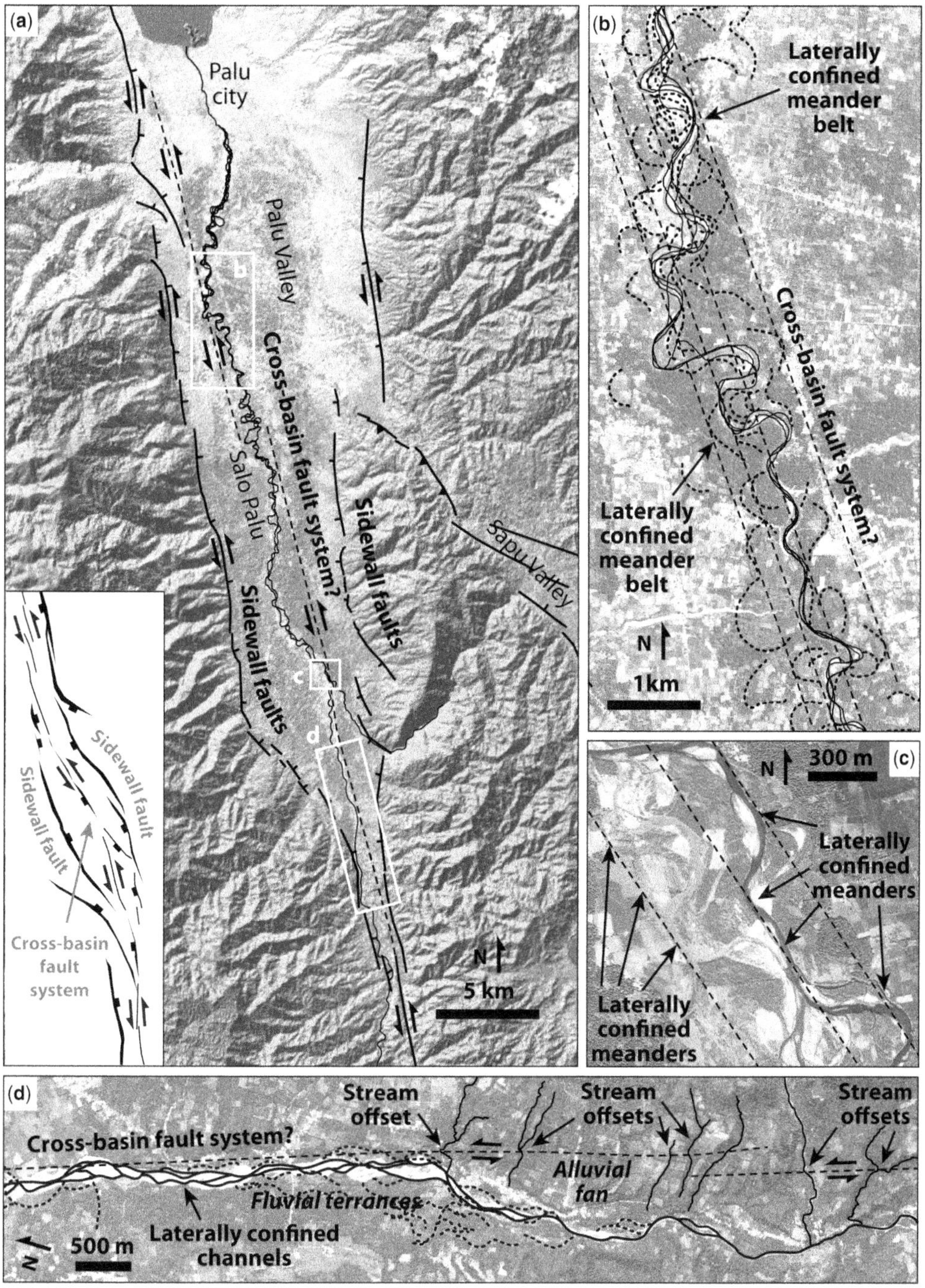

Fig. 6. Evidence of a cross-basin fault system within the Palu valley Quaternary fill. (**a**) Overview ASTER digital elevation model draped with ESRI imagery layer. Illumination from NW. Palu River channels traced from six separate images from 2003 to 2015. Inset shows fault pattern developed in an analogue model of a releasing bend, modified after Wu *et al.* (2009), reflected and rotated to mimic the Palu valley. Sidewall faults and cross-basin fault system are highlighted in the model and on the satellite imagery. (**b**, **c**) Laterally confined meander belts, interpreted as representing minor subsidence within the cross-basin fault system. (**d**) Laterally confined river channels directly along-strike from a Palu–Koro Fault strand seen to offset alluvial fans in the south of the valley. (**c**, **d**, **e**) show ESRI imagery.

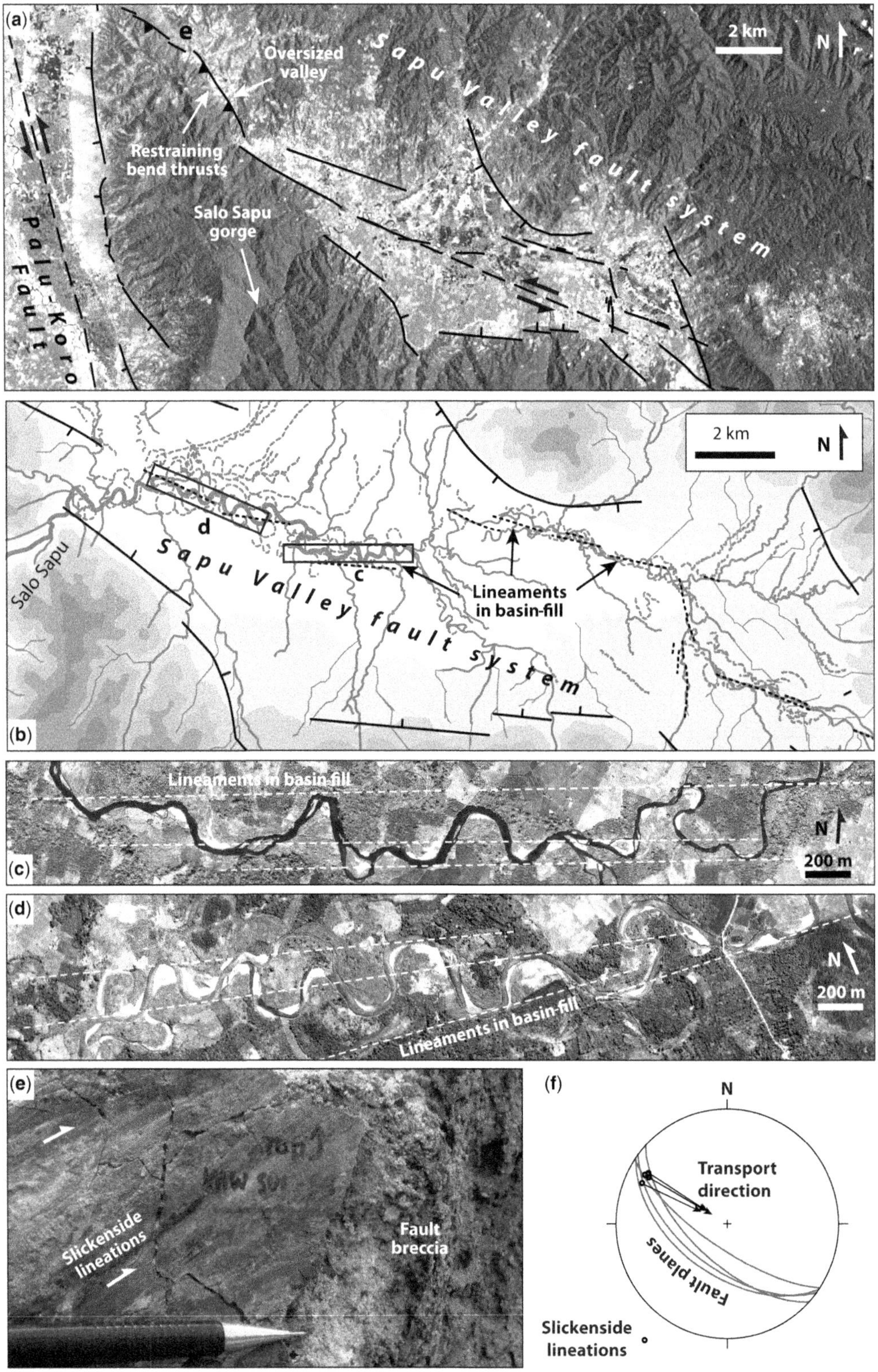
(a)
e
Oversized valley
Restraining bend thrusts
Salo Sapu gorge
Palu-Koro Fault
Sapu Valley fault system
2 km
N
(b)
2 km
N
d
c
Salo Sapu
Sapu Valley fault system
Lineaments in basin-fill
(c)
Lineaments in basin-fill
N
200 m
(d)
N
200 m
Lineaments in basin-fill
(e)
Slickenside lineations
Fault breccia
(f)
N
Transport direction
Fault planes
Slickenside lineations

Quaternary valley-fill resulting from penetration by strike-slip strands.

The valley's eastern sidewall fault is generally much more segmented and strongly eroded than in the west, with gentle slopes and irregular mountain fronts (Fig. 6a). South of the intersection with the Sapu valley fault system, S_{mf} values are 1.19–1.59 and V_f averages of 0.55 indicate rapid to moderate tectonic activity. North of the Sapu valley intersection, S_{mf} is 2.30 and the average V_f is 0.80, indicating slow tectonic activity.

Further south along the Palu–Koro Fault, the Gimpu basin exists at a small releasing step-over and the Leboni basin occupies a releasing bend near the southern termination of the fault (Fig. 4). The Palu–Koro Fault bounding these flat-topped Quaternary basins has S_{mf} values of 1.11 and 1.12, respectively, and similarly low V_f values of 0.56 and 0.89, indicating rapid to moderate tectonic activity.

Sapu valley fault system

A complex NW–SE-trending, 75 km long fault system cuts across crystalline basement between Palu valley and the Tokorondo Mountains in the east (Fig. 4). The fault system is dominated by a double bend: a releasing bend forming the intermontane Sapu valley (*c.* 600 m elevation) and a restraining bend associated with uplift at the head of the valley (Fig. 5a). Both bends are consistent with an overall left-lateral shear sense for the fault system. Anecdotal reports from residents of the valley (various, pers. comm. 2009) suggest that earthquakes are frequent and well known, although there is little instrumental seismicity and no record of historical earthquakes.

Sapu valley is an irregular rhomboidal basin bounded by normal faults trending NNE–SSE and east–west (Fig. 7a). Many of the faults are arcuate, convex into the basin. Their range front slopes are generally gentle, but S_{mf} values of 1.09–1.45 and an average V_f of 0.40 suggest rapid to moderate tectonic activity (Fig. 7b). A conspicuous feature of the basin floor is the strong confinement of river channels to narrow linear meander belts (Fig. 7b), as discussed earlier for the Palu River. Both modern and abandoned channels have linear meander belt margins and square longitudinal sections parallel to the projected trace of the fault system through the valley, implying fault penetration through the Quaternary basin-fill (Fig. 7c, d). In the same way as for the Palu valley, this evidence supports a cross-basin fault system that accommodates most of the strike-slip strain, whereas the prominent sidewall faults are dominantly extensional structures. The cross-basin fault system is buried by fluvial sediments, but co-seismic subsidence, or changes in permeability, cementation or compaction caused by periodic surface rupture through the Quaternary basin-fill continue to influence the meander patterns.

At the head of the valley the entire fault system curves to a more NNW–SSE trend – a restraining geometry under sinistral shear. A broad, oversized valley in the west is presently at 700 m elevation (Fig. 7a), i.e. 100 m above the modern Sapu valley floor. Exhumed brittle SW-dipping reverse-sinistral faults in mica schists along the uplifted valley support long-lived uplift at this restraining bend (Fig. 7e, f). At the foot of the westernmost oblique-reverse fault, S_{mf} is 1.08, suggesting maximal tectonic activity (Fig. 5a).

Drainage networks extracted from SRTM data show that there is presently a drainage divide separating the Salo Wuno and Salo Sapu catchment basins at the position of the thrust-related uplift and oversized valley (Fig. 5a). Water presently exits Sapu valley via a narrow, steep-sided gorge (Fig. 7a). The extreme steepness and geomorphic immaturity of that gorge suggests that it has recently captured the Sapu valley drainage, perhaps in response to tectonic uplift of its former well-established route to the NW via Salo Wuno. It is likely that the Sapu valley was internally drained for some time after uplift in the NW and may have contained an intermontane lake similar to Lake Lindu to the south (Fig. 4), explaining the flat base of the Sapu valley.

Four lines of evidence suggest the Sapu valley fault system has been active during the Quaternary: (1) control of the modern river meander belts by a cross-basin fault system that traverses the Quaternary basin-fill; (2) youthful geomorphic expression of the Salo Sapu gorge where it has recently captured the Salo Sapu drainage in response to tectonic uplift in the NW; (3) rapid to moderate tectonic activity along the transtensional segment sidewall faults, indicated by geomorphic indices; and (4) maximal tectonic activity along the transpressional

Fig. 7. Details of the Sapu valley fault system. (**a**) ESRI imagery of the Sapu and central Palu valleys showing major structural and geomorphic features, particularly the releasing–restraining double bend and re-routing of axial drainage from the NW valley to the Salo Sapu gorge. (**b**) Detail of the Sapu valley showing drainage and highlighting fault control of the axial river. Location shown on Figure 5a. (**c, d**) Laterally confined meander belts and lineaments, interpreted as representing minor subsidence within the cross-basin fault system. (**e**) Lineated slickenside surface from an exhumed fault core within the Sapu restraining bend. Location shown in Figure 7a. (**f**) Lower hemisphere stereographic projection of fault planes (great circles) and slickenside lineations (points) from the fault shown in Figure 7e. ESRI imagery.

segment's reverse faults implicated in uplifting the oversized palaeovalley in the east, indicated by geomorphic indices.

Matano Fault

The Matano Fault passes from southern central Sulawesi through the island's SE arm to Tolo Bay (Fig. 4). It is typically shown to mark the southern edge of the Sula Block, linking to the Palu–Koro Fault to the west and the North Sulawesi Trench to the north (e.g. Hamilton 1979; Rangin *et al.* 1999). A hard linkage between either the Lawanopo or Matano and Palu–Koro faults is a requirement of many rigid-block models for Sulawesi (e.g. Bellier *et al.* 2006; Socquet *et al.* 2006). However, Silver *et al.* (1983*b*) noted that the nature of the connection was not known. Modern satellite imagery shows a highly segmented and discontinuous westernmost Matano Fault curving towards the Palu–Koro Fault, but the two structures remain largely isolated either side of the Gunung Balease massif (Fig. 4).

In the east, the Matano Fault passes into the northern Banda Sea. Some workers link it to the Tolo Thrust (sometimes referred to as the Hamilton Thrust or the East Sulawesi Trench) (Fig. 1), an ESE-verging thrust zone NE of Buton. Silver *et al.* (1983*b*) suggest that the Matano and Palu–Koro faults act as a trench–trench transform between the north Sulawesi subduction and the Tolo Thrust. This thrust has been considered to accommodate convergence between the Makassar block and the Banda Sea block (e.g. Socquet *et al.* 2006). However, recent work suggests that the Tolo Thrust is a gravity-driven feature at the foot of a series of slumps (Rudyawan 2011), rather than a structure bounding a tectonic block (e.g. Silver *et al.* 1983*b*; Rangin *et al.* 1999).

Geological offsets (e.g. Ahmad 1978) and stream offsets (e.g. Hamilton 1979) across the Matano Fault confirm that it is a left-lateral structure and that it has been active during the Quaternary (Bellier *et al.* 2006). Laterally offset streams are routinely used to assess the shear sense and Quaternary activity of strike-slip faults, usually in arid environments (e.g. Sieh & Jahns 1984), but also in humid, forested environments (e.g. Lacassin *et al.* 1998; Wang *et al.* 2014). Nonetheless, such observations must be interpreted cautiously, as stream offset may result from stream diversion along a fault and capture by another downstream reach, as well as by the genuine tectonic displacement of a single stream (Wallace 1990). No study has used such offsets to evaluate Quaternary slip rates along the Matano Fault.

The Matano Fault is highly segmented and lacks a single through-going strand (Fig. 8a). Several linear basins (e.g. Pansu, Matano and Mahalona) lie within or adjacent to the fault zone, often at step-overs between strands. Each basin is 4–6 km wide and 20–30 km long. The Matano basin hosts Lake Matano, which, at 590 m deep (Haffner *et al.* 2001), is the deepest lake in Indonesia and the tenth deepest lake in the world. A fault passing from the northern margin of the Pansu Basin is very prominent as it cuts through ultramafic rocks in the SW corner of Lake Matano, just south of Desa Matano (Fig. 8a). The fault then steps to the left to another very prominent fault in the NW of the lake, from where it passes across the northern margin of the Mahalona Basin. Rapid subsidence in the lake and earthquake focal mechanisms recording east–west extension close to the lake probably result from this releasing geometry (McCaffrey & Sutardjo 1982). Two major pop-ups associated with the uplift, thrusting and exhumation of metamorphic rocks and serpentinite at restraining bends occur east of the Mahalona Basin and west of the Pansu Basin (Fig. 8a).

A number of consistent left-lateral stream offsets, evidence of stream capture across two fault strands west of Pansu Basin (Fig. 8b) and steep-sided, narrow fault valleys (Fig. 8c) suggest youthful fault activity. Geomorphic indices of oblique basin-bounding faults range from S_{mf} 1.06–1.28, average V_f 0.69 (Pansu Basin), S_{mf} 1.02–1.17, average V_f 0.78 (Matano Basin), S_{mf} 1.19, V_f 0.45 (Mahalona Basin) to S_{mf} 1.08–1.9, average V_f 0.51 (eastern termination splay) and indicate mostly rapid to moderate tectonic activity.

On 15 February 2011, a shallow focus M_w 6.1 earthquake near the western end of Lake Matano (NEIC) had a focal mechanism consistent with left-lateral slip along the Matano Fault. The earthquake caused damage to concrete walls and buildings, including a newly built hospital in the Mahalona valley (Fig. 8d). The earthquake's location suggested that the prominent fault segment that links the NE corner of Lake Matano with the Mahalona Basin failed (Fig. 8e). ‘Surface cracks’ were reported by local people at the eastern end of the basin but, although we visited the area in October 2011, a surface rupture could not be located. Close to the lake, very high resolution satellite imagery recently made available (Bing Maps) shows three clear lineaments cutting across boggy ground and low-lying forest (Fig. 8f) along-strike from a Matano Fault strand that offsets drainage to the left. Although it is not possible to confirm that they represent the 2011 surface rupture, these lineaments appear to be tectonic in origin and are clearly very young. Linking these lineaments with the reported surface cracks in the east, along a topographically clearly defined fault strand, yields a postulated surface rupture length of >39 km, which is longer than expected for a M_w 6.1 earthquake from empirical relationships (Wells & Coppersmith 1994).

Fig. 8. Details of the Matano Fault. (**a**) Map of the Matano Fault, Lake Towuti and the northern part of the Lawanopo Fault. Base map is ASTER digital elevation model draped with ESRI imagery. Location shown in Figure 4. (**b**) Systematic stream offsets along strands of the Matano Fault west of the Pansu Basin. (**c**) Deep, steep-sided valley marking the westernmost Matano Fault NE of Gunung Balease. (**d**) Hospital in the Mahalona valley damaged during the 15 February 2011 M_w 6.1 earthquake. Location shown in Figure 8e. (**e**) Detail of the eastern Matano and Mahalona valleys, showing features related to the 2011 earthquake and inferred surface rupture extent. ASTER base map. (**f**) Imagery from Bing Maps showing strong topographic lineaments in low ground in the NE corner of Lake Matano, inferred to represent recent (2011?) surface ruptures. Location shown in Figure 8e. (**b**) and (**f**) images © 2016 DigitalGlobe.

Lawanopo Fault and Lake Towuti

The Lawanopo Fault (Fig. 4) consists of several straight NW-trending fault segments that cross Sulawesi's SE arm south of the Matano Fault. The Lawanopo Fault is used in preference to the Matano Fault by Socquet *et al.* (2006) as the southern margin of the 'East Sula Block'. However, discontinuous and eroded fault traces along strands of the Lawanopo Fault system suggest that it has been mostly inactive during the Quaternary (Bellier *et al.* 2006; Natawidjaja & Daryono 2014). Nonetheless, recent earthquakes close to Kendari may indicate that at least some strands of the Lawanopo Fault system remain active. An M_w 7.5 earthquake in the Banda Sea 170 km SE of Kendari on 19 October 2001 had a strike-slip focal mechanism and may have originated on the projected offshore trace of the Lawanopo Fault (Yeats 2010).

Like the Matano Fault, the Lawanopo Fault is highly segmented and there is no through-going strand at the surface (Fig. 4). Mountain front sinuosity values on the few segments associated with adjacent basins range from 1.21 to 1.75 and valley depth to width ratios average 0.55–0.83, indicating moderate to slow tectonic activity.

Lake Towuti, the largest of the Malili lakes, occupies an intermontane basin at 318 m elevation and has a maximum water depth of 203 m (Haffner *et al.* 2001). The basin lies in the wedge between the Matano and Lawanopo faults and is itself cut by linear fault strands that internally deform the wedge (Fig. 4). Two prominent curvilinear faults lie along the south and east of the lake (Fig. 8a). The closest, trending NE–SW and downthrown to the NW, forms the linear eastern lake boundary and is marked by a number of fans prograding into the lake. Its high mountain front sinuosity (2.04) and valley depth to width ratio (1.22) suggest slow tectonic activity. However, a large earthquake along this >25 km long structure could cause a substantial tsunami or seiche in the lake. The second fault, to the east, is longer still (>55 km) and highly continuous. It intersects the Lawanopo Fault at a small angle and may directly transfer slip away from that structure. Mountain front sinuosity ranges from 1.03 to 1.15, suggesting maximal to rapid tectonic activity, although the valley floors are rather rounded (average V_f 0.49). Lake Towuti would rapidly fill with sediment if it were not actively subsiding, therefore the bounding normal faults must be considered to be active during the Quaternary.

Kolaka Fault

The Kolaka Fault (Simandjuntak *et al.* 1984, 1994; Surono 1994) (Fig. 9a) lies along the southern margin of the Mengkoka mountains and is sub-parallel to the Lawanopo Fault to its north. It is equivalent to the Mendoke Fault of Bellier *et al.* (2006). Hamilton (1979) interpreted the fault as a SW-dipping thrust and Bellier *et al.* (2006) considered the fault as a pre-Early Pleistocene strike-slip continuation of the Palu–Koro Fault, but there is little evidence to support either hypothesis. One strand of the Kolaka Fault is sealed by 4.4 $\pm$ 0.2 Ma dacites, potentially placing a limit on the timing of faulting (White *et al.* 2014).

The fault is composed of several NE–SW-trending, gently arcuate segments up to 45 km long in map view. Along the Bone Bay coast and at Kolaka town the downthrown side is to the south and the easternmost segment is downthrown to the north (Fig. 9a). The polarity shift occurs across a 10 km wide relay straddling the Anggowala mountains. The orientation of these two apparently normal fault systems is kinematically consistent with sinistral slip along the overall Kolaka trend.

Geomorphic indices are highest closest to Kolaka town, where S_{mf} values of 1.22–1.30 and V_f values of 0.23–1.68 suggest that there is rapid to slow active dip-slip across the fault, which has a clear surface expression and is marked by triangular facets (Fig. 9c). Along-strike to the NW a series of linear valleys and low ridges near Lasusua may be a continuation of the Kolaka Fault (Fig. 9b). An absence of fault scarps or clearly displaced features makes fault activity hard to evaluate, but meander confinement within a linear graben across the Lasusua alluvial plain and asymmetrical subsidence highlighted by the river's proximity to the bounding fault suggests recent fault activity (Fig. 9d).

Faults downthrown to the WSW at the western end of the Kolaka Fault have very low S_{mf} and V_f values (1.05 and 1.25, respectively), deeply incised streams and well-developed triangular facets, suggesting Quaternary dip-slip. These faults face into Bone Bay and may be related to basin-bounding extensional structures accommodating subsidence in the bay (Camplin & Hall 2014).

Balantak Fault

A prominent ENE-trending linear structure, the Balantak Fault, lies at the eastern end of Sulawesi's east arm and separates the Batui thrust system in the south from mountainous highlands in the north (Fig. 10a). It has been considered to be part of the Batui thrust system (Silver *et al.* 1983*b*), but its remarkably straight outcrop, field observations (Simandjuntak 1986) and along-strike alternation between local uplift and subsidence suggest that it is a steep, possibly strike-slip, fault.

Onshore, where the fault bends gently to the right, small, apparently Quaternary basins are developed (Fig. 10b). There is uplift where the fault bends

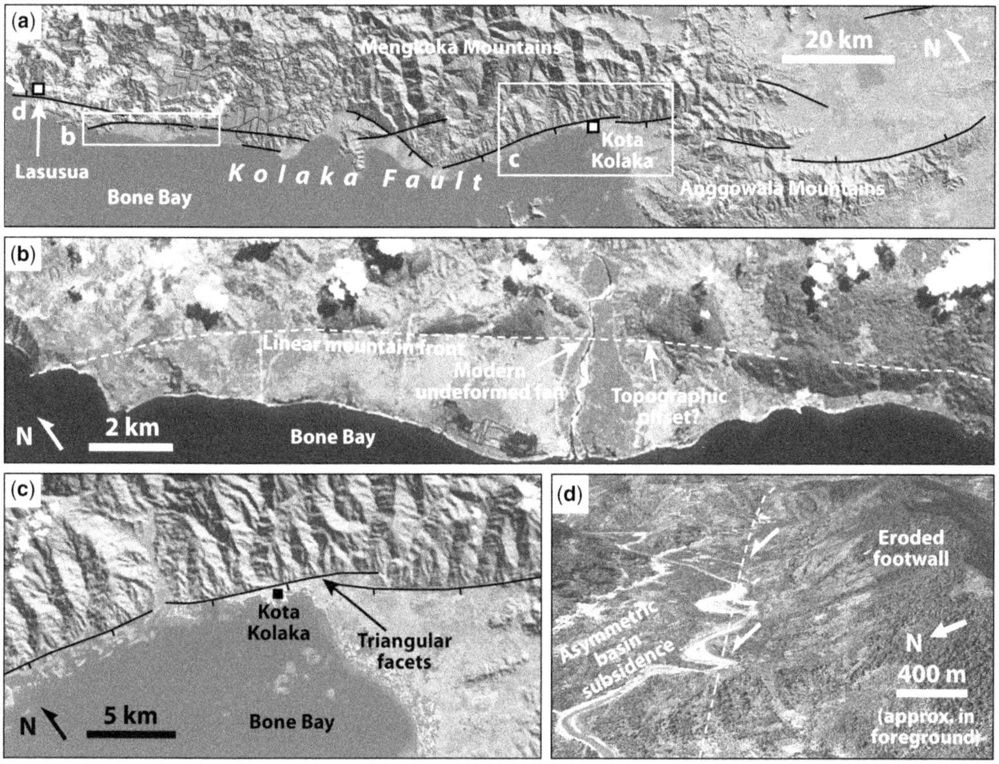

Fig. 9. Details of the Kolaka Fault. (**a**) Overview map of the main Kolaka Fault segments. Base map is ASTER digital elevation model draped with ESRI imagery. Location shown in Figure 1. (**b**) Straight segment of the Kolaka Fault associated with linear ridges and valleys. ESRI imagery. (**c**) Linear fault-bounded mountain front and triangular facets indicating Quaternary fault activity at Kolaka town. (**d**) Asymmetrical axial drainage at a splaying fault segment near the western fault termination, indicating Quaternary subsidence along the bounding fault system.

gently to the left. Both observations are kinematically compatible with a dextral shear sense. One of the zones of Quaternary subsidence is shown in Figure 10c. A basin-bounding fault at a small clockwise angle from the regional Balantak Fault trend is crossed by streams that show no systematic offset, suggesting dominant dip-slip. To the north, a prominent lineament crosses the basin, expressed by lines of vegetation and slightly darker (moister?) soil. This lineament's parallelism with the Balantak Fault to the east and its negligible topographic relief suggests it is the through-going strike-slip fault strand. Although stream avulsion across the flat-topped basin is too dynamic to preserve meaningful offsets, the clear expression of the fault in the young sediments suggests the Balantak Fault has been active during the Quaternary.

The Balantak Fault's termination system offshore to the east of Poh Head is composed of left-stepping segments separated by folds and thrusts (Fig. 10d). Contraction between left-stepping main segments, an apparently antithetic sinistral fault and the orientation of folds and thrusts are all kinematically compatible with dextral shear along the Balantak Fault (Watkinson *et al.* 2011). Earthquakes located onshore and west of Poh Head also suggest right-lateral and reverse slip parallel to the Balantak Fault (Fig. 10a). However, a swarm of offshore earthquakes between Peleng and Taliabu to the east have focal mechanisms that support sinistral slip along the Balantak trend. This apparent contradiction is discussed in Watkinson *et al.* (2011). Here we conclude that the geological and geomorphic evidence supports long-term Quaternary dextral slip. Further work is required to understand the significance of a small number of contradictory seismological signals in the area.

The Balantak Fault is almost continuous for 54 km from Balantak town in the east to Poh Bay in the west, where it probably continues just offshore for another >30 km. Extending to include the dextral fault system offshore to the SE makes the

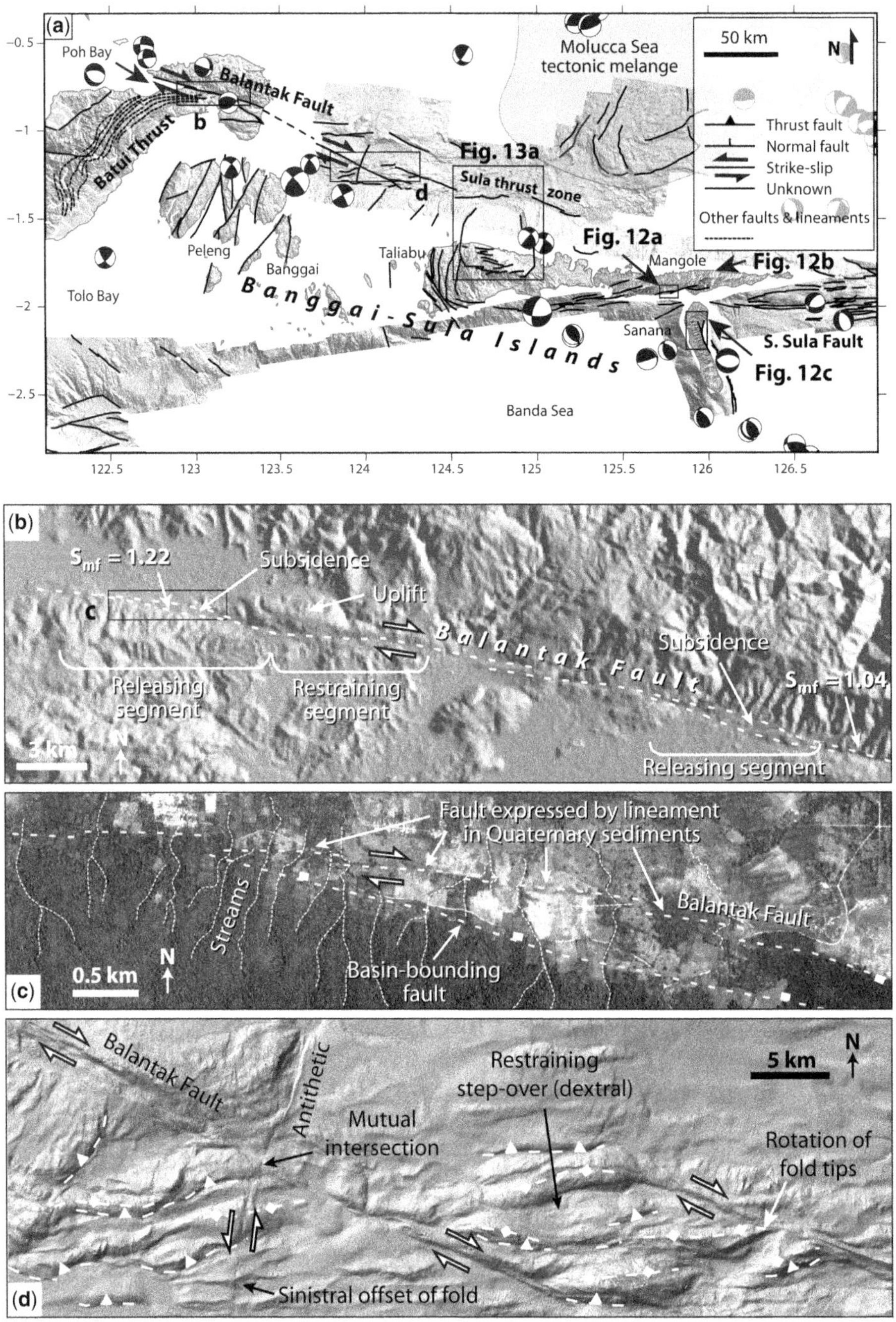

Fig. 10. (**a**) East arm of Sulawesi and Banggai–Sula Islands digital elevation model (SRTM), multibeam bathymetry, CMT catalogue earthquakes <35 km depth and structures showing geomorphic evidence of Quaternary tectonic activity. After Watkinson *et al.* (2011). Location shown in Figure 1. (**b**) Subsidence and uplift associated with releasing and restraining segments of the onshore Balantak Fault. ASTER digital elevation model base map. (**c**) Detail of bounding fault system of a Balantak Fault releasing segment, showing a north-dipping normal fault and sub-parallel lineament in agricultural land, inferred to represent a through-going strike-slip strand. ESRI imagery base map. (**d**) Detail of the offshore Balantak Fault expressed in multibeam imagery (illumination from NW) showing evidence of dextral shear.

fault up to 250 km long. The onshore fault scarp has exceptionally low S_{mf} values, from 1.04 to 1.22 (Fig. 10b), with correspondingly low average V_f values of 0.36, suggesting maximal to moderate tectonic activity.

Gorontalo Fault

The Gorontalo Fault (Katili 1973) (Fig. 11a) has been considered to be one of the major block-bounding structures of Sulawesi (e.g. Socquet *et al.* 2006; Molnar & Dayem 2010). Geodetic modelling suggests a 11 mm a^{-1} dextral slip rate and 10 km locking depth; however, because the observation points are widely spaced, it remains possible that global positioning system (GPS) data record rotation of the entire north arm of the island rather than discrete slip across a fault (Socquet *et al.* 2006). There is little modern shallow seismicity in the Gorontalo area, suggesting that the fault is inactive or remains locked (Fig. 11a).

The fault is composed of several branching segments, including major *c.* 30 km long segments south and north of Gorontalo city (Fig. 11b). Limboto Lake lies in the 7 km wide step-over between these two segments, indicating local transtension. The fault is expressed by highly eroded scarps passing along the Tomini Bay coast and bounding the Gorontalo/Limboto depression. Geomorphic indices suggest that the segments experience slow to minimal tectonic activity, with S_{mf} values ranging from 1.83 to 2.36 and an average V_f of 1.28. Although there is considerable human development within the Gorontalo/Limboto depression, which may obscure neotectonic activity, there appears to be little evidence of deformation within the Quaternary sediment-fill, except for the presence of Limboto Lake subsidence at the releasing step-over.

Western Tomini Bay bounding faults

A series of faults along the margin of Tomini Bay shows evidence of recent activity. The faults are arcuate and generally mark the boundary between mountainous ground along Sulawesi's narrow 'neck' and Tomini Bay, which is up to 2 km deep and contains a sedimentary succession up to 10 km thick (Jablonski *et al.* 2007; Pholbud *et al.* 2012). Extension and mantle decompression across the bay are associated with Plio-Pleistocene volcanism in the Togian Islands and possibly with modern volcanism at Una Una volcano (Cottam *et al.* 2011), supporting recent extensional faulting and lithospheric thinning both onshore and offshore (Pholbud *et al.* 2012).

The northernmost bounding fault bounds the 2.5 km high Molino Metamorphic Complex (Fig. 11c), a suite of quartzo-feldspathic mica schists and gneisses that may be an exhumed metamorphic core complex (van Leeuwen & Muhardjo 2005). The faults dip north and south on the north and south sides of the complex, respectively, and have crystalline basement in their footwalls. The southern segment has a curvilinear trace >75 km long with extremely low S_{mf} values (1.05) and well-developed triangular facets at the end of V-shaped valleys with V_f values of 0.33–0.64 within an uplifted footwall block. On this basis, combined with no evidence of strike-slip, it is interpreted as a normal fault.

Further SW, the Tomini Bay bounding faults are crossed by a number of fan deltas prograding into the bay. These are surprisingly short (<3 km), given the potential upstream sediment source, suggesting rapid and recent hanging wall subsidence (Fig. 11d, e). Segments further south along the 'neck' have higher S_{mf} values (1.66) and the fan delta lobe length increases to >10 km, suggesting less significant recent subsidence (Fig. 11f).

At the southern end of the neck, a NE-dipping fault system, including the Tambarama Fault (Pholbud *et al.* 2012), forms an apparently continuous arcuate trace at Parigi (Fig. 4), marking the boundary between the Palu Metamorphic Complex onshore (van Leeuwen & Muhardjo 2005) and Tomini Bay subsidence offshore. S_{mf} values are generally high (2.77–3.25), although a short northern segment is less sinuous at 1.32. A well-developed apron of fan deltas extends 6 km from the mountain front.

Maluku and North Maluku

Maluku and North Maluku are composed of numerous islands affected by disparate neotectonic processes. In the north, Halmahera (Fig. 1) and the Sangihe Arc are involved in the active collision of two accretionary complexes above the subducted Molucca Sea slab, where the Sangihe forearc is being thrust eastwards over the Halmahera forearc (e.g. Silver & Moore 1978; Hamilton 1979; Hall 1987; Hall *et al.* 1995). The entire system accommodates 80 mm a^{-1} of the 105 mm a^{-1} Philippine Sea plate–Sundaland convergence (Rangin *et al.* 1999). Splays of the left-lateral Sorong Fault pass through and to the south of Halmahera and Bacan, where there is abundant modern seismicity (e.g. Ali & Hall 1995; Hall *et al.* 1995) (Fig. 1).

South of Bacan, islands with Australian continental basement, such as the Banggai–Sula Islands and Obi, are bounded by strands of the Sorong Fault and were for a long time considered to have been translated from New Guinea along a 1900 km long Sorong Fault passing from northern Papua New Guinea towards Sulawesi (e.g. Visser & Hermes 1962; Audley-Charles *et al.* 1972; Hamilton 1979; Pigram *et al.* 1985; Garrard *et al.* 1988; Hutchison

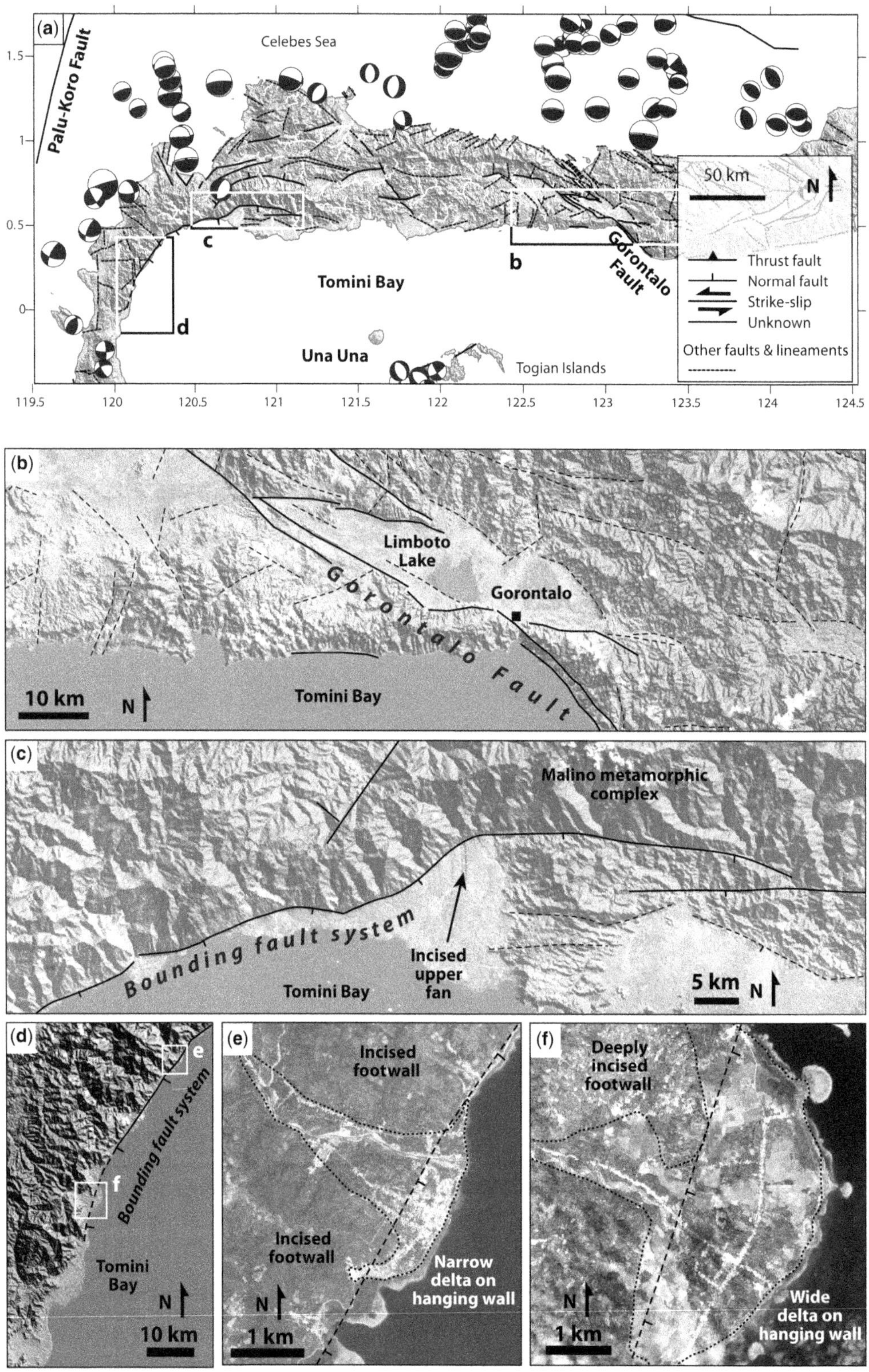
(a)
Palu-Koro Fault
Celebes Sea
Tomini Bay
Una Una
Togian Islands
Gorontalo Fault
50 km
N
Thrust fault
Normal fault
Strike-slip
Unknown
Other faults & lineaments
(b)
Limboto Lake
Gorontalo
Gorontalo Fault
Tomini Bay
10 km
(c)
Malino metamorphic complex
Bounding fault system
Incised upper fan
Tomini Bay
5 km
(d)
Bounding fault system
Tomini Bay
10 km
(e)
Incised footwall
Incised footwall
Narrow delta on hanging wall
1 km
(f)
Deeply incised footwall
Wide delta on hanging wall
1 km

1989). New interpretations based on evidence of extreme crustal extension and mantle exhumation, mantle tomography and geodynamic models (e.g. Spakman & Hall 2010; Hall 2011; Spencer 2010, 2011; Pownall *et al.* 2014) suggest that those islands, together with others along the northern Banda Arc such as Buru and Seram, were part of a continental spur that was fragmented during Miocene–Pliocene times by lower crustal delamination driven by Banda Sea rollback.

Quaternary extension in Maluku appears to be as important as it is in Sulawesi, despite an overall collisional tectonic setting. Young metamorphic core complexes exhumed in Seram (Pownall *et al.* 2013) and possibly Buru (Roques 1999) are associated with low-angle and steep normal faults. A significant component of the seismic moment release in Maluku is by normal and strike-slip earthquakes, alongside important thrusting in the Molucca Sea and north Seram (e.g. Rangin *et al.* 1999). Sinistral transpression through Seram accommodates Australia–Pacific convergence and links into the Tarera–Aiduna Fault of West Papua (e.g. Rangin *et al.* 1999; Stevens *et al.* 2002; Teas *et al.* 2009).

Banggai–Sula Islands

The Banggai–Sula Islands (Fig. 10a) occupy a fragment of continental crust of Australian affinity that has collided with the east arm of Sulawesi (e.g. Audley-Charles *et al.* 1972; Hamilton 1979; Pigram *et al.* 1985; Garrard *et al.* 1988). The South Sula–Sorong Fault was interpreted by Hamilton (1979) to follow the break in slope south of Taliabu and pass between Mangole and Sanana. North of the Banggai–Sula Islands the North Sula–Sorong Fault (e.g. Hamilton 1979; Norvick 1979; Silver *et al.* 1983*b*; Sukamto & Simandjuntak 1983), previously considered to pass from the Bird's Head, past Obi and along the north margin of the Banggai–Sula Islands towards Sulawesi's east arm, cannot be detected in new geophysical data and must lie below the Molucca Sea collision complex to the north (Ferdian *et al.* 2010; Watkinson *et al.* 2011).

Despite the density of deformation in the area, there is very little shallow seismicity immediately north of the Banggai–Sula margin (Engdahl *et al.* 1998; Rangin *et al.* 1999; Beaudouin *et al.* 2003), indicating that there are few active structures, that deformation is largely aseismic or that the main faults have interseismic periods that exceed instrumental records. This is a marked contrast with the abundant shallow seismicity associated with the Molucca Sea collisional zone further north. However, a number of focal mechanisms north and south of the islands indicate that there is some residual left-lateral slip on east–west- to NW–SE-trending faults (Fig. 10a).

Mangole Island appears to be bordered along its north and south sides by several linear east–west-trending normal faults, indicated by straight traces and well-developed triangular facets (Fig. 12a, b). Mountain front sinuosity values range from 1.11 to 1.57 and V_f is from 0.44 to 0.55, suggesting that some of the structures have been active during the Quaternary. Sanana Island, topographically orthogonal to Mangole, is bounded by NNW–SSE-trending faults that can be traced offshore in multibeam bathymetry. The most prominent fault, on the east coast, forms a well-defined scarp >20 km long, dipping and downthrown to the east, making it likely to be a normal fault (Fig. 12c). Triangular facets, hanging valleys, deeply incised streams (Fig. 12d) and an absence of subaerial prograding fan delta tops wider than *c.* 400 m suggest rapid recent eastwards subsidence along the fault, supported by S_{mf} values of 1.27–1.34.

Taliabu Island (Fig. 10a) is cut by a number of east–west- and north–south-trending Quaternary faults. The north–south-trending faults in the west have a particularly fresh geomorphic expression. A north coast bedding-parallel dip slope dips 6° into the Molucca Sea (Fig. 13a). Offshore to the north a planar detachment surface 34 km wide exactly corresponds to the Taliabu dip slope onshore and represents a submarine slope failure (Watkinson *et al.* 2011). Both onshore and offshore slopes appear to be part of a single large glide surface of a mega-debris slide that translated much of north Taliabu at least 37 km north into the Molucca Sea, probably causing a significant tsunami.

Fig. 11. (**a**) North arm of Sulawesi digital elevation model (SRTM), CMT catalogue earthquakes <35 km depth and structures showing geomorphic evidence of Quaternary tectonic activity. Rivers marked in white. Location shown in Figure 1. (**b**) Detail of the onshore Gorontalo Fault and associated basins. ASTER digital elevation model draped with ESRI imagery layer. (**c**) Fault system bounding the Malino metamorphic complex showing remarkably straight and steep mountain front and well-developed triangular facets. (**d**) Overview of fan deltas prograding into western Tomini Bay across the bounding fault system. (**e**) Narrow fan delta clearly cut by the basin-bounding fault, indicating rapid subsidence. Google Earth imagery. Image © 2016 DigitalGlobe. Data SIO, NOAA, U.S. Navy, NGA, GEBCO. Image (2016) Terrametrics. Image Landsat. (**f**) Wide fan delta further south indicating a slower rate of hanging wall subsidence. Google Earth imagery. Image © 2016 DigitalGlobe. Data SIO, NOAA, U.S. Navy, NGA, GEBCO. Image (2016) Terrametrics. Image Landsat.

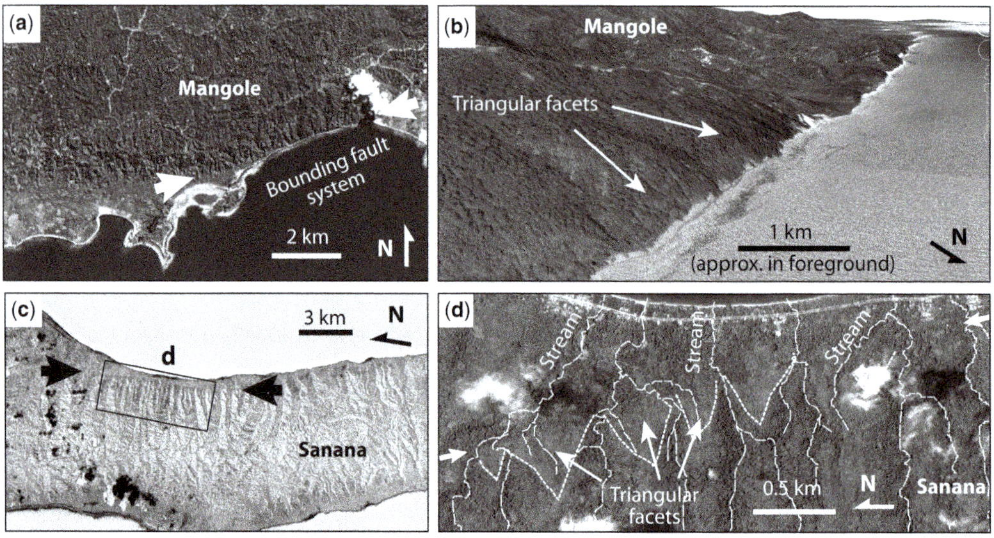

Fig. 12. Details of Quaternary faults in the Banggai–Sula Islands. For locations, see Figure 10a. (**a**) Part of the linear normal fault system bounding the southern margin of Mangole Island. Google Earth image. Data SIO, NOAA, U.S. Navy, NGA, GEBCO, Image Landsat. Image © 2016 TerraMetrics. (**b**) Triangular facets along the north coast of Mangole Island. Oblique view in Google Earth. Data SIO, NOAA, U.S. Navy, NGA, GEBCO. Image © DigitalGlobe. Image © 2016 TerraMetrics. Image Landsat. (**c**) Fault control along the eastern coast of Sanana Island. Image from Google Earth (greyscale inverted for clarity). Image Landsat. Image © DigitalGlobe. Image © 2016 TerraMetrics. Data SIO, NOAA, U.S. Navy, NGA, GEBCO. (**d**) Detail of the Sanana fault, showing the extremely linear mountain front, narrow V-shaped valleys and triangular facets. Image © DigitalGlobe.

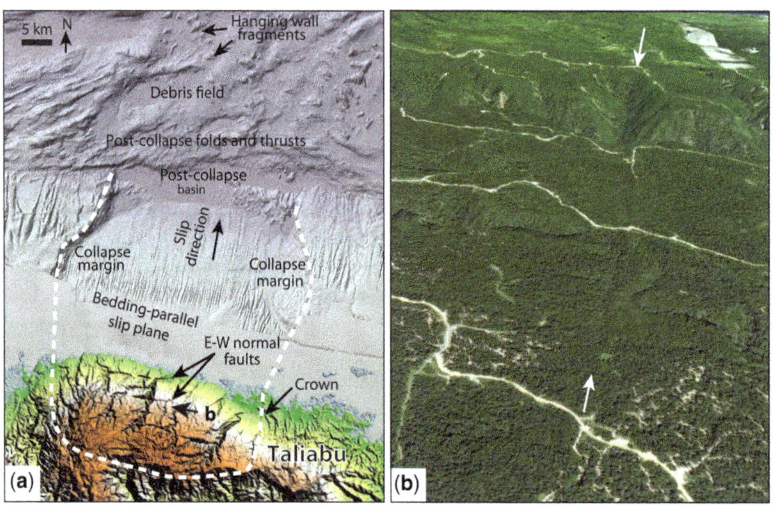

Fig. 13. (**a**) Mega-debris slide and post-collapse normal faults, north coast of Taliabu. Base map is SRTM topography onshore and multibeam bathymetry offshore. Location shown in Figure 10a. (**b**) Oblique perspective view from Google Earth of one of the normal faults on the north slope of Taliabu. White arrows mark fault tips. View to the west. Field of view is *c.* 1 km in foreground. View location and direction indicated by arrow in Figure 13a. Image Landsat. Image © DigitalGlobe. Data SIO, NOAA, U.S. Navy, NGA, GEBCO.

The north Taliabu dip slope is truncated by several prominent east–west-trending faults that dip steeply north. The geomorphic expression is very fresh (Fig. 13b). The footwall crests are only slightly eroded and, in most cases, the drainage runs parallel to fault scarps and has not cut across them, except for a few prominent high-order streams. The faults displace the dip slope and must therefore post-date the mega-debris slide. Although we have no absolute constraint on the timing of the slide, reef build-ups are conspicuously poorly developed along the section of coast at the foot of the dip slope, but are extensive along the coast and small islands on either side. The slide must have happened recently enough that corals have been unable to fully recolonize the new coastline, suggesting that the post-slide normal faults are late Quaternary and probably still active.

Sorong Fault from Obi to Waigeo

Westward splaying segments of the Sorong Fault emanate from the western Bird's Head and pass close to the islands of Salawati, Misool, Obi, Bacan, south Halmahera and Waigeo (e.g. Katili 1975; Hamilton 1979; Ali & Hall 1995) (Fig. 1). Although there is debate about whether the Sorong Fault onshore West Papua is tectonically active (discussed later in this paper), at the latitude of Obi there is 19 ± 8 mm a^{-1} left-lateral displacement between Ternate and the Bird's Head that may be accommodated by one or more strands of the Sorong Fault (Bock *et al.* 2003). Seismicity is limited in the islands immediately west of the Bird's Head, but intense seismicity occurs around Obi, Bacan and south Halmahera (Rangin *et al.* 1999), which may be where sinistral strain is transferred from Seram into the Molucca Sea.

Seram fold–thrust belt

Between northern Seram and the Bird's Head is a broad zone of transpression linked to convergence between Australia and the Pacific plate (Fig. 1). A deep bathymetric trough, the Seram Trough, lies 150 km north of Seram Island and curves around the Banda Sea, linking to the Timor Trough and ultimately the Java Trench. The Seram Trough has been interpreted as a subduction trench (e.g. Hamilton 1979), a foredeep ahead of a fold–thrust belt (e.g. Audley-Charles 1986) and a hinge zone marking the northern limit of delaminated and subducted lower continental crust (Spakman & Hall 2010).

Convergence across the Seram Trough is presently 20 mm a^{-1} (Rangin *et al.* 1999; Stevens *et al.* 2002) and is associated with intense seismicity generated by shallow thrust faulting (McCaffrey 1989; Engdahl *et al.* 1998) mainly concentrated along the northern edge of Seram (Fig. 14a) and entirely in the western part of the fold belt (Teas *et al.* 2009).

Seram is centred on a belt of high mountains (>3 km elevation), which include tracts of continental metamorphic rocks, ultramafic rocks and the Earth's youngest exposed ultra-high-temperature granulites, exhumed since 16 Ma (Pownall *et al.* 2014). The Plio-Pleistocene Wahai and Fufa formations onlap the elevated pre-Pliocene succession, forming low plains along the northern coast (Pairault *et al.* 2003), and are themselves overlain by modern alluvial and reef deposits. There is evidence of active contraction within these plains.

On the north coast of Seram, onshore fold growth affects the modern drainage, suggesting that the folds have been active during the Quaternary (Fig. 15a). Three large rivers draining the northern slopes of the Kobipoto Mountains are deflected from a linear route to the coast by two sets of segmented east–west- to NW–SE-trending hills. Progressive migration of the rivers away from the hilltops is recorded by a trail of abandoned and filled river channels left behind by the deflected river, expressed by oxbow-shaped fields and areas of vegetation (Fig. 15b, c). Larger hills, like that in the centre of Figure 15a, cause more deflection than smaller folds, like that in the east which only deflects Wai (stream) Kobi slightly. In all cases the abandoned channels are located upslope of the modern river, suggesting that progressive uplift is forcing river avulsion. This tendency for the hills to grow symmetrically from a central axis, their elongate morphology and their asymmetry (steep northern slopes, shallow southern slopes) support the interpretation that they are the surface expression of shallow, north-vergent fault propagation folds above south-dipping thrusts (Fig. 15d).

Abandoned meander channels and point bars on the coastal plain in the central part of Figure 15d are not associated with any obvious modern river, but seem to originate at the foot of the central frontal thrust. Abandoned remnants of a comparably large river can also be observed in an uplifted valley immediately to the south, and directly north of a fourth major north-flowing river, which presently abruptly curves around the eastern tip of the fault before joining Wai Musi. It is interpreted that the abandoned channels here represent a river that flowed directly north before the fold developed. An uplifted valley across the mid-point of the fold shows that the river attempted to downcut as the fold grew, but was ultimately thwarted by a high uplift rate and swung east to be captured by Wai Musi. Deep lateral incision by the captured river into the back limb of the fold (Fig. 15b) suggests that the fold growth, and presumably underlying thrust activity, is ongoing.

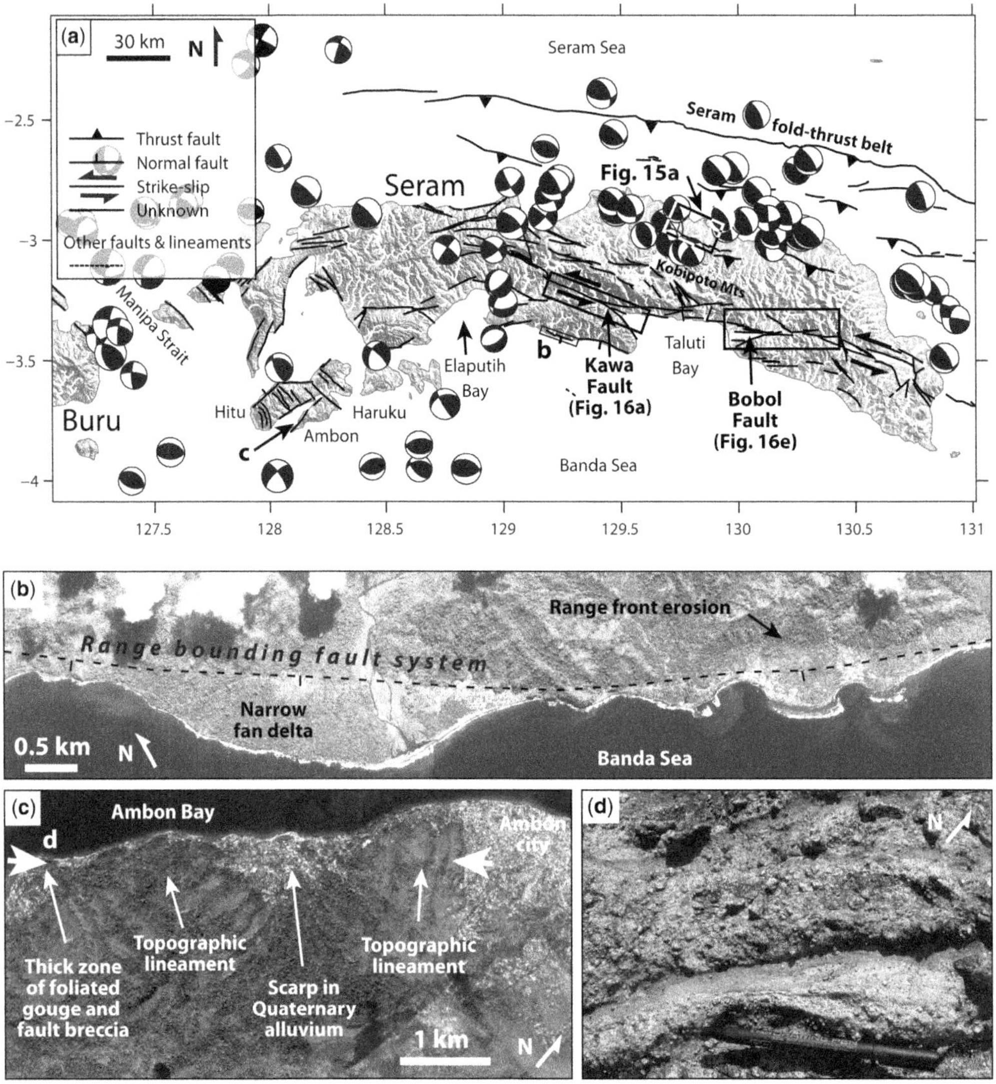

Fig. 14. (**a**) Seram digital elevation model (SRTM), CMT catalogue earthquakes <35 km depth and structures showing geomorphic evidence of Quaternary tectonic activity. Rivers marked in white. Offshore structures from Teas *et al.* (2009). Location shown in Figure 1. (**b**) Normal faults along the south coast of Seram, marked by a linear mountain front and a prominent lineament crossing a narrow fan delta. Google Earth imagery. Data SIO, NOAA, U.S. Navy, NGA, GEBCO. Image © 2016 DigitalGlobe. Image © 2016 TerraMetrics. (**c**) Possible Quaternary fault SW of Ambon, marked by a lineament that crosses volcanic hills and Quaternary drift. Google Earth imagery. Image © 2016 DigitalGlobe. (**d**) Example of foliated gouge from a thick fault zone located where the lineament illustrated in Figure 14c reaches the coast. Pen is 14 cm long.

A series of abandoned channels east of Wai Musi, the easternmost of which link to Wai Kobi, indicates that river itself may previously have been a tributary to Wai Kobi, before being deflected to the west and ultimately cut off from the trunk stream, presumably by uplift above the eastern frontal thrust.

Such evidence of recent hanging wall uplift and tectonic folding, together with the low relief of the range front, leads to the conclusion that the faults are youthful, low-angle, south- to SW-dipping thrusts supported by focal mechanisms along the north coast (Fig. 14a). Uplifted coastal terraces in the foreland of the onshore thrusts and a conspicuously wide coastal plain (Fig. 15d) suggest additional young uplift north of the onshore thrusts, perhaps in response to a third set of active faults just offshore. This is consistent with modern thrust activity within the broad fold–thrust belt offshore (e.g. Engdahl *et al.* 1998; Teas *et al.* 2009) and a 1629 mega-thrust earthquake probably originating in the Seram Trough (Liu & Harris 2013).

Kawa Fault

The Kawa Fault (Pownall *et al.* 2013) lies in the prominent ESE-WNW-trending deep linear valley that passes through central Seram (Fig. 14a) and is occupied by the Kawa River. The fault broadly separates upper greenschist to mid-amphibolite facies Tehoru Formation rocks in the south from generally higher grade metamorphic rocks of the Saku and Taunusa complexes in the north (Germeraad 1946; Tjokrosapoetro *et al.* 1993; Pownall *et al.* 2013). The Kawa Fault coincides with the position of strongly mylonitic garnet-bearing Tehoru Formation schists with a steeply dipping foliation considered by Linthout *et al.* (1991) to record dextral shear, but now recognized to have been intensely folded and possibly originating in a low-angle normal fault, resulting in complexly re-oriented kinematic indicators (Pownall *et al.* 2013).

A brittle fault zone up to 2 km wide (Pownall *et al.* 2013) overprints the mylonitic rocks and controls the modern topography (Fig. 16a, b). Fault strands are generally parallel to the mylonitic foliation and contain abundant serpentinite slivers and smears. Mid-way along the fault is a prominent right step associated with uplift and a major drainage divide, pointing to local transpression due to left-lateral slip. Stream offsets measured from Landsat and Google Earth imagery along the fault (Fig. 16a) range from 66–605 m of left-lateral offset (22 measurements) to 62–334 m of right-lateral offset (five measurements). Most measurements have a high uncertainty, increased by Seram's extremely humid climate and thick forest cover. Nonetheless, some measurements – for example, the 268 and 253 m left-lateral offsets (e.g. Fig. 16c) – are considered to be robust because: (1) they lie on fault segments that are well defined (narrow linear valleys with other independent evidence of a fault origin such as triangular facets and steps/bends with the corresponding uplift/subsidence appropriate to the sense of river offset); (2) there is no evidence of stream capture; and (3) upstream and downstream valleys have a similar geomorphic character. A left-lateral shutter ridge displacement and a NW–SE-trending fold within the Kawa River delta (Fig. 16a) support recent sinistral shear. A few earthquakes close to the western end of the fault yield CMT solutions suggesting dextral slip along NW–SE-trending planes, whereas one close to the westernmost splay indicates sinistral slip (Fig. 14a).

The fault zone splays as it enters Taluti Bay in the east (Fig. 16a). The splay strands are associated with well-developed triangular facets that record Quaternary normal faulting (Fig. 16d). The two major splays have low S_{mf} values of 1.10 in the north and 1.33 in the south and an average V_f of 0.27, indicating rapid to moderate tectonic activity. The Kawa River flows hard against the southern splay, suggesting active subsidence along that segment, despite its higher S_{mf} indicating slower tectonic activity than in the north. However, the river's position may also be influenced by landslips, debris flows and anticline growth in the northern part of its valley. In the west, the fault splays north of Elaputih Bay, attaining a total onshore length of 90 km, or 120 km including a possible splay fault along the north coast of Taluti Bay (Fig. 14a).

Although the fault zone is thickly forested, numerous landslip scars can be recognized along the fault, indicating recent seismicity (Fig. 16a). In addition, the eastern termination is characterized by a series of discontinuous tilted blocks suggestive of slope failure along the southern margin of the Manusela Mountains (Fig. 16a). An M 7.8 earthquake in 1899 triggered landslides that caused a 12 m high local tsunami at Tehoru (http://www.ngdc.noaa.gov; Brune *et al.* 2010), although it is unclear whether the source was the Kawa Fault or a more distant earthquake. However, all evidence points to the Kawa Fault being active during the Quaternary and capable of generating large earthquakes.

Other active faults of Seram

Along-strike from the Kawa Fault on the east side of Taluti Bay, a fault zone occupies the valleys of Wai Masumang and Wai Bobol and is here termed the Bobol Fault (Fig. 14a). It is highly segmented, although with a total onshore length of 100 km and possible along-strike continuity with the Kawa Fault, it is a significant structure. Four large basins are developed along its length, each bounded by ESE–WNW- to SE–NW-trending normal faults. The mountain front sinuosity along these structures ranges from 1.26 in the central section to 1.99 in the west and the average V_f is 1.66, indicating moderate to slow tectonic activity. There are a number of stream offsets both across the basin-bounding faults and across parallel faults in adjacent mountains (Fig. 16e). Convincing displacements are all left-lateral and range from 310 m to 2.06 km (Fig. 16f). Most strike-slip fault segments within the fault zone are parallel to the Kawa Fault and the two fault systems appear to be tectonically related and part of a broader zone of active left-lateral shear linking to the Tarera–Aiduna Fault in West Papua.

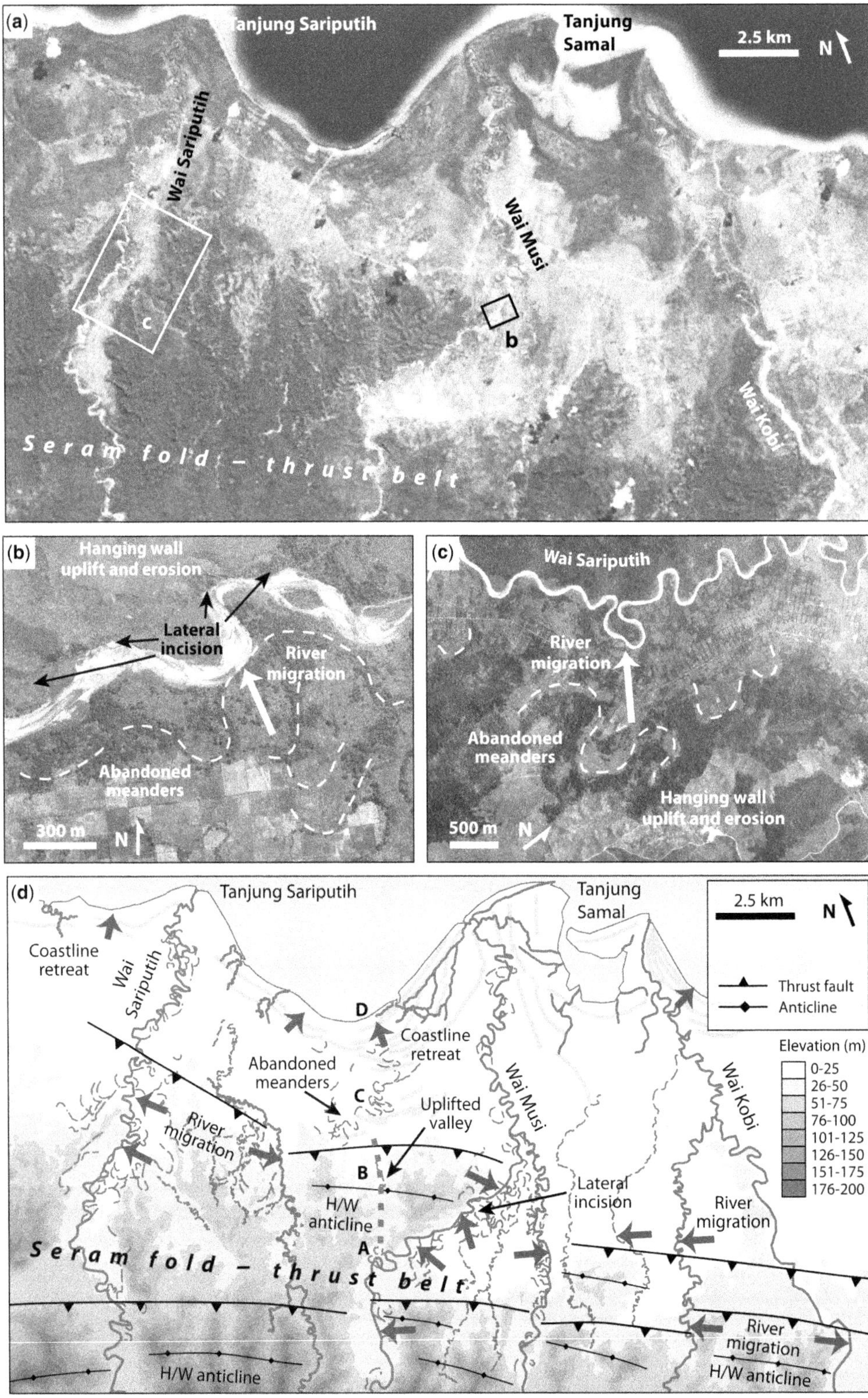
(a)
Tanjung Sariputih
Tanjung Samal
2.5 km
N
Wai Sariputih
Wai Musi
Wai Kobi
c
b
Seram fold – thrust belt
(b)
Hanging wall uplift and erosion
Lateral incision
River migration
Abandoned meanders
300 m
N
(c)
Wai Sariputih
River migration
Abandoned meanders
Hanging wall uplift and erosion
500 m
N
(d)
Tanjung Sariputih
Tanjung Samal
2.5 km
N
Thrust fault
Anticline
Elevation (m)
0-25
26-50
51-75
76-100
101-125
126-150
151-175
176-200
Coastline retreat
Wai Sariputih
D
Coastline retreat
Abandoned meanders
C
Uplifted valley
Wai Musi
Wai Kobi
River migration
B
Lateral incision
River migration
H/W anticline
A
Seram fold – thrust belt
River migration
H/W anticline
H/W anticline

The southern margin of Seram is locally formed by linear mountain fronts flanked by narrow fan deltas not more than 1 km wide. The steep, linear aspect and high topographic relief of the mountain fronts and the topographic lineaments that cross the fans parallel to the mountain front (Fig. 14b) suggest that the mountain front is defined by Quaternary normal faults. However, the coastal range is deeply eroded, with S_{mf} values of 1.84–2.08 and an average V_f of 1.62, indicating slow tectonic activity. Earthquake focal mechanisms towards the west of the coastal fault system in the region of Elaputih Bay support shallow focus, broadly south-directed, steep normal faulting (Fig. 14a).

A number of other small suspected normal fault systems occur around the SW coast of Seram, including those bounding the Ambon Islands. One fault along the northern coast of Hitu (Fig. 14a) is particularly steep and straight, with an S_{mf} of 1.16 and well-developed triangular facets along its 16 km long trace. A NE–SW-trending lineament that passes through Ambon city marks the southern coast of Ambon Bay (Fig. 14c) and is associated with a zone of fault breccia and foliated gouge several metres thick (Fig. 14d). An M 7.6 earthquake occurred on 8 October 1950 close to the south coast of Ambon (Bath & Duda 1979), although it is unlikely that such an event could have been caused by the relatively short, dominantly normal faults visible onshore.

Buru

Buru consists of a presumed Palaeozoic continental metamorphic basement flanked by a Mesozoic sedimentary succession (Tjokrosapoetro *et al.* 1993), both of which are probably continuous with similar units in Seram (e.g. Pigram & Panggabean 1984; Linthout *et al.* 1989). Young K–Ar ages of 4–5 Ma (Linthout *et al.* 1989) and an apatite fission track central age of 2.5 $\pm$ 0.5 Ma suggest late Neogene exhumation, possibly accommodated by low-angle normal faults (Roques 1999) as similarly postulated for western Seram (Pownall *et al.* 2013).

Intense shallow seismicity associated with Seram terminates abruptly in Manipa Strait, east of Buru (Fig. 17a). A broad belt of earthquakes in Manipa Strait have CMT solutions indicating either NNE–SSW dextral events or WNW–ESE sinistral events, including a 14 March 2006 M_w 6.7 earthquake 25 km offshore. Most earthquakes have a component of reverse slip; others are pure thrust earthquakes with a NW–SE trend.

Most of Buru's sparse population lives in the NE of the island, including the major town, Namlea. A 5–10 km wide system of NW–SE-trending faults cuts through the town, across Kayeli Bay, and defines the coastline (Fig. 17b). The faults are expressed in remote sensing data by linear hills and sag ponds at releasing right step-overs, notably at Jikumerasa (Fig. 17c). Fault strands that cut through basement metamorphic rocks and alluvial fans show consistent stream offsets and pass directly into Quaternary alluvium and control modern river channels (Fig. 17d). Stream offsets of up to 85 m across individual strands are mostly right-lateral; where they are left-lateral there is clear evidence for stream capture. Variations in offset sense and amount are to be expected – streams are dynamic and are not passively offset like pre-kinematic geological markers. The process of offset, beheading and capture, leading to stream offsets of zero or opposite to the fault's shear sense, is well documented and widely observable in active faults worldwide (e.g. Wallace 1968; Sieh & Jahns 1984; Huang 1993; Walker & Allen 2012). All these features imply Quaternary NW–SE-trending dextral fault activity in NE Buru, despite the apparent discordance with the few earthquake focal mechanisms recorded.

A broad fault zone 65 km long almost bisects Buru from the NE to SW (Figs 17a & 18a). Identified as left-lateral on early geological maps (e.g. Tjokrosapoetro *et al.* 1981), little else is known about the fault zone, here termed the Rana Fault. Danau (lake) Rana, in the centre of Buru, occupies an intermontane basin within a right step-over between two segments of the Rana Fault, suggesting that the fault is dextral. West of Wadule, Wa (river) Geren is abruptly diverted 90° from a broad oversized valley, which would have taken it to the coast in the NE of the island, into a narrow and steep-sided canyon (Fig. 18a) that links with Wa Apu and empties into Kayeli Bay further south

Fig. 15. Evidence of Quaternary thrusting along the north coast of Seram. (**a**) ESRI image showing a number of NE-flowing rivers flowing around linear elevated and forested regions. Location shown in Figure 14a. (**b**, **c**) Migrating rivers marked by filled channels and oxbow lakes and incision into uplifting regions. Google Earth imagery. Image © 2016 DigitalGlobe. (**d**) Interpretation of Figure 15a. Thick arrows indicate progressive migration of river channels; short arrows show coastline regression. Abandoned channels at points A, B and C are interpreted to represent the previous route of a river that entered the sea at D north of a meander plain at C, but was cut off by thrust hanging wall (HW) uplift at B and was forced to divert east from point A to join Wai Musi, leaving previous channels abandoned. Other rivers show lateral migration away from the growing tips of thrusts in response to hanging wall fold growth. See text for further details.

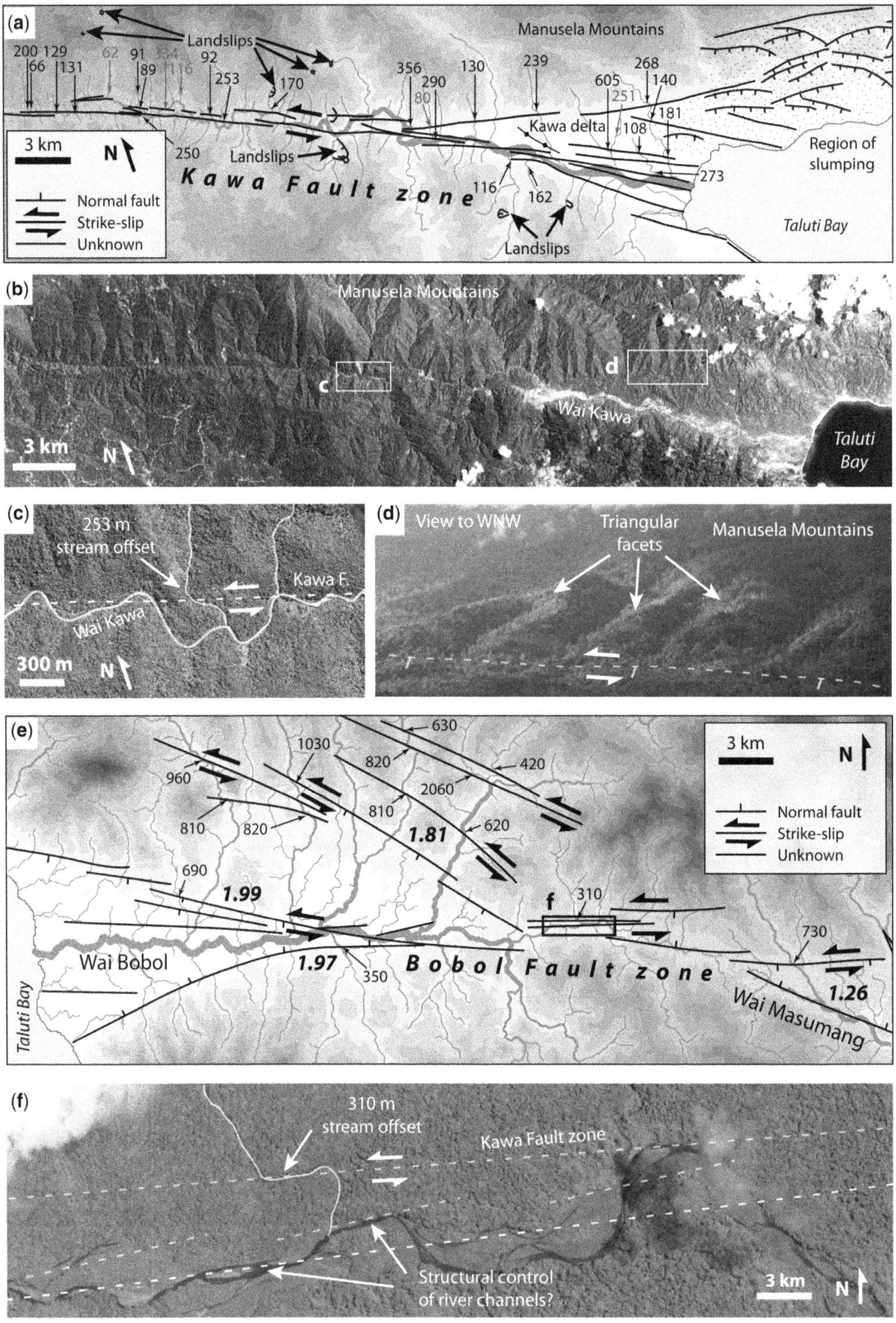

Fig. 16. Strike-slip faults of southern Seram. (**a**) Overview map of the Kawa fault zone showing Quaternary fault strands, rivers, river offsets (in metres) and landslips. Left-lateral offsets in black, right-lateral offsets in grey. See Figure 14a for location. (**b**) ESRI image of the Kawa fault zone highlighting its clear geomorphic expression and

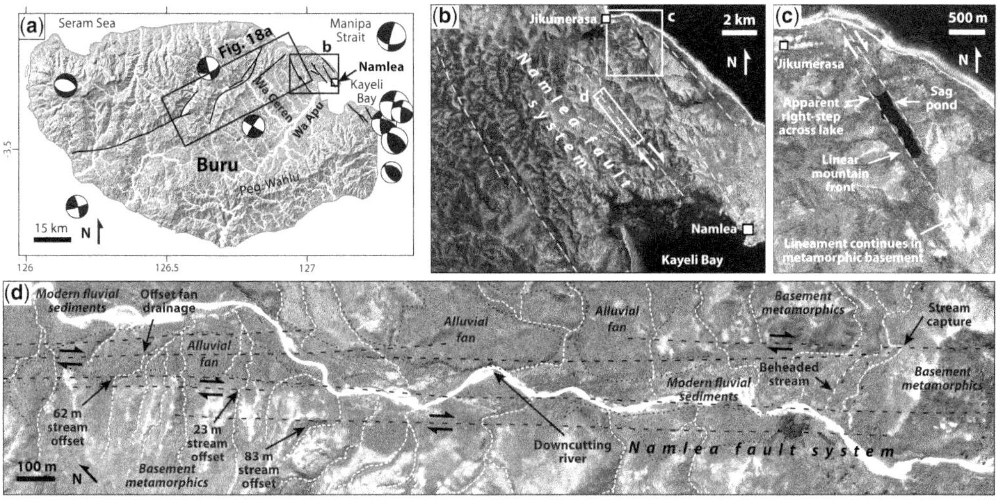

Fig. 17. Quaternary fault features in Buru. (**a**) Digital elevation model (SRTM), CMT catalogue earthquakes <35 km depth and structures that show geomorphic evidence of Quaternary tectonic activity. Rivers marked in white. Location shown in Figure 1. (**b**) Overview of topographic lineaments passing through Namlea and NE Buru. ESRI imagery. (**c**) Detail showing possible sag pond developed at the releasing step-over between right-stepping fault segments. (**d**) Evidence of strike-slip faulting along the Namlea lineament trend. Lineaments pass from basement rock through Quaternary drift and are associated with systematic right-lateral stream offsets. Image © 2016 DigitalGlobe.

(Fig. 17a). This pronounced capture of a northern drainage basin by a relatively minor tributary of Wa Apu appears to have been triggered by uplift at a left bend in the Rana Fault immediately east of the capture point (Fig. 18a), again suggesting Quaternary dextral shear.

Upstream of the Wa Geren stream capture, the Rana Fault has exceptionally fresh geomorphic expression (Fig. 18b, c), with pronounced triangular facets and very low S_{mf} values from 1.01 to 1.18 along the southern valley slope and a correspondingly low average V_f of 0.25, all suggesting a maximal to rapid tectonic rate. There are a number of beheaded and offset streams along the southern valley slope, although there is no consistent tectonic lateral offset. The axial river has migrated systematically eastwards in two places, leaving behind abandoned channels uplifted up to 10 m above the modern river channel (Fig. 18c). The uplift defines a pair of low-amplitude right-stepping en echelon periclines, consistent with Quaternary right-lateral shear. There is abundant evidence of revegetated landslip scars in the surrounding hills close to the fault.

A *c.* 10 m high scarp along the base of alluvial fans in the valley, visible in high-resolution DigitalGlobe satellite imagery from Google Earth, has the appearance of a normal fault surface rupture (Fig. 18d, e). The valley is relatively thinly vegetated and the scarp, discontinuous over *c.* 7.5 km, is well preserved. Although in places it is parallel to the modern river valley, the linear scarp also crosses higher ground, proving that it is not simply an erosional feature. By analogy with proved historical earthquake surface ruptures with a similar topographic expression – for example, the 1857 Lone Pine earthquake (Beanland & Clark 1994) and the 1609 Hongyazi earthquake (Xu *et al.* 2010) – the Buru scarp may have formed during the last few hundred years. The entire 10 m throw could have developed during a single M 7.5 earthquake, according to empirical relationships (Wells & Coppersmith 1994), or during a number of smaller events, similar to the Star Valley Fault at Afton, Wyoming,

Fig. 16. (*Continued*) thick forest cover. (**c**) Representative stream offset across the main Kawa Fault strand, image from Google Earth. (**d**) View into the Kawa Fault from the Wai Kawa delta showing the linear mountain front and triangular facets developed along the northern strand of the Taluti Bay splay. Image © 2016 DigitalGlobe. (**e**) Overview map of the Bobol fault zone showing Quaternary fault strands, rivers and left-lateral river offsets (in metres). Bold italic numbers are S_{mf} values. See Figure 14a for location. (**f**) Representative stream offset across the main Kawa Fault strand, also showing fault control of river channels. ESRI imagery.

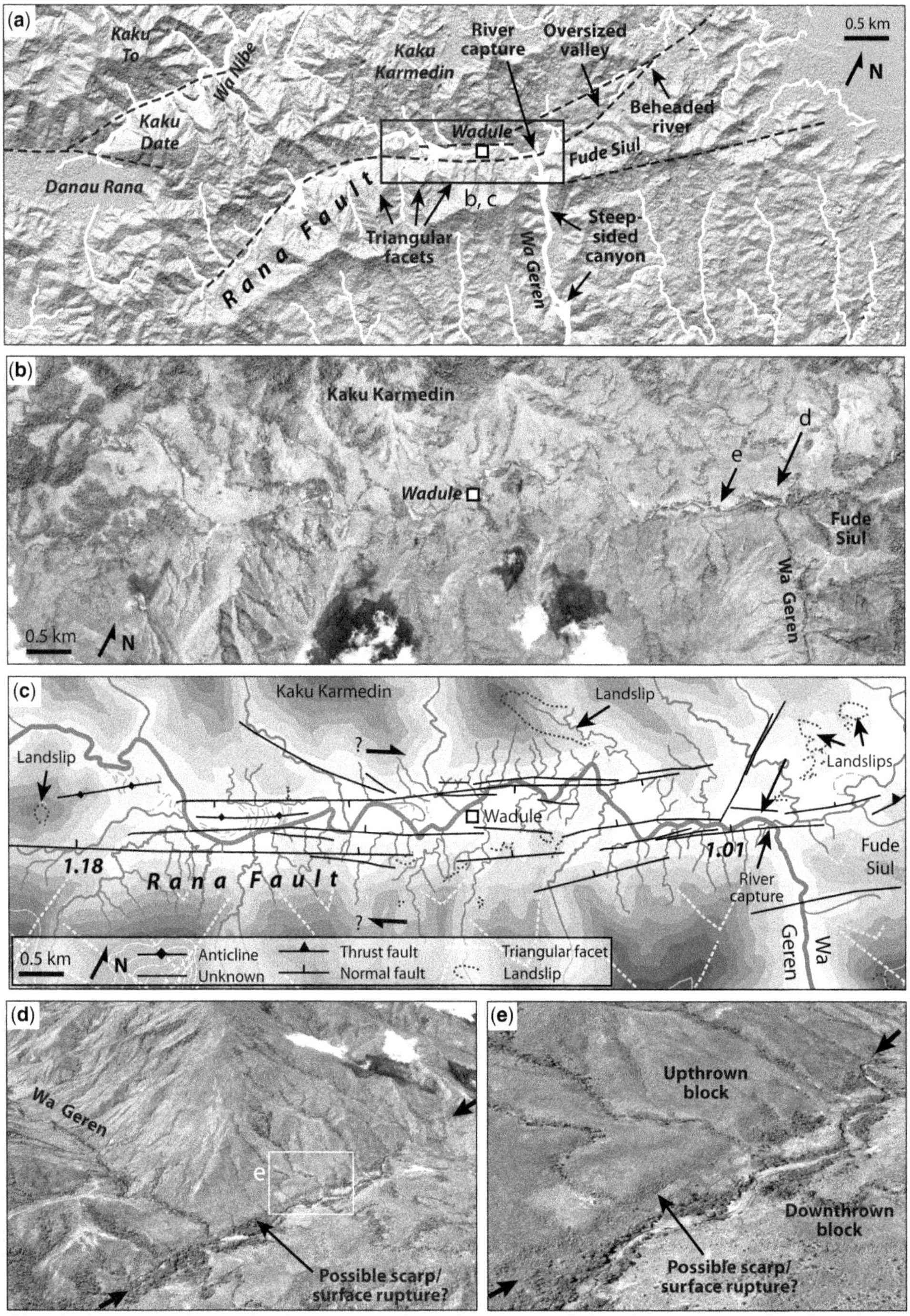

Fig. 18. Rana Fault, central Buru. (**a**) Overview of the Rana Fault, ASTER GDEM base map. Rivers marked in white, with white arrows showing the flow direction of major rivers discussed in the text. Location shown in Figure 17a. (**b**) ESRI image of the central part of the Rana Fault. (**c**) Interpretation of the image in (b), showing evidence of Quaternary fault activity. (**d**) Possible fault scarp along the foot of triangular facets marking the Rana Fault. (**e**) Detail of the possible fault scarp, showing steep dip, fresh geomorphic expression and straight trace. (d) and (e) are oblique views from Google Earth. Image © 2016 DigitalGlobe.

where an 11 m high scarp formed during three late Quaternary earthquakes (Piety *et al.* 1992). This is perhaps a more likely scenario given the relatively short length of the Rana Fault.

Elsewhere in Buru the geomorphic expression of other steep normal faults suggests rapid to moderate tectonic activity. Faults associated with the Rana Lake basin have S_{mf} values of 1.33–1.49 (Fig. 2j). Short fault segments in the SE of the island have S_{mf} values of 1.23 and 1.44, whereas those on the extreme east coast are more eroded, with S_{mf} values of 1.99 and 2.14 (Fig. 2k), indicating that they have been less active during the Quaternary.

Papua and West Papua

Oblique convergence at an angle of *c.* 60° between Australia and the Pacific is accommodated across Papua and West Papua in a complex zone of strain partitioning between shortening and left-lateral shear (e.g. Abers & McCaffrey 1988; McCaffrey 1996). West of about 138° E shortening is largely accommodated on a variety of structures in the New Guinea Trench and Manokwari Trough, in the Mamberamo fold–thrust belt and in the central Highlands to the south (e.g. Milsom *et al.* 1992; Puntodewo *et al.* 1994; Stevens *et al.* 2002). The largest earthquake to occur in eastern Indonesia since 1938 was the tsunamigenic 17 February 1996 M_w 8.2 Biak earthquake, which was also the largest thrust event worldwide since 1977 (Henry & Das 2002) and may have been associated with the 1979 M 7.9 Yapen earthquake (Okal 1999).

Left-lateral strain of up to 80 mm a^{-1} resulting from oblique Australia–Pacific convergence is accommodated across a 300 km wide zone of sinistral shear (Stevens *et al.* 2002) focused on the Yapen Fault system in the north and stepping across Cenderawasih Bay to the Tarera–Aiduna Fault system in the south, largely bypassing the antecedent Sorong Fault in West Papua (e.g. Puntodewo *et al.* 1994; McCaffrey 1996; Stevens *et al.* 2002; Bock *et al.* 2003). Left-lateral shear is passed from the Tarera–Aiduna Fault westwards into Maluku via the highly transpressive Seram fold–thrust belt (Teas *et al.* 2009).

As in Sulawesi, Maluku and North Maluku, extension is important within the overall convergent orogen. Cenderawasih Bay and the adjacent Waipoga Basin contain thick sediment piles (e.g. Dow & Sukamto 1984; Pubellier *et al.* 1999; Charlton 2010) and metamorphic core complex exhumation at the Wandamen Peninsula (e.g. Bailly *et al.* 2009) indicates extreme lithospheric stretching. Although extension may be related to processes within the wide left-lateral shear zone (Stevens *et al.* 2002), lessons from Sulawesi suggest that

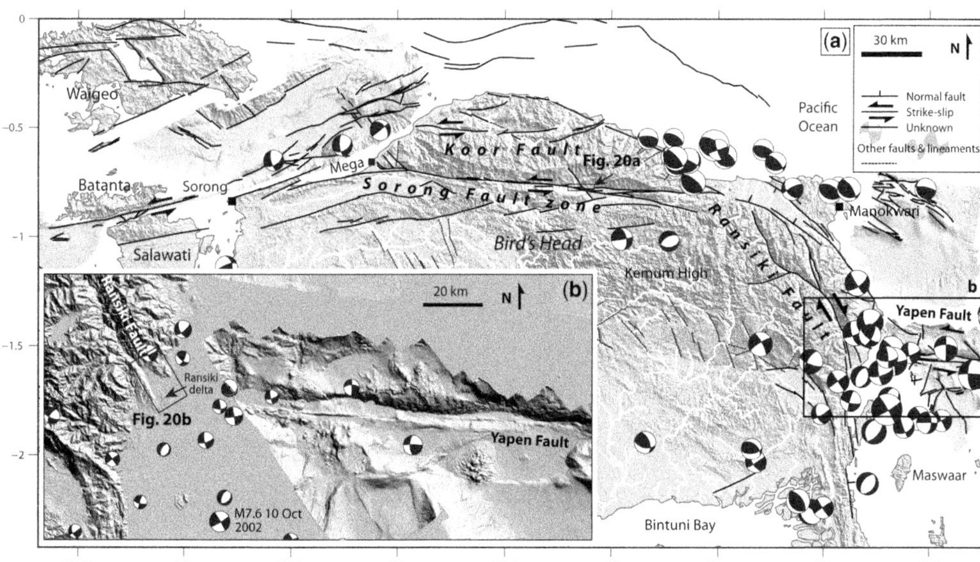

Fig. 19. (**a**) Bird's Head (West Papua) digital elevation model (SRTM), multibeam bathymetry, CMT catalogue earthquakes <35 km depth and structures showing evidence of Quaternary tectonic activity. Rivers marked in white. Location shown in Figure 1. Offshore structures north of the Koor Fault from Milsom *et al.* (1992). (**b**) Detail of the intersection between the Ransiki and Yapen faults, south of Manokwari. Eastern limit of image is to the east of the main map. SRTM onshore, multibeam bathymetry offshore.

far-field subduction-related mechanisms may also be significant.

Sorong Fault in West Papua

The Sorong Fault in West Papua is marked by a 15 km wide zone of pronounced linear ridges and valleys trending ENE from northern Salawati through Sorong city and into the deep valley cutting across the northernmost mainland towards Manokwari in the east (Fig. 19a). Hamilton (1979) questioned whether this structure was significant in post-Miocene tectonics, pointing out that parts of it were covered by post-Miocene strata, and it is now generally considered to be inactive (e.g. Puntodewo *et al.* 1994; Decker *et al.* 2009; Charlton 2010).

There is little significant seismicity along much of the fault and geodetic measurements suggest that both sides of the fault are broadly moving together and with the Pacific (e.g. Puntodewo *et al.* 1994; Stevens *et al.* 2002), with slight residual left-lateral motion between the Sorong and Fakfak GPS stations possibly accommodated on the Sorong Fault or the Koor Fault to the north (Bock *et al.* 2003). However, the Sorong GPS station is south of important strands of the Sorong Fault, which lie offshore to the north and come onshore at Mega, and the station is certainly south of the Koor Fault, leaving substantial uncertainty in the amount of present day left-lateral strain accommodated across this zone. The April 1937 M 6.9 and April 1944 M 7.2 and 7.4 earthquakes relocated by Okal (1999) were located on the onshore Sorong Fault 50–100 km west of Manokwari and had focal mechanisms indicating left-lateral shear. Apparent right-lateral motion between the Sorong and Biak GPS stations, taken to lie on opposite sides of the Sorong Fault (Puntodewo *et al.* 1994), is complicated by other structures such as the Ransiki and Yapen faults, which also lie between the stations.

Numerous convincing left-lateral stream offsets of up to 300 m are documented in the central part of the fault valley (Dow & Sukamto 1984) (Fig. 20a). Similar sized displacements of Wallace Creek crossing the San Andreas Fault have been dated to 13 259 years (Sieh & Jahns 1984). It is unclear how long such offsets can be preserved in the landscape of an environment like West Papua, but it is unlikely they are pre-Quaternary. Given that few such offsets are preserved in the more obviously active faults of eastern Indonesia, such as the Palu–Koro and Matano faults, the Sorong Fault examples must reflect relatively recent and significant strike-slip. Mountain front sinuosity along those segments of the fault associated with vertical motions is also conspicuously low, ranging from 1.16 to 1.17 along segments NNE of Sorong city to 1.14 along the central section, where Dow &

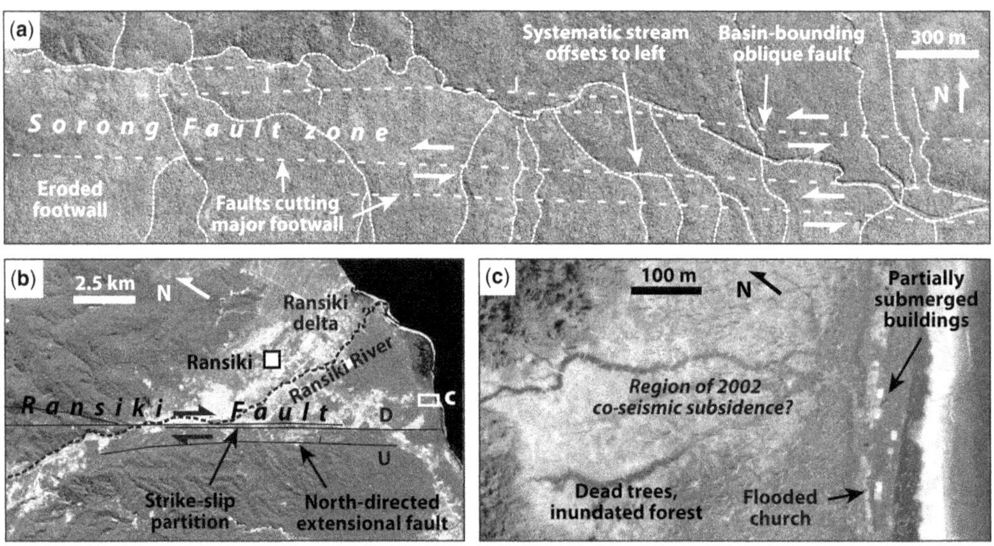

Fig. 20. Evidence of Quaternary fault activity in the Bird's Head. (**a**) Section of the onshore Sorong Fault showing a basin-bounding normal fault in the north and two strike-slip fault strands offsetting streams to the left in the south. Location shown in Figure 19a. ESRI imagery. (**b**) Ransiki delta at the southern end of the Ransiki Fault showing prominent western normal fault. Location shown in Figure 19b. (**c**) Region of co-seismic subsidence showing flooded forest and buildings adjacent to the western normal fault. Google Earth Imagery. Image © 2016 DigitalGlobe.

Sukamto (1984) measured displaced streams and where triangular facets and shutter ridges are well developed. In the east, S_{mf} values of 1.20 and 1.33 also suggest active tectonics. Faults adjacent to flat-topped Quaternary basins associated with Sorong Fault releasing geometries are interpreted to be dominantly normal faults (Fig. 20a) and these structures have generally higher S_{mf} values, including 1.60, 1.61, 1.74 and 2.79. The average V_f value for all these fault segments is 1.15, consistent with moderate to slow tectonic activity.

Koor Fault and Ransiki Fault

The Koor Fault is an east–west-trending structure 20–30 km north of the Sorong Fault (Fig. 19a), which lies within a boundary zone between the oceanic Pacific plate and continental crust in the south (Dow & Sukamto 1984). The NNW-trending Ransiki Fault (Fig. 19a) has been viewed as a dextral shear zone linking the easternmost Sorong Fault and the Yapen Fault (e.g. Robinson & Ratman 1978; Milsom *et al.* 1992; Charlton 2010).

Like the Sorong Fault in West Papua, both the Koor and Ransiki faults have been considered to be inactive (e.g. Hamilton 1979; Puntodewo *et al.* 1994). However, a shallow M 7.6 earthquake on 10 October 2002 at the southern end of the Ransiki Fault (Fig. 19b) had a focal mechanism and aftershock distribution consistent with dextral slip along the Ransiki Fault (NEIC), although the possibility of sinistral slip along a NE–SW-trending splay of the Yapen Fault cannot be excluded. Topographic and bathymetric data from the intersection (Fig. 19b) could be interpreted to show the two structures curving gently into each other, leading to the possibility of contraction in the Ransiki area.

Mountain front sinuosity measured along two splays of the southern Ransiki Fault yields values of 2.64 for a clearly inactive, *c.* north–south-trending southwestern strand, and 1.06 for the linear fault bounding the southern margin of Ransiki delta (Figs 2p & 19b). The very low S_{mf} and the asymmetrical position of the Ransiki River close to the fault scarp (Fig. 20b) support recent extensional activity along the fault. A 2 m high co-seismic surface rupture formed close to the fault scarp during the 2002 earthquake and was associated with subsidence of the delta that flooded a low-lying church (D. Gold, pers. comm. 2013), visible in satellite imagery to be coincident with a large region of flooded forest (Fig. 20c).

Yapen Fault

The Yapen Fault (Fig. 21a) is a highly linear east–west-trending structure that crosses the 320 km wide northern Cenderawasih Bay and is similar in character to the Sorong Fault in West Papua (e.g. Hamilton 1979; Dow & Sukamto 1984). In the east, the Yapen Fault vanishes into the Mamberamo delta (Fig. 21a), where it forms a subtle linear valley delineated by active mud volcanoes (Dow & Sukamto 1984) and may dissipate into the Mamberamo fold–thrust belt (Puntodewo *et al.* 1994). In the west, the Yapen Fault has an unclear termination, variously interpreted as being dextrally offset from the Sorong Fault along the Ransiki Fault (Puntodewo *et al.* 1994; Charlton 2010), linking/terminating against the Ransiki Fault (Milsom *et al.* 1992) and unconnected to inactive Ransiki/onshore Sorong faults, but transferring strain south to the Wandamen fault system (Bailly *et al.* 2009).

Geodetic measurements indicate a fast left-lateral slip rate of 46 ± 12 mm a^{-1} across the Yapen Fault (Bock *et al.* 2003), expressed by intense seismicity and focal mechanisms indicating left-lateral slip along east–west-trending subvertical planes (e.g. Okal 1999; Stevens *et al.* 2002). The 12 September 1979 M 7.9 tsunamigenic earthquake on the south coast of Yapen island (Fig. 21a) was associated with sinistral slip along a ESE–WNW-trending plane focused at a depth of 5 km and probably caused 2 m of displacement (Okal 1999).

The Randaway Fault Zone (Dow & Hartono 1982) is a set of NW–SE-trending faults onshore Yapen that link to strands of the Yapen Fault in the north (Fig. 21b). Interpreted as post-Plio-Pleistocene normal faults, they have previously been used to support a period of right-lateral shear along the Yapen Fault zone (Charlton 2010). However, we saw no geomorphic evidence of significant normal faulting along the Randaway trend – instead we saw a small linear basin and lake near the northern tip of the Randaway Fault at a left step-over and a deeply incised stream offset to the left by almost 1 km – both evidence of Quaternary sinistral shear (Fig. 21b).

Although the north coast of Yapen is remarkably straight and clearly fault-controlled, the main fault mostly lies just offshore to the north, meaning that geomorphic indices could not be usefully measured along the Yapen Fault. Multibeam bathymetry east of the island shows the Yapen Fault expressed by a straight, narrow lineament marked by pressure ridges and parallel to a prominent set of curvilinear normal faults (Fig. 21c). Splays of the fault curving to the WSW delimit at least two rhomboidal pull-apart basins. At the western limit of the multibeam data a splay appears to enter a third pull-apart basin, which is associated with a prominent north–south-trending sidewall fault. It is significant that this structure is parallel to, and 60 km north, of the Wandamen Peninsula – perhaps support for the southwards transfer of sinistral shear from the Yapen

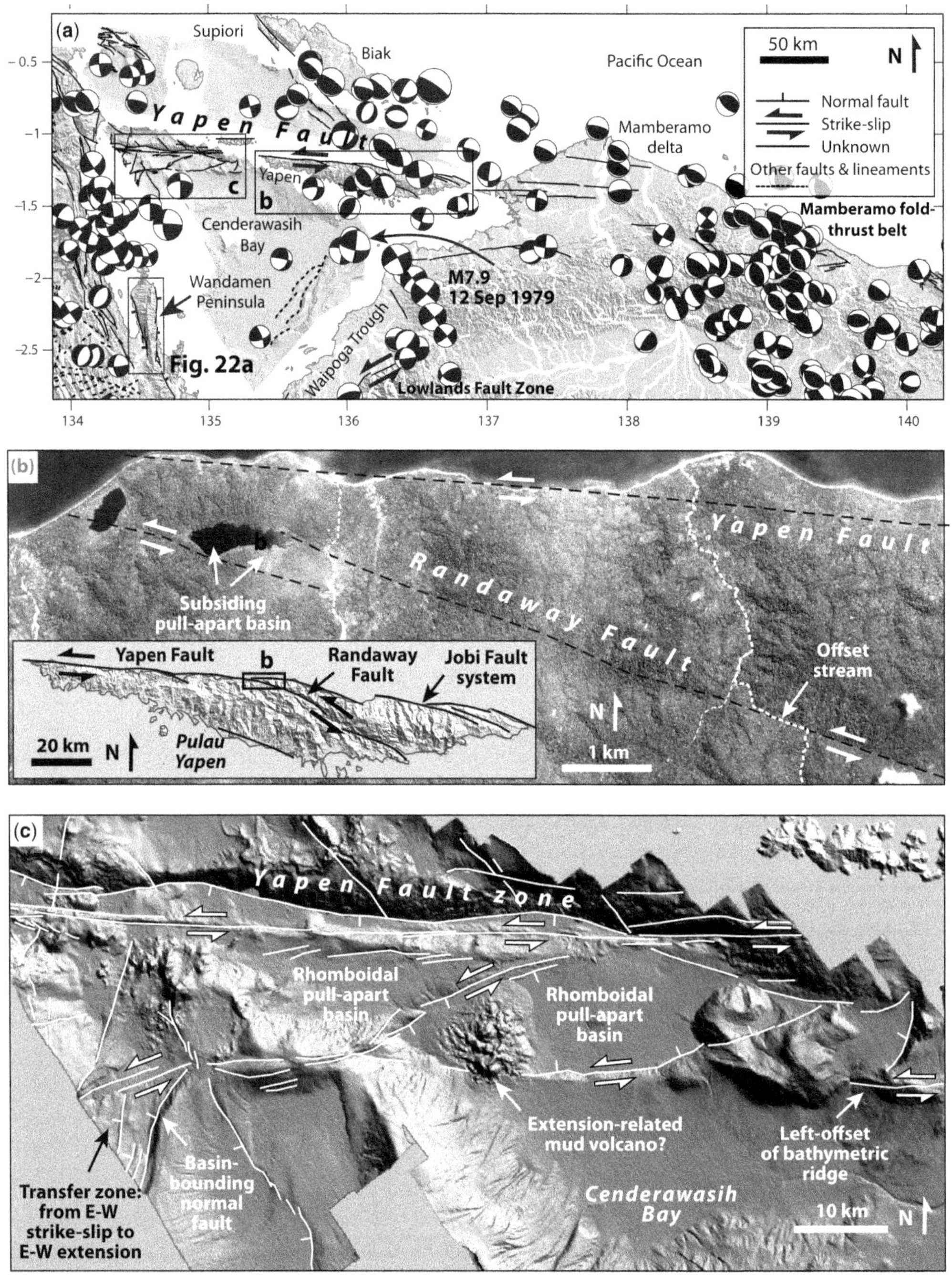

Fig. 21. Northern Papua and Cenderawasih Bay digital elevation model (SRTM), multibeam bathymetry, CMT catalogue earthquakes <35 km depth and structures showing geomorphic evidence of Quaternary tectonic activity. Rivers marked in white. Location shown in Figure 1. (**b**) Expression of the Yapen and Randaway faults along the northern coast of Pulau Yapen, showing evidence for Quaternary sinistral slip along the Randaway Fault. Inset shows the topography and major structures of Yapen. ESRI imagery. (**c**) Multibeam bathymetry detail showing the Yapen Fault to the west of Pulau Yapen; the southern strands appear to transfer to north–south extension via a series of pull-apart basins.

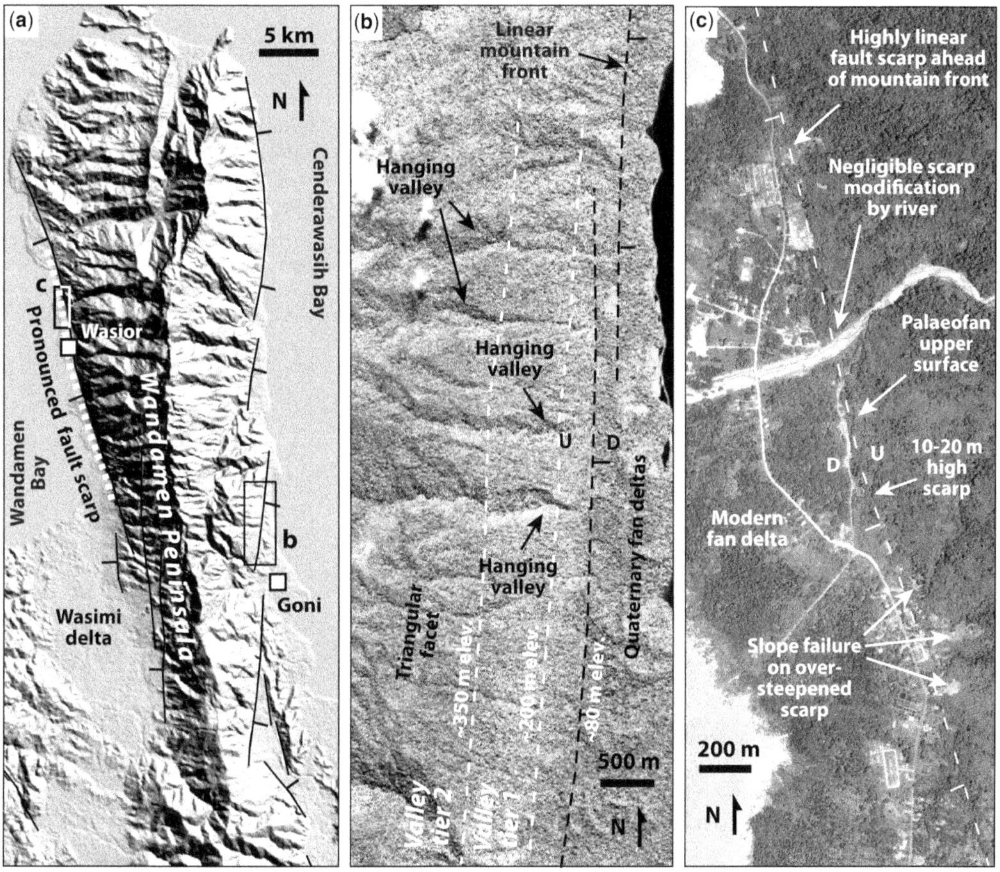

Fig. 22. Evidence of Wandamen Peninsula Quaternary fault activity. (**a**) Overview digital elevation model (SRTM) showing bounding normal faults. Location shown in Figure 21a. (**b**) Pronounced triangular facets and hanging valleys along the eastern bounding fault system. ESRI imagery. (**c**) Inferred Quaternary fault trace across the top of alluvial fans crossing the western fault system. Google Earth imagery. Image © 2016 DigitalGlobe.

Fault via a region of east–west extension, as proposed by Bailly *et al.* (2009).

Mamberamo fold–thrust belt

The Mamberamo fold–thrust belt (Fig. 21a) probably accommodates some Australia–Pacific shortening in eastern Papua and lies north of the Highlands thrust belt of central New Guinea (e.g. Dow & Sukamto 1984). Unlike the complex oblique convergence and strain partitioning further west, the belt contains relatively simple NW-trending active structures oriented normal to convergence (McCaffrey 1996). Despite intense and widespread seismicity, less than 15 mm a^{-1} of shortening occurs across the Mamberamo belt, leaving much of the remaining 45 mm a^{-1} Australia–Pacific convergence and 100 mm a^{-1} of left-lateral motion to offshore structures to the north and the Highlands thrust belt to the south (Puntodewo *et al.* 1994; McCaffrey 1996; Bock *et al.* 2003) (Fig. 1).

Wandamen Peninsula faults

The Wandamen Peninsula projects into Cenderawasih Bay from the eastern edge of the Lengguru fold belt, and is bounded on the east and west sides by north–south-trending faults (Fig. 22). We refer here specifically to these faults, not to the Wandamen Fault Zone of Dow & Sukamto (1984) that connects the Sorong Fault with the Tarera–Aiduna fault system via the Ransiki Fault.

The peninsula is considered to represent the exhumed internal zone of the Lengguru fold belt and is composed of an amphibolite–eclogite grade metamorphic dome rising to >2 km elevation (Robinson *et al.* 1990; Bailly *et al.* 2009; Charlton 2010), which may be a metamorphic core complex (e.g. Hill *et al.* 2002). Seismicity and GPS vectors either side of Cenderawasih Bay (Stevens *et al.* 2002) suggest active extension accommodated on north–south-trending structures close to the Wandamen Peninsula, which may connect to the western releasing termination array of the Yapen Fault in the north (Fig. 21c).

Normal faults bounding the peninsula are expressed by curvilinear en echelon segments up to 20 km long trending north–south to NNW–SSE. These make up the east and west detachment systems of Bailly *et al.* (2009). Triangular facets, hanging valleys and V-shaped valleys are common and indicate rapid tectonic activity (Fig. 22b). Two tiers of hanging valleys on the eroded scarp of the eastern fault system are defined by changes in valley width or orientation at common elevations along the scarp. They probably record variations in the tectonic rate or climate during exhumation of the fault surface. Mountain front sinuosity values of four segments on the east side are uniform at 1.25, 1.28 and 1.29, with one more eroded segment of 1.72. Fan deltas are well developed at relays between the fault segments, notably at Goni and another smaller delta 21 km further north (Fig. 22a). As well as localizing sediment transport, the relays are likely to be sites of active displacement minima, allowing subaerial delta progradation on the hanging wall.

On the west of the peninsula, S_{mf} values range from 1.05 to 1.43, indicating maximal to rapid tectonic activity. A 21 km long section of the western fault system passing through Wasior shows evidence of recent normal faulting (Fig. 22c). Upper modern fan deltas are abruptly terminated by a linear scarp, above which are narrow truncated palaeofans. Rivers vertically incised into footwall palaeofans show little evidence of lateral erosion and small landslides are localized along the over-steepened scarp. The scarp is marked by a linear change in topography, lines of vegetation and often an abrupt change from meandering rivers upstream to anastomosing rivers downstream of the scarp. A southern continuation of the Wandamen fault system bounds the eastern margin of the Wasimi delta and has an S_{mf} value of 2.33, indicating slow to minimal tectonic activity.

Other circum-Cenderawasih Bay structures

The locus of active Australia–Pacific left-lateral strain partitioning shifts from the Yapen Fault system to the Tarera–Aiduna Fault system across Cenderawasih Bay, defining a 300 km wide shear zone that involves a complex array of Quaternary faults within the two bounding strike-slip zones (e.g. Stevens *et al.* 2002; Bock *et al.* 2003). Along the eastern margin of Cenderawasih Bay, the NE-trending

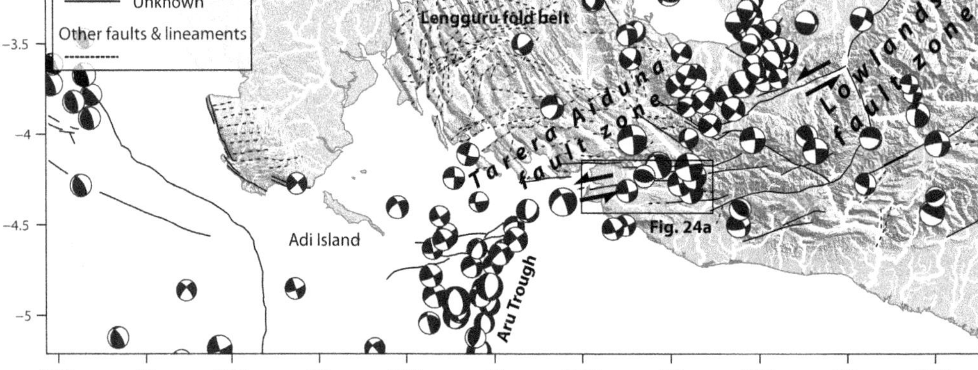

Fig. 23. Southern West Papua and Cenderawasih Bay digital elevation model (SRTM), multibeam bathymetry, CMT catalogue earthquakes <35 km depth and structures showing geomorphic evidence of Quaternary tectonic activity. Rivers marked in white. Location shown in Figure 1. Offshore structures from Teas *et al.* (2009).

Lowlands Fault Zone (bounding the Waipoga Trough of Visser & Hermes 1962) and the Paniai Fault Zone are associated with thrust and left-lateral strike-slip earthquakes (Fig. 23), offset drainage and high fault scarps, indicating modern tectonic activity (Pubellier *et al.* 1999; Stevens *et al.* 2002). The faults have a soft linkage with the Tarera–Aiduna Fault system in the south and splays curve into parallelism with the Yapen Fault and Mamberamo fold–thrust belt in the north.

The Lengguru fold belt (Visser & Hermes 1962) lies SW of Cenderawasih Bay and the Wandamen Peninsula, east of Bintuni Bay and the Bomberai Peninsula, and is bounded by the Tarera–Aiduna fault system in the south (Fig. 23). Compressional deformation terminated during the Pleistocene (Decker *et al.* 2009) and the belt is presently largely inactive, except for a few earthquakes related to gravitational collapse (Bailly *et al.* 2009), often with a left-lateral component related to residual Tarera–Aiduna strain.

Tarera–Aiduna Fault

The Tarera–Aiduna Fault (Visser & Hermes 1962) is an east–west-trending left-lateral shear zone that forms the southern boundary of the Lengguru fold belt and passes offshore to the west, north of the Aru Trough (Fig. 23). The Tarera–Aiduna Fault *sensu stricto* is part of a wide system of faults that pass, via a diffuse zone of sinistral transpression, into the Seram fold–thrust belt in the west (Teas *et al.* 2009). The fault system is at least 130 km long onshore (Fig. 24) and is expressed by straight lineaments clearly visible on satellite imagery (Hamilton 1979) and a set of en echelon folds (Katili 1986). Including possible soft linkage to Seram via sinistral transpression within the Seram fold–thrust belt, imaged in multibeam bathymetric data (Teas *et al.* 2009), the whole fault system may be >700 km long. Geodetic measurements show high relative motion between the Bird's Head north of the Tarera–Aiduna Fault and GPS stations south of the fault, such as Aru and Timika (Bock *et al.* 2003). Earthquake focal mechanisms showing sinistral slip along east–west-trending vertical planes (e.g. Seno & Kaplan 1988) suggest that the motion onshore is seismic and occurs along a broad zone (Fig. 23). Seismicity is largely absent west of the Bomberai peninsula, suggesting either a wide zone of aseismic deformation linking the Tarera–Aiduna Fault with the Seram sinistral transpression (Teas *et al.* 2009), a region of seismic deformation with recurrence times longer than the instrumental record, or no structural connection between the two regions.

The onshore Tarera–Aiduna Fault has a geomorphic expression typical of a major strike-slip fault zone (Fig. 24a). In the west it passes across a low-lying mangrove plain with minimal topographic relief. It is possible to trace several fault strands from linear features revealed by abandoned river channels and coastline segments (Fig. 24b). Its central section is expressed by a series of linear ridges of moderate relief bounding a wide rhomboidal basin (Fig. 24c), across which the captured Aru River passes into the Uruma River in the south. The river is abruptly deflected as it crosses two prominent fault strands, with 65–75 m left-lateral displacement, which may reflect recent Tarera–Aiduna Fault slip (Fig. 24d, e), although this offset is rather speculative.

An asymmetrical graben developed at the eastern termination of the Tarera–Aiduna Fault is bounded by NE–SW-trending normal faults (Fig. 24f). Rivers pressed hard against the NW-dipping bounding faults and a SE-dipping set of antithetic faults indicate active subsidence. The easternmost Tarera–Aiduna Fault itself has a significant dip-slip component, forming the northern margin of an 800 m high ridge. The Tarera–Aiduna Fault and the eastern bounding normal fault have S_{mf} values of 1.08 and 1.21, respectively, indicating that they are both active. Bounding faults along the northern margin of the rhomboidal basin, including segments corresponding to the Aria River Fault of Hamilton (1979), have S_{mf} values of 1.63, 1.91 and >4.00, pointing to slow to inactive tectonics.

Discussion

Challenges

The identification of Quaternary/modern fault activity in eastern Indonesia has historically proved difficult (e.g. Hamilton 1979; Dow & Sukamto 1984; Puntodewo *et al.* 1994; Socquet *et al.* 2006; Bailly *et al.* 2009; Teas *et al.* 2009). In part, this is because eastern Indonesia cannot be well described in terms of rigid plate tectonics, involving instead diffuse boundaries and boundary linkages, lithospheric strength heterogeneity and lower crustal flow (Hall 2011). All the fault zones in the region that are relatively well-constrained by geodetic data display strain gradients that can be explained in terms of multiple fault strands, distributed deformation or elastic strain surrounding a locked fault (e.g. Walpersdorf *et al.* 1998; Rangin *et al.* 1999; Stevens *et al.* 2002; Bock *et al.* 2003; Socquet *et al.* 2006). Poor historical earthquake records and few palaeoseismic data mean that it is difficult to distinguish between these options and so attention is naturally focused on geomorphologically prominent faults and lineaments or structures with instrumentally recorded seismicity. Faults or segments of fault systems with recurrence intervals greater than

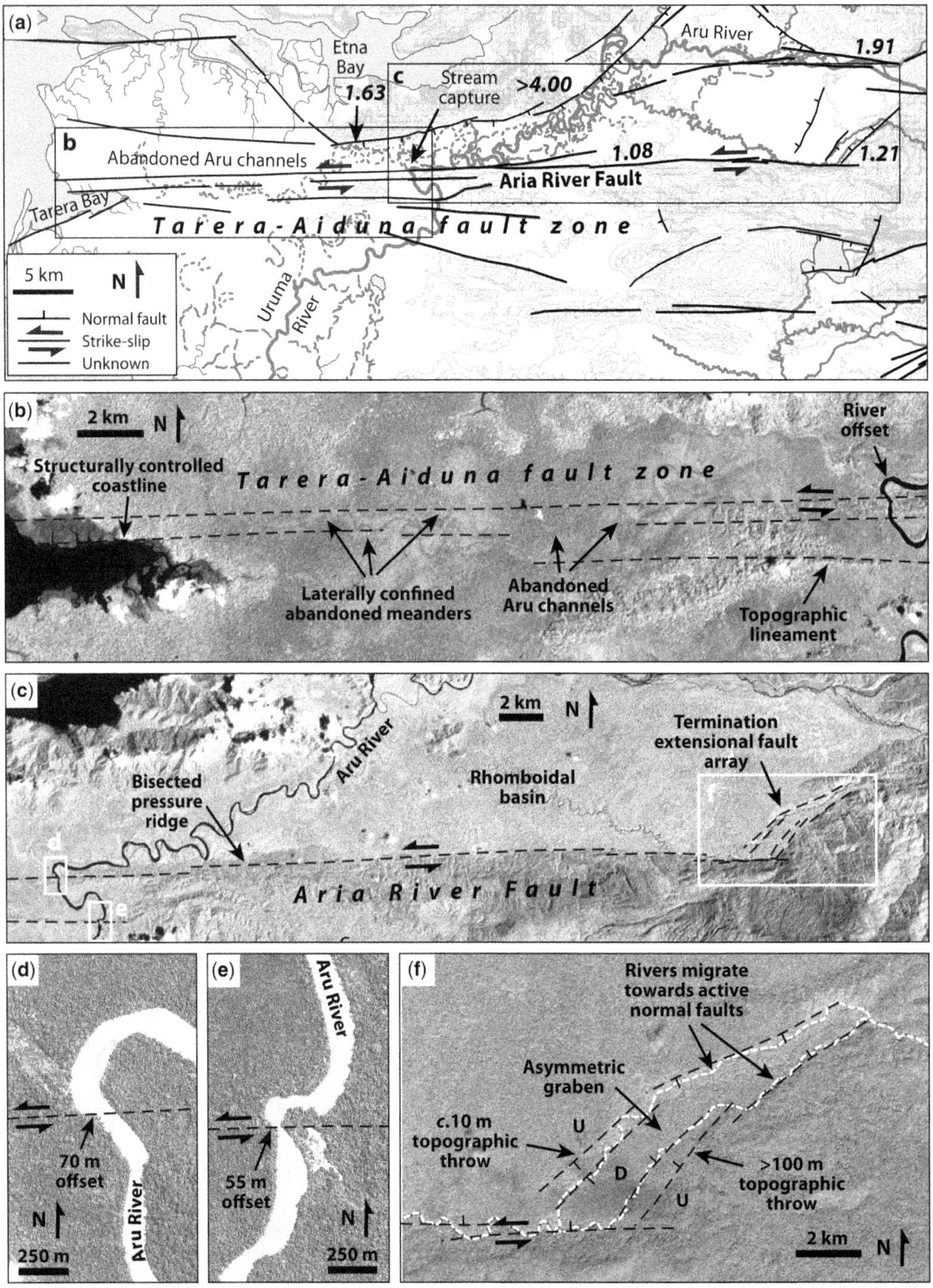

Fig. 24. (**a**) Map of the onshore Tarera–Aiduna Fault showing structures with geomorphic evidence of Quaternary tectonic activity. Bold italic numbers are S_{mf} values. Location shown in Figure 23. (**b**) Detail from greyscale Landsat TM 432 image showing linear confinement of abandoned River Aru channels, indicating strike-slip strands across the plain. (**c**) Major strand of the Tarera–Aiduna Fault bounding a steep-sided ridge and rhomboidal basin. (**d**, **e**) Possible river offset across the Tarera–Aiduna Fault. (**f**) Termination extensional fault array developed at the eastern end of the main Tarera–Aiduna Fault strand. (d, e, f) from ESRI imagery.

the short period of instrumental or even historical seismic records inevitably remain undocumented.

Additional challenges to the identification of Quaternary faults include thick forest over most of the islands (e.g. Pubellier *et al.* 1999), the abundance of important structures located entirely offshore and not readily available for study (e.g. Silver *et al.* 1983*b*; Henry & Das 2002; Teas *et al.* 2009; Liu & Harris 2013), the rapid erosion of tectonic landforms in the humid environment, the rapid burial of co-seismic features by a high sediment flux (e.g. Suggate & Hall 2003) and the high density of active and inactive structures within a large region (e.g. Puntodewo *et al.* 1994; Stevens *et al.* 2002).

Of the 27 fault systems described here, none can be confidently described as inactive during the Quaternary. Eleven show evidence of 'maximal' tectonic activity according to the classification summarized in McCalpin (2009) and a further five show evidence of 'rapid' tectonic activity (Table 3). It is important to note that the Quaternary faults discussed here are not exhaustive – there are numerous other active faults in the region, in addition to major offshore seismic sources, such as the Molucca Sea collision complex, the Banda Sea and Molucca Sea subducted slabs, and ongoing subduction of the Celebes Sea (e.g. Cardwell & Isacks 1978; Silver & Moore 1978; Cardwell *et al.* 1980; Silver *et al.* 1983*a*; Engdahl *et al.* 1998), which also need to be taken into account in any hazard analysis.

Quaternary fault geometry and earthquakes

The largest earthquakes in eastern Indonesia have been thrust and mega-thrust events, including those of the Seram Trough (1629, $M > 8.5$), the Banda Sea (1938, $M > 8.0$) and Biak (1996, M_w 8.2) (e.g. Wichmann 1918; Henry & Das 2002; Okal & Reymond 2003; Liu & Harris 2013). However, many major historical earthquakes in the studied region have occurred on strike-slip faults, including the Sorong Fault (1944, M 7.5), the Yapen Fault (1979, M 7.9) the Ransiki Fault (2002, M 7.6) and perhaps the Kawa Fault (1899, M 7.8) (e.g. Okal 1999; Brune *et al.* 2010; NEIC). Sixteen of the studied faults are dominantly strike-slip and an additional five may have a substantial strike-slip component (Table 3). As they are often long, straight, geometrically simple and subvertical, strike-slip faults are capable of generating large, shallow and damaging earthquakes – for example, the 1906 M 7.7 San Francisco earthquake (e.g. Wald *et al.* 1993), the 2001 M_w 7.8 Kunlun Shan earthquake (e.g. Lin *et al.* 2003) and the 2002 M_w 7.9 Denali earthquake (e.g. Haeussler *et al.* 2004).

A crucial barrier to the propagation of lateral ruptures and hence earthquake magnitude, even on straight strike-slip faults, is the presence of discontinuities or step-overs (e.g. Segal & Pollard 1980; Sibson 1985; Barka & Kadinsky-Cade 1988). The majority of historical strike-slip earthquake ruptures were arrested by step-overs wider than 3–5 km (Lettis *et al.* 2002; Wesnousky 2006). For example, the 1999 M_w 7.1 Düzce earthquake ruptured a 40 km segment of the North Anatolian Fault (Aydın & Kalafat 2002) and terminated in the >4 km wide Eften releasing bend in the west and the 4–5 km wide Bakacak releasing step-over in the east (Duman *et al.* 2005). Straight, continuous faults are therefore capable of generating larger earthquakes than curved or segmented faults, of generating ruptures that penetrate below the seismogenic layer (King & Wesnousky 2007) and of sustained super-shear rupture propagation, causing enhanced ground motion (Robinson *et al.* 2010). Eastern Indonesia's major strike-slip faults show a variety of levels of segmentation, which may be viewed as an indication of their structural maturity, with high cumulative displacements empirically known to remove fault zone complexities (e.g. Wesnousky 1988; Stirling *et al.* 1996; King & Wesnousky 2007). Other properties such as block rotation and pre-existing weaknesses may complicate this simple relationship.

The Matano Fault is an example of a structurally immature fault zone. Its onshore length of 195 km is punctuated by three major basins, each one 4–6 km wide, and two major restraining bends. The resultant maximum potential rupture length is 90 km (Table 3). Empirical rupture length–magnitude relationships (Wells & Coppersmith 1994) suggest a potential M 7.4 earthquake for such a rupture length. Uncertainties in this estimate include the unknown ability of a rupture to bypass the relatively gentle restraining bend east of the Mahalona Basin, the possibility of a through-going strike-slip fault at seismogenic depths below Lake Matano, the effect on fault strength of widespread serpentinite smears along the fault zone and the unknown length to which the fault continues offshore to the east.

The Sorong Fault in West Papua, part of the fault system at the southern end of the Philippine Sea plate, is a much more established fault zone with a long history of slip (e.g. Ali & Hall 1995), reflected in an apparent absence of step-overs >1 km and a continuous, straight onshore length of 420 km equating to a potential $M > 8.0$ earthquake if the entire linked system failed. The M_w 7.9 Yapen earthquake of 1979 ruptured an unknown length of the potentially 420 km long quasi-continuous Yapen Fault (Okal 1999), showing that such a scenario is possible. Despite evidence that most left-lateral strain is focused south of the Sorong Fault in West Papua, a conservative slip rate estimate of 2 mm a^{-1} could accumulate 2 m of elastic

Table 3. *Summary of observations made from Quaternary faults in eastern Indonesia, with hypothetical earthquake magnitudes, styles and tsunami risk*

Fault	Typical segment length (km)	Maximum observed total length (km)	Step-over/relay width (km)	Potential rupture length (km)*	Attributable seismicity	Notable historical events	S_{mf} range	V_f range	Tectonic activity class[†]	Potential earthquake magnitude[‡]	Potential earthquake style	Associated tsunami?
Malino boundary	25–75	130	1.2	130	Y		1.05–1.66	0.22–1.01	Maximal to slow	7.6	Normal	Y
Gorontalo	30	95	7	35	N		1.83–2.36	0.88–1.69	Slow to minimal	6.9	Strike-slip	Y
Palu–Koro	10–35	220	<1	135	Y	M_w7.7, 1996	1.08–2.30	0.24–0.89	Maximal to slow	7.6	Strike-slip	Y
Parigi boundary	10–45	95	3	80	Y		1.32–3.25	0.50–1.45	Minimal	7.3	Normal	Y
Sapu valley	5–20	75	? <2	75	N		1.08–1.45	0.40	Maximal to moderate	7.3	Strike-slip	N
Balantak	54	250	10 (offshore)	54	Y		1.04–1.22	0.25–0.47	Maximal to moderate	7.1	Strike-slip	Y
Matano	10–60	195	6	90	Y	M_w 6.1, 2011	1.02–1.9	0.23–0.78	Maximal to slow	7.4	Strike-slip	Y (lake)
Lawanopo and Kendari	10–45	200	7	70	?	M_w 7.5, 2001?	1.21–1.75	0.55–0.83	Moderate to slow	7.2	Strike-slip	N
Towuti bounding	25–55	55	<1	55	N		1.03–2.04	0.41–1.22	Maximal to slow	7.1	Normal	Y (lake)
Kolaka	5–45	175	10	50	Y		1.05–1.64	0.23–1.68	Maximal to slow	7.0	Strike-slip	Y
Mangole	20	135	2	135	Y		1.12–1.57	0.49–0.55	Rapid to slow	7.6	Normal/strike-slip	Y
Sanana	5–20	60	?	60	Y		1.27	0.44	Rapid	7.2	Normal/strike-slip	Y
Rana	10	65	3–4	>40	N	S_{fc} rupture?	1.01–1.96	0.23–1.53	Maximal to slow	>6.9	?Strike-slip	N
East Buru	10	48	<1	48	Y					7.0	Strike-slip	Y
Southern Seram	5–15	60	2	60	Y	?M 7.6, 1950	1.84–2.08	1.36–1.88	Slow	7.2	Normal	Y

Kawa	15–40	120	2	120	Y		1.10–1.33	0.26–0.28	Rapid to moderate	7.5	Strike-slip	Y
Bobol	10–15	100	2.5	100	N		1.26–1.99	1.04–2.60	Moderate to minimal	7.4	Strike-slip	Y
Combined Kawa–Bobol	10–40	240	2.5	240	Y	?M 7.8, 1899	1.10–1.99	0.26–2.66	Rapid to minimal	7.8	Strike-slip	Y
Seram FTB	5–15	135	>2	?135	Y					7.6	Thrust	Y
Sorong (West Papua)	45–75	420	<1	420	Y	M 7.4, 1944	1.14–2.85	0.27–8.68	Rapid to minimal	>8.0	Strike-slip	Y
Koor	15–35	100	6	75	Y					7.3	Strike-slip	Y
Ransiki	20–50	100	<1	100	Y	M 7.6, 2002	1.06–2.64		Maximal to minimal	7.4	Strike-slip	Y
Yapen	30–50	420	2–3	420	Y	M 7.9, 1979				>8.0	Strike-slip	Y
Mamberamo	10–20	180	<2	?180	Y	Numerous				7.7	Thrust	Y
Wandamen boundary	6–20	55	2	55	Y	S_{fc} rupture?	1.05–2.33		Maximal to slow	7.1	Normal	Y
Lowlands	30–70	220	?	?	Y					?	Normal/strike-slip	N
Paniai	?	150	?	?	Y					?	Normal/strike-slip	N
Tarera–Aiduna	30–60	130	7	90	Y		1.08–4.58		Maximal to minimal	7.4	Strike-slip	Y

*Maximum length of segment(s) separated by step-overs <3 km wide.
†Using scheme modified after McCalpin (2009).
‡Based on potential rupture length estimate and empirical length–magnitude relationships from Wells & Coppersmith (1994).

displacement across the northern Bird's Head (similar to the 1979 Yapen earthquake release; Okal 1999) in 1000 years. Even if all of this occurred west of the 1937 and 1944 earthquakes, and assuming complete stress release during those events, the remaining *c.* 200 km western portion of the fault could still generate a M > 7.7 earthquake.

The apparently very young and highly segmented Tarera–Aiduna fault zone and structures in the near-offshore Seram fold–thrust belt (Teas *et al.* 2009) and onshore Seram (Kawa and Bobol faults) could be part of a single soft-linked fault system and seem to partition much of the present day left-lateral motion between Australia and the Bird's Head. This fault system may thus be taking over the Pre-Pleistocene role of the Sorong Fault. Although there is not yet a through-going fault on the scale of the Sorong Fault, the individual components of the Tarera–Aiduna Fault and left-lateral faults in Seram are each capable of generating M > 7 earthquakes. Geomorphic observations suggest that they have all been active during the Quaternary, even if some segments (e.g. the Bobol Fault) lack instrumental seismicity records. A major uncertainty in assessing the Tarera–Aiduna fault system is the type and degree of linkage along its segments. The longest segment onshore with geomorphic evidence for rapid activity is 60 km long and may be traced, via an abrupt releasing bend 3 km wide, another 30 km to the east. Assuming rupture is not terminated by the bend, an M 7.4 earthquake is possible on this 90 km long segment. It is reasonable to assume that the fault passes some distance offshore before the next terminating step-over, so the maximum magnitude is likely to be larger.

In a similar manner, the maximum potential magnitudes for observed quasi-continuous segments of the Kawa and Bobol faults of southern Seram are 7.5 and 7.4, respectively, but a continuous rupture linking across Taluti Bay could achieve a length of 240 km and an earthquake magnitude of 7.8. The 1899 M 7.8 event, which caused slope failure north of Tehoru and a tsunami around Taluti Bay, could have originated from such a rupture.

The Palu valley has previously been considered to represent a pull-apart basin between two strands of the Palu–Koro Fault (Bellier *et al.* 2001, 2006; Beaudouin *et al.* 2003; Socquet *et al.* 2006). The width between the two strands would be about 6 km, ample to terminate earthquake rupture, limiting the maximum length and magnitude of Palu–Koro Fault earthquakes to the segments north and south of Palu valley. However, the possibility of a continuous, buried cross-basin fault system within the Palu valley as proposed here has significant implications for seismic hazard assessment in the densely populated valley. A continuous cross-basin fault within the Palu valley, as seen in analogue models (e.g. Wu *et al.* 2009) and natural strike-slip basins (e.g. the Clonard Basin, Haiti; Mann *et al.* 1995), means that the Palu–Koro Fault may be straighter and more continuous than previously suggested and palaeoseismic trenches across the border faults may not record major historical strike-slip earthquakes. The postulated buried and locked section alone is 50 km long and is thus capable of generating an M 7.0 earthquake. The total onshore length of the Palu–Koro Fault between Leboni valley and Palu city, lacking step-overs wider than 1 km and bends greater than 5°, is 135 km. As such, the Palu–Koro Fault must qualify as a 'fault superhighway', potentially capable of sustained super-shear rupture speeds (Robinson *et al.* 2010) and earthquakes up to M 7.6.

Other smaller structures that are geologically less significant because they are either not associated with instrumental seismicity (e.g. the Sapu valley fault system), have very low geomorphic tectonic activity indices (e.g. the Gorontalo Fault) or are composed of short and discontinuous fault segments (e.g. the Namlea fault system) are of particular importance from a hazard analysis perspective because of their proximity to large population centres with little to no earthquake resistance. Similarly, structures such as the Kolaka Fault, which has its most geomorphologically youthful segment bounding steep uplifted topography immediately adjacent to Kolaka town, may also be associated with secondary seismic hazards such as landslides. Large earthquakes along many of the faults, particularly the Palu–Koro, Matano and Balantak faults and the Molino, Towuti and Wandamen Peninsula boundary faults, may also trigger local tsunami, as has been already demonstrated in Palu and Taluti bays (e.g. Prasetya *et al.* 2001; Brune *et al.* 2010).

Conclusions

Neotectonic deformation in eastern Indonesia is rarely focused on discrete shear zones bounding rigid blocks, although this is often how it is interpreted. The pattern of seismicity and the broad distribution of Quaternary faults suggests that the region is more closely approximated by continuum mechanics than by rigid microplates (e.g. Thatcher 1995). All of the studied faults show geomorphic evidence of Quaternary tectonic activity, even in areas where high strain rates are not inferred from geodetic measurements (e.g. Buru, south Seram and northern West Papua).

The zone of left-lateral deformation that includes the Yapen Fault, the Tarera–Aiduna Fault and strike-slip associated with the Seram fold–thrust belt is perhaps the most active onshore/nearshore fault system of eastern Indonesia as recorded by instrumental seismicity and geodetics. However, in terms of seismic risk, the Palu–Koro Fault is considered to be the most significant structure due to its proximity to Palu city, the possibility of a cross-basin fault system close to the city, the fault's unpredictability as a result of its poorly known seismic history, and the fault's potential to cause large, shallow focus, super-shear earthquakes. Additional factors increasing the risk of the Palu–Koro Fault include the possibility of liquefaction in the deep Quaternary sedimentary basin on which Palu city is built and the low-lying city's vulnerability to tsunami travelling down the narrow Palu Bay.

The Sorong Fault in West Papua should be viewed as the wildcard of eastern Indonesian active tectonics. Although GPS measurements appear to show little sinistral strike-slip motion, or even a degree of dextral slip, station locations in Sorong and Biak cannot resolve the complexity of the Sorong–Yapen–Ransiki Fault and may omit shear to the north. Convincing left-lateral stream offsets and low mountain front sinuosity values show that the fault has been active during the Quaternary. A dearth of seismicity, rather than indicating that the fault is benign, may instead indicate that it is locked and accumulating elastic strain. Magnitude 6.9, 7.2 and 7.4 earthquakes in 1937 and 1944, located on the fault west of Manokwari, prove that the fault is capable of generating large earthquakes. The Sorong Fault's contribution to the seismic hazard of West Papua should not be underestimated, particularly given its proximity to large towns such as Sorong and Manokwari.

There is great potential for the palaeoseismic study of some of the faults discussed in this paper to confirm Quaternary activity and to provide more detailed answers to questions about seismic hazards, particularly characteristic earthquake sizes and recurrence intervals. It is recommended that trenching work is carried out across possible surface ruptures identified along the Matano, Balantak, Rana, Ransiki and Wandamen Peninsula faults. Geophysical studies to image shallow fault strands in the Quaternary sedimentary-fill of several strike-slip basins, including the Palu, Sapu and Mahalona valleys, would help to confirm the existence of cross-basin strike-slip fault systems that may pose a previously unrecognized seismic hazard.

We are grateful to TGS (http://www.tgs.com), who provided the multibeam data, to John Decker, Phil Teas and the crew of the Shakti dive boat for support and valuable discussions during Banda Sea fieldwork. Alfend Rudyawan, Dave Gold, Ferry Yulian, Ega Abang Surya Nugraha, Jonathan Pownall and Benjamin Sapiie are thanked for their assistance and contributions to the ideas discussed here. Two anonymous reviewers are thanked for detailed and constructive reviews, which considerably improved the final paper. Recent fieldwork was funded by the SE Asia Research Group, Royal Holloway University of London.

Correction notice: In the revised paper, the meaning was changed in the opening sentence.

References

ABERS, G. & MCCAFFREY, R. 1988. Active deformation in the New Guinea Fold-and-Thrust Belt: seismological evidence for strike-slip faulting and basement-involved thrusting. *Journal of Geophysical Research*, **93**, 13332–13354.

AHMAD, W. 1978. Geology along the Matano Fault Zone, East Sulawesi, Indonesia. *In*: WIRYOSUJONO, S. & SUDRADJAT, A. (eds) *Proceedings: Regional Conference on the Geology and Mineral Resources of South East Asia*, Jakarta, Indonesia, 4–7 August 1975. Indonesian Association of Geologists, 143–150.

ALI, J.R. & HALL, R. 1995. Evolution of the boundary between the Philippine Sea Plate and Australia: palaeomagnetic evidence from eastern Indonesia. *Tectonophysics*, **251**, 251–275.

AUDLEY-CHARLES, M.G. 1986. Timor-Tanimbar Trough: the foreland basin to the evolving Banda orogen. *In*: ALLEN, P.A. & HOMEWOOD, P. (eds) *Foreland Basins*. International Association of Sedimentologists, Special Publications, **8**, 91–102.

AUDLEY-CHARLES, M.G., CARTER, D.J. & MILSOM, J. 1972. Tectonic development of Eastern Indonesia in relation to Gondwanaland dispersal. *Nature*, **239**, 35–39.

AYDIN, A. & KALAFAT, D. 2002. Surface ruptures of the 17 August and 12 November 1999 Izmit and Duzce earthquakes in northwestern Anatolia, Turkey: their tectonic and kinematic significance and the associated damage. *Bulletin of the Seismological Society of America*, **92**, 95–106.

BAILLY, V., PUBELLIER, M., RINGENBACH, J.-C., DE SIGOYER, J. & SAPIN, F. 2009. Deformation zone 'jumps' in a young convergent setting; the Lengguru fold-and-thrust belt, New Guinea Island. *Lithos*, **113**, 306–317.

BARKA, A. & KADINSKY-CADE, K. 1988. Strike-slip fault geometry in Turkey and its influence on earthquake activity. *Tectonics*, **7**, 663–684.

BATH, M. & DUDA, S.J. 1979. *Some Aspects of Global Seismicity*. Seismological Institute, Uppsala, 1–79.

BEANLAND, S. & CLARK, M. 1994. *The Owens Valley Fault Zone, Eastern California, and Surface Faulting Associated with the 1872 Earthquake*. US Geological Survey Bulletin, **1982**. US Government Printing Office, Washington DC.

BEAUDOUIN, T. 1998. *Tectonique active et sismotectonique du système de failles décrochantes de Sulawesi Central*. Thèse de Doctorat, Universitá Paris-Sud.

BEAUDOUIN, T., BELLIER, O. & SEBRIER, M. 2003. Present-day stress and deformation fields within the Sulawesi Island area (Indonesia): geodynamic implications. *Bulletin de la Société géologique de France*, **174**, 305–317.

BELLIER, O., BEAUDOUIN, T. *ET AL*. 1998. Active faulting in central Sulawesi (eastern Indonesia). *In*: WILSON, P.

& Michel, G.W. (eds) *The Geodynamics of S, SE Asia (GEODYSSEA Project)*. GeoForshingsZentrum, Potsdam, 276–312.

Bellier, O., Sebrier, M. et al. 2001. High slip rate for a low seismicity along the Palu-Koro active fault in central Sulawesi (Indonesia). *Terra Nova*, **13**, 463–470.

Bellier, O., Sébrier, M., Seward, D., Beaudouin, T., Villeneuve, M. & Putranto, E. 2006. Fission track and fault kinematics analyses for new insight into the Late Cenozoic tectonic regime changes in West-Central Sulawesi (Indonesia). *Tectonophysics*, **413**, 201–220.

Bock, Y., Prawirodirdjo, L. et al. 2003. Crustal motion in Indonesia from global positioning system measurements. *Journal of Geophysical Research*, **108**, 2367.

Brune, S., Babeyko, A.Y., Ladage, S. & Sobolev, S.V. 2010. Landslide tsunami hazards in the Indonesian Sunda Arc. *Natural Hazards Earth Systems Science*, **10**, 589–604.

Bull, W.B. 1978. *Geomorphic Tectonic Activity Classes of the South Front of the San Gabriel Mountains, California*. Final Report, US Geological Survey, Contract No. 14-08-0001-G-394.

Bull, W.B. 2007. *Tectonic Geomorphology of Mountains: a New Approach to Paleoseismology*. Blackwell Publishing, Malden, MA.

Bull, W.B. & McFadden, L.D. 1977. Tectonic geomorphology north and south of the Garlock fault, California. *In*: Doehering, D.O. (ed.) *Geomorphology in Arid Regions. Proceedings at the Eighth Annual Geomorphology Symposium*. State University of New York, Binghamton, NY, 115–138.

Burchfiel, B.C., Royden, L.H. et al. 2008. A geological and geophysical context for the Wenchuan earthquake of 12 May 2008, Sichuan, People's Republic of China. *Geological Society of America Today*, **18**, 4–11.

Camplin, D.J. & Hall, R. 2014. Neogene history of None Gulf, Sulawesi, Indonesia. *Marine and Petroleum Geology*, **57**, 88–108.

Cardwell, R.K. & Isacks, B.L. 1978. Geometry of the subducted lithosphere beneath the Banda Sea in eastern Indonesia from seismicity and fault plane solutions. *Journal of Geophysical Research*, **83**, 2825–2838.

Cardwell, R.K., Isacks, B.L. & Karig, D.E. 1980. The spatial distribution of earthquakes, focal mechanism solutions, and subducted lithosphere in the Philippine and northeastern Indonesian islands. *In*: Hayes, D.E. (ed.) *The Tectonic, Geologic Evolution of Southeast Asian Seas, Islands*. American Geophysical Union, Geophysical Monograph, **23**, 1–35.

Charlton, T.R. 2010. The Pliocene–Recent anticlockwise rotation of the Bird's Head, the opening of the Aru Trough–Cenderawasih Bay sphenochasm, and the closure of the Banda double arc. *In*: *Proceedings, Indonesian Petroleum Association, 34th Annual Convention and Exhibition*. IPA10-G-008.

Chen, W.-P. & Hsu, L. 2013. Historic seismicity near the source zone of the great 2008 Wenchuan earthquake: implications for seismic hazards. *Tectonophysics*, **584**, 114–118.

Cottam, M.A., Hall, R., Forster, M.A. & Boudagher-Fadel, M.K. 2011. Basement character and basin formation in Gorontalo Bay, Sulawesi, Indonesia: new observations from the Togian Islands. *In*: Hall, R., Cottam, M.A. & Wilson, M.E.J. (eds) *The SE Asian Gateway: History, Tectonics of the Australia–Asia Collision*. Geological Society, London, Special Publications, **355**, 177–202, https://doi.org/10.1144/SP355.9

Decker, J., Bergman, S.C., Teas, P.A., Baillie, P. & Orange, D.L. 2009. Constraints on the tectonic evolution of the Bird's Head, West Papua. *In*: *Proceedings, Indonesian Petroleum Association, 33rd Annual Convention and Exhibition*, IPA09-G-139.

Dehbozorgi, M., Pourkermani, M., Arian, M., Matkan, A.A., Motamedi, H. & Hosseiniasl, A. 2010. Quantitative analysis of relative tectonic activity in the Sarvestan area, central Zagros, Iran. *Geomorphology*, **121**, 329–341.

DeMets, C., Gordon, R., Argus, D. & Stein, S. 1994. Effects of recent revisions to the geomagnetic reversal time scale on estimates of current plate motions. *Geophysical Research Letters*, **21**, 2191–2194.

Dow, D.B. & Hartono, U. 1982. The nature of the crust underlying Cenderawasih (Geelvink) Bay, Irian Jaya. *Proceedings of the Indonesian Petroleum Association*, **11**, 203–210.

Dow, D.B. & Sukamto, R. 1984. Western Irian Jaya: the end-product of oblique plate convergence in the Late Tertiary. *Tectonophysics*, **106**, 109–139.

Duman, T.Y., Emre, O., Dogan, A. & Ozalp, S. 2005. Step-over and bend structures along the 1999 Duzce Earthquake surface rupture, North Anatolian Fault, Turkey. *Bulletin of the Seismological Society of America*, **95**, 1250–1262, https://doi.org/10.1785/0120040082

Engdahl, E.R., van der Hilst, R. & Buland, R. 1998. Global teleseismic earthquake relocation with improved travel times and procedures for depth determination. *Bulletin of the Seismological Society of America*, **88**, 722–743.

Eshghi, S. & Zare, M. 2003. Bam (SE Iran) Earthquake of 26 December 2003, Mw 6.5. A Preliminary Reconnaissance Report, www.iiees.ac.ir/en/wp-content/uploads/2010/12/Bam_report_zare.pdf [last accessed 29 October 2016].

Ferdian, F., Hall, R. & Watkinson, I. 2010. A structural re-evaluation of the north Banggai-Sula area, eastern Indonesia. *In*: *Proceedings, Indonesian Petroleum Association, 34th Annual Convention*, IPA07-G-009, 1–20.

Fu, B., Ninomiya, Y., Lei, X., Toda, S. & Awata, Y. 2004. Mapping active fault associated with the 2003 M_w 6.6 Bam (SE Iran) earthquake with ASTER 3D images. *Remote Sensing of Environment*, **92**, 153–157.

Garrard, R.A., Supandjono, J.B. & Surono, 1988. The geology of the Banggai-Sula microcontinent, eastern Indonesia. *In*: *Proceedings, Indonesian Petroleum Association, 17th Annual Convention*, 23–52.

Germeraad, J.H. 1946. Geology of central Seran. *In*: Rutten, L. & Hotz, W. (eds) *Geological, Petrographical, and Palaeontological Results of Explorations Carried Out September 1917–June 1919, Island of Ceram*. De Bussy, Amsterdam.

Haeussler, P.J., Schwartz, D.P., Dawson, T.E., Stenner, H.D., Lienkaemper, J.J., Sherrod, B. & Personius, S.F. 2004. Surface rupture and slip distribution of the Denali and Totschunda faults in the 3 November 2002 M 7.9 earthquake, Alaska. *Bulletin of the Seismological Society of America*, **94**, S23–S52.

Haffner, G.D., Hehanussa, P.E. & Hartoto, D. 2001. The biology and physical processes of large lakes of Indonesia: Lakes Matano and Towuti. *In*: Munawar, M. & Hecky, R.E. (eds) *The Great Lakes of the World (GLOW): Food-Web, Health, Integrity, Ecovision*. Ecovision World Monograph Series. Backhuys Publishers, Leiden, 183–192.

Hall, R. 1987. Plate boundary evolution in the Halmahera region, Indonesia. *Tectonophysics*, **144**, 337–352.

Hall, R. 1996. Reconstructing Cenozoic SE Asia. *In*: Hall, R. & Blundell, D.J. (eds) *Tectonic Evolution of SE Asia*. Geological Society, London, Special Publications, **106**, 153–184, https://doi.org/10.1144/GSL.SP.1996.106.01.11

Hall, R. 2002. Cenozoic geological and plate tectonic evolution of SE Asia and the SW Pacific: computer-based reconstructions, model and animations. *Journal of Asian Earth Sciences*, **20**, 353–434.

Hall, R. 2011. Australia–SE Asia collision: plate tectonics and crustal flow. *In*: Hall, R., Cottam, M.A. & Wilson, M.E.J. (eds) *The SE Asian Gateway: History, Tectonics of the Australia–Asia Collision*. Geological Society, London, Special Publications, **355**, 75–109, https://doi.org/10.1144/SP355.5

Hall, R. 2012. Late Jurassic–Cenozoic reconstructions of the Indonesian region and the Indian Ocean. *Tectonophysics*, **570**, 1–41.

Hall, R., Fuller, M., Ali, J.R. & Anderson, C.D. 1995. The Philippine Sea plate: magnetism and reconstructions. *In*: Taylor, B. & Natland, J.H. (eds) *Active Margins, Marginal Basins: a Synthesis of Western Pacific Drilling Results*. American Geophysical Union Monograph, **88**, 371–404.

Hamilton, W. 1979. *Tectonics of the Indonesian Region*. US Geological Survey, Professional Papers, **1078**.

Harris, R.A. 1992. Peri-collisional extension and the formation of Oman type ophiolites in the Banda Arc and Brooks Range. *In*: Parson, L.M., Murton, B.J. & Browning, P. (eds) *Ophiolites, their Modern Oceanic Analogues*. Geological Society, London, Special Publications, **60**, 301–325, https://doi.org/10.1144/GSL.SP.1992.060.01.19

Hennig, J., Advokaat, E., Rudyawan, A. & Hall, R. 2014. Large sediment accumulations and major subsidence offshore; rapid uplift on land: consequences of extension of Gorontalo Bay and Northern Sulawesi. *In*: *Proceedings, Indonesian Petroleum Association, 38th Annual Convention and Exhibition*, IPA14-G-304.

Henry, C. & Das, S. 2002. The M_w 8.2, 17 February 1996 Biak, Indonesia, earthquake: rupture history, aftershocks, and fault plane properties. *Journal of Geophysical Research*, **107**, https://doi.org/10.1029/2001jb000796

Hill, K.C., Hoffman, N., Lunt, P. & Paul, R. 2002. Structure and hydrocarbons in the Sareba block, 'Bird's Neck', West Papua. *In*: *Proceedings, Indonesian Petroleum Association, 28th Annual Convention*, IPA01-G-115, 227–248.

Holzer, L.T. & Savage, J.C. 2013. Global earthquake fatalities and population. *Earthquake Spectra*, **29**, 155–175.

Huang, W. 1993. Morphologic patterns of stream channels on the active Yishi Fault, southern Shandong Province, Eastern China: implications for repeated great earthquakes in the Holocene. *Tectonophysics*, **219**, 283–304.

Hutchison, C.S. 1989. *Geological Evolution of South-East Asia*. Clarendon Press, Oxford.

Jablonski, D., Priyono, P., Westlake, S. & Larsen, O.A. 2007. Geology and exploration potential of the Gorontalo basin, central Indonesia – eastern extension of the North Makassar Basin? *In*: *Proceedings, Indonesian Petroleum Association, 31st Annual Convention and Exhibition*. IPA07-G-083.

Katili, J.A. 1970. Additional evidence of transcurrent faulting in Sumatra and Sulawesi. *Bandung National Institute of Geology and Mining Bulletin*, **3**, 15–28.

Katili, J.A. 1973. On fitting certain geological and geophysical features of the Indonesian island arc to the new global tectonics. *In*: Coleman, P.J. (ed.) *The Western Pacific: Island Arcs, Marginal Seas, Geochemistry*. Western Australia University Press, Perth, 287–305.

Katili, J.A. 1975. Volcanism and plate tectonics in the Indonesian island arcs. *Tectonophysics*, **26**, 165–188.

Katili, J.A. 1986. Geology and hydrocarbon potential of the Arafura Sea. *In*: Halbouty, M.T. (ed.) *Future Petroleum Provinces of the World*. AAPG Memoirs, **40**, 487–501.

Kavalieris, I., van Leeuwen, T.M. & Wilson, M. 1992. Geological setting and styles of mineralization, north arm of Sulawesi, Indonesia. *Journal of Southeast Asian Earth Sciences*, **7**, 113–130.

Keller, E.A. 1986. Investigation of active tectonics: use of surficial Earth processes. *In*: Wallace, R.E. (ed.) *Active Tectonics*, Studies in Geophysics. National Academy Press, Washington, DC, 136–147.

King, G.C.P. & Wesnousky, S.G. 2007. Scaling of fault parameters for continental strike-slip earthquakes. *Bulletin of the Seismological Society of America*, **97**, 1833–1840.

Lacassin, R., Replumaz, A. & Leloup, P.H. 1998. Hairpin river loops and slip-sense inversion on Southeast Asian strike-slip faults. *Geology*, **26**, 703–706.

Lettis, W., Bachhuber, J. et al. 2002. Influence of releasing step-overs on surface fault rupture and fault segmentation: examples from the 17 August 1999 Izmit earthquake on the North Anatolian Fault, Turkey. *Bulletin of the Seismological Society of America*, **92**, 19–42.

Lin, A., Kikuchi, M. & Fu, B. 2003. Rupture segmentation and process of the 2001 M_w 7.8 central Kunlun earthquake, China. *Bulletin of Seismological Society of America*, **93**, 2477–2492.

Linthout, K., Helmers, H., Sopaheluwakan, J. & Surya Nila, E. 1989. Metamorphic complexes in Buru and Seram, northern Banda Arc. *Netherlands Journal of Sea Research*, **24**, 345–356.

Linthout, K., Helmers, H. & Andriessenm, P.A.M. 1991. Dextral strike-slip in Central Seram and 3–4.5 Ma Rb/Sr ages in pre-Triassic metamorphics related to Early Pliocene counterclockwise rotation of the Buru-Seram microplate (E. Indonesia). *Journal of Southeast Asian Earth Sciences*, **6**, 335–342.

LIU, Z.Y.-C. & HARRIS, R.A. 2013. Discovery of possible mega-thrust earthquake along the Seram Trough from records of 1629 tsunami in eastern Indonesian region. *Natural Hazards*, **72**, 1311–1328.

MANN, P. 2007. Global catalogue, classification and tectonic origins of restraining- and releasing bends on active and ancient strike-slip fault systems. *In*: CUNNINGHAM, W.D. & MANN, P. (eds) *Tectonics of Strike-Slip Restraining and Releasing Bends*. Geological Society, London, Special Publications, **290**, 13–142, https://doi.org/10.1144/SP290.2

MANN, P., TAYLOR, F.W., LAWRENCE EDWARDS, R. & TEH-LUNG, Ku. 1995. Actively evolving microplate formation by oblique collision and sideways motion along strike-slip faults: an example from the northeastern Caribbean plate margin. *Tectonophysics*, **246**, 1–69.

MCCAFFREY, R. 1989. Seismological constraints and speculations on Banda arc tectonics. *Netherlands Journal of Sea Research*, **24**, 141–152.

MCCAFFREY, R. 1996. Slip partitioning at convergent plate boundaries of SE Asia. *In*: HALL, R. & BLUNDELL, D.J. (eds) *Tectonic Evolution of Southeast Asia*. Geological Society, London, Special Publications, **106**, 3–18, https://doi.org/10.1144/GSL.SP.1996.106.01.02

MCCAFFREY, R. & SUTARDJO, R. 1982. Reconnaissance microearthquake survey of Sulawesi, Indonesia. *Geophysical Research Letters*, **9**, 793–796.

MCCALPIN, J.P. (ed.) 2009. *Palaeoseismology*. 2nd ed. International Geophysics Series, **95**. Academic Press, Burlington, USA.

MICHETTI, A.M., AUDEMARD, F.A. & MARCO, S. 2005. Future trends in palaeoseismology: integrated study of the seismic landscape as a vital tool in seismic hazard analyses. *Tectonophysics*, **408**, 3–21.

MILSOM, J., MASSON, D., NICHOLS, G., SIKUMBANG, N., DWIYANTO, B., PARSON, L. & KALLAGHER, H. 1992. The Manokwari Trough and the western end of the New Guinea Trench. *Tectonics*, **11**, 145–153.

MOLNAR, P. & DAYEM, K.E. 2010. Major intercontinental strike-slip faults and contrasts in lithospheric strength. *Geosphere*, **6**, 444–467.

NATAWIDJAJA, D.H. & DARYONO, M.R. 2014. The Lawanopo Fault, Central Sulawesi, East Indonesia. *In*: *Proceedings, 4th International Symposium on Earthquake and Disaster Mitigation 2014 (ISEDM 2014)*. 11–12 November 2014, Bandung, Indonesia, https://doi.org/10.1063/1.4915009

NATIONAL EARTHQUAKE INFORMATION CENTRE (NEIC). http://earthquake.usgs.gov/contactus/golden/neic.php [last accessed 29 October 2016].

NATIONAL GEOPHYSICAL DATA CENTER/WORLD DATA SERVICE. Global Significant Earthquake Database. National Geophysical Data Center, NOAA, https://doi.org/10.7289/V5TD9V7K [last accessed 29 October 2016].

NORVICK, M.S. 1979. The tectonic history of the Banda arcs, eastern Indonesia: a review. *Journal of the Geological Society, London*, **136**, 519–527, https://doi.org/10.1144/gsjgs.136.5.0519

OKAL, E.A. 1999. Historical seismicity and seismotectonic context of the great 1979 Yapen and Biak, Irian Jaya earthquakes. *Pure and Applied Geophysics*, **154**, 633–675.

OKAL, E.A. & REYMOND, D. 2003. The mechanism of great Banda sea earthquake of 1 February 1938: applying the method of preliminary determination of focal mechanism to a historical event. *Earth and Planetary Science Letters*, **216**, 1–15.

PAIRAULT, A.A., HALL, R. & ELDERS, C.F. 2003. Tectonic Evolution of the Seram Trough, Indonesia. *In*: *Proceedings, Indonesian Petroleum Association, 29th Annual Convention*, 355–370.

PAVLIDES, S.B. 1989. Looking for a definition of neotectonics. *Terra Nova*, **1**, 233–235, https://doi.org/10.1111/j.1365-3121.1989.tb00362.x

PELINOVSKY, E., YULIADI, D., PRASETYA, G. & HIDAYAT, R. 1997. The 1996 Sulawesi tsunami. *Natural Hazards*, **16**, 29–38.

PHOLBUD, P., HALL, R., ADVOKAAT, E., BURGESS, P. & RUDYAWAN, A. 2012. A new interpretation of Gorontalo Bay, Sulawesi. *In*: *Proceedings, Indonesian Petroleum Association, 36th Annual Convention*, IPA12-G-029, 197–224.

PIETY, L., SULLIVAN, J.T. & ANDERS, M.H. 1992. Segmentation and paleoseismicity of the Grand Valley Fault, southeastern Idaho and western Wyoming. *In*: LINK, P.K., KUNTZ, M.A. & PLATT, L.B. (eds) *Regional Geology of Eastern Idaho and Western Wyoming*. Geological Society of America, Memoirs, **179**, 155–182.

PIGRAM, C.J. & PANGGABEAN, H. 1984. Rifting of the eastern margin of the Australian continent and the origin of some microcontinents in Indonesia. *Tectonophysics*, **107**, 331–353.

PIGRAM, C.J. SURONO & SUPANDJONO, J.B. 1985. Origin of the Sula Platform, eastern Indonesia. *Geology*, **13**, 246–248.

POWNALL, J.M., HALL, R. & WATKINSON, I.M. 2013. Extreme extension across Seram and Ambon, eastern Indonesia: evidence for Banda slab rollback. *Solid Earth*, **4**, 277–314.

POWNALL, J.M., HALL, R., ARMSTRONG, R.A. & FORSTER, M.A. 2014. Earth's youngest known ultrahigh-temperature granulites discovered on Seram, eastern Indonesia. *Geology*, **42**, 279–282.

PRASETYA, G.S., DE LANGE, W.P. & HEALY, T.R. 2001. The Makassar Strait Tsunamigenic Region, Indonesia. *Natural Hazards*, **24**, 295–307.

PUBELLIER, M., DESFONTAINES, B., CHOROWICZ, J., RUDANT, J.-P. & PERMANA, H. 1999. Active denudation morphostructures from SAR ERS-1 images (SW Irian Jaya). *International Journal of Remote Sensing*, **20**, 789–800.

PUNTODEWO, S.S.O., MCCAFFREY, R. *ET AL*. 1994. GPS measurements of crustal deformation within the Pacific–Australia plate boundary zone in Irian Jaya, Indonesia. *Tectonophysics*, **237**, 141–153.

QUIGLEY, M., VAN DISSEN, R. *ET AL*. 2012. Surface rupture during the 2010 M_w 7.1 Darfield (Canterbury) earthquake: implications for fault rupture dynamics and seismic-hazard analysis. *Geology*, **40**, 55–58.

RAMÍREZ-HERRERA, M.T. 1998. Geomorphic assessment of active tectonics in the Acambay Graben, Mexican Volcanic Belt. *Earth Surface Processes and Landforms*, **23**, 317–332.

Rangin, C., Le Pichon, X. *et al.* 1999. Plate convergence measured by GPS across the Sundaland/Philippine Sea plate deformed boundary: the Philippines and eastern Indonesia. *Geophysical Journal International*, **139**, 296–316.

Robinson, D.P., Das, S. & Searle, M.P. 2010. Earthquake fault superhighways. *Tectonophysics*, **493**, 236–243.

Robinson, G.P. & Ratman, N. 1978. The stratigraphic and tectonic development of the Manokwarai area, Irian Jaya. *BMR Journal of Australian Geology & Geophysics*, **3**, 19–24.

Robinson, G.P., Ryburn, R.J., Harahap, B.H., Tobing, S.L. & Achdan, A. 1990. *Geology of the Steenkool Sheet Area, Irian Jaya. 1:250,000 Scale.* Geological Research and Development Centre, Bandung.

Rockwell, T.K., Keller, E.A. & Johnson, D.L. 1984. Tectonic geomorphology of alluvial fans and mountain fronts near Ventura, California. *In*: Morisawa, M. (ed.) *Tectonic Geomorphology. Proceedings of the 15th Annual Geomorphology Symposium.* Allen and Unwin, Boston, MA, 183–207.

Roques, D. 1999. *The Metamorphic Core of Buru.* Southeast Asia Research Group, London, Report No. 204.

Royden, L.H. 1993. Evolution of retreating subduction boundaries formed during continental collision. *Tectonics*, **12**, 629–638.

Rudyawan, A. 2011. *Tectonostratigraphy and structural style of the South Banggai–Sula Block and NW Banda Basin, Indonesia.* MSc thesis, University of London.

Segal, P. & Pollard, D.D. 1980. Mechanics of discontinuous faults. *Journal of Geophysical Research*, **85**, 4337–4350.

Seno, T. & Kaplan, D.E. 1988. Seismotectonics of western New Guinea. *Journal of Physics of the Earth*, **36**, 107–124.

Sibson, R.H. 1985. Stopping of earthquake ruptures at dilational fault jogs. *Nature*, **316**, 248–251.

Sieh, K.E. & Jahns, R.H. 1984. Holocene activity of the San Andreas fault at Wallace Creek, California. *Journal of Geophysical Research*, **95**, 883–896.

Silva, P.G., Goy, J.L., Zazo, C. & Bardají, T. 2003. Fault-generated mountain fronts in southeast Spain: geomorphologic assessment of tectonic and seismic activity. *Geomorphology*, **50**, 203–225.

Silver, E.A. & Moore, J.C. 1978. The Molucca Sea collision zone, Indonesia. *Journal of Geophysical Research*, **83**, 1681–1691.

Silver, E.A., McCaffrey, R. & Smith, R.B. 1983*a*. Collision, rotation and the initiation of subduction in the evolution of Sulawesi, Indonesia. *Journal of Geophysical Research*, **88**, 9407–9418.

Silver, E.A., McCaffrey, R., Joyodiwiryo, Y. & Stevens, S. 1983*b*. Ophiolite emplacement by collision between the Sula Platform and the Sulawesi Island Arc, Indonesia. *Journal of Geophysical Research*, **88**, 9419–9435.

Simandjuntak, T.O. 1986. *Sedimentology and tectonics of the collision complex in the East Arm of Sulawesi, Indonesia.* PhD thesis, University of London.

Simandjuntak, T.O., Surono & Sukido. 1984. *Geological Report of the Kolaka Quadrangle, 394 Sulawesi, Scale 1:250,000.* Open File Report. Indonesian Geological Research and Development 395 Centre, Bandung.

Simandjuntak, T.O., Surono & Sukido. 1994. *Geological Map of the Kolaka Quadrangle, Sulawesi, 1:250,000 scale.* Geological Research and Development Centre, Bandung.

Socquet, A., Vigny, C., Chamot-Rooke, N., Simons, W., Rangin, C. & Ambrosius, B. 2006. India and Sunda plates motion and deformation along their boundary in Myanmar determined by GPS. *Journal of Geophysical Research*, **111**, https://doi.org/10.1029/2005JB003877

Sohoni, P.S., Malik, J.N., Merh, S.S. & Karanth, R.V. 1999. Active tectonics astride Katrol Hill Zone, Kachchh, Western India. *Journal of the Geological Society of India*, **53**, 579–586.

Spakman, W. & Hall, R. 2010. Surface deformation and slab–mantle interaction during Banda Arc subduction rollback. *Nature Geoscience*, **3**, 562–566.

Spencer, J.E. 2010. Structural analysis of three extensional detachment faults with data from the 2000 Space-Shuttle Radar Topography Mission. *Geological Society of America Today*, **20**, 4–10.

Spencer, J.E. 2011. Gently dipping normal faults identified with Space Shuttle radar topography data in central Sulawesi, Indonesia, and some implications for fault mechanics. *Earth and Planetary Science Letters*, **308**, 267–276.

Stein, S., Geller, R.J. & Liu, M. 2012. Why earthquake hazard maps often fail and what to do about it. *Tectonophysics*, **562–563**, 1–25.

Stevens, C.W., McCaffrey, R. *et al.* 1999. GPS evidence for rapid rotations about a vertical axis in a collisional setting: the Palu fault of Sulawesi, Indonesia. *Geophysical Research Letters*, **26**, 2677–2680.

Stevens, C.W., McCaffrey, R., Bock, Y., Genrich, J.F., Pubellier, M. & Subarya, C. 2002. Evidence for block rotations and basal shear in the world's fastest slipping continental shear zone in NW New Guinea. *In*: Stein, S. & Freymueller, J. (eds) *Plate Boundary Zones.* American Geophysical Union, Geodynamics Series, **30**, 87–99.

Stirling, M.W., Wesnousky, S.G. & Shimazaki, K. 1996. Fault trace complexity, cumulative slip, and the shape of the magnitude-frequency distribution for strike-slip faults: a global survey. *Geophysical Journal International*, **124**, 833–868.

Suggate, S. & Hall, R. 2003. Predicting sediment yields from SE Asia: a GIS approach. *In*: *Proceedings, Indonesian Petroleum Association, 29th Annual Convention*, IPA03-G-015, 289–304.

Sukamto, R. & Simandjuntak, T.O. 1983. Tectonic relationship between geologic provinces of western Sulawesi, eastern Sulawesi and Banggai-Sula in the light of sedimentological aspects. *Bulletin Geological Research and Development Centre, Bandung*, **7**, 1–12.

Surono 1994. Stratigraphy of the southeast Sulawesi continental terrane, Eastern Indonesia. *Journal of Geology and Mineral Resources*, **4**, 4–11.

Teas, P.A., Decker, J., Orange, D. & Baillie, P. 2009. New insight into structure and tectonics of the Seram Trough from SEASEEPTM high resolution bathymetry. *In*: *Proceedings, Indonesian Petroleum Association, 33rd Annual Convention and Exhibition*, IPA09-G-091.

THATCHER, W. 1995. Microplate v. continuum descriptions of active tectonic deformation. *Journal of Geophysical Research*, **100**, 3885–3894.

TJOKROSAPOETRO, S., BUDHITRISNA, T. & RUSMANA, D. 1981. *Geological Map of the Buru Island Quadrangle, Maluku, 1:250,000 Scale*. Geological Research and Development Centre, Bandung.

TJOKROSAPOETRO, S., ACHDAN, A., SUWITODIRDJO, K., RUSMANA, E. & ABIDIN, H.Z. 1993. *Geological Map of the Masohi Quadrangle, 2200 Maluku, 1:250000*. Geological Research and Development Centre, Bandung.

USGS EARTHQUAKE HAZARDS PROGRAM. Significant Earthquakes, http://earthquake.usgs.gov/earthquakes/browse/significant.php [last accessed 29 October 2016].

VAN LEEUWEN, T.M. & MUHARDJO, 2005. Stratigraphy and tectonic setting of the Cretaceous and Paleogene volcanic-sedimentary successions in northwest Sulawesi, Indonesia implications for the Cenozoic evolution of Western and Northern Sulawesi. *Journal of Asian Earth Sciences*, **25**, 481–511.

VAN LEEUWEN, T., ALLEN, C.M., KADARUSMAN, A., ELBURG, M., PALIN, J.M., MUHARDJO & SUWIJANTO 2007. Petrologic, isotopic, and radiometric age constraints on the origin and tectonic history of the Malino Metamorphic Complex, NW Sulawesi, Indonesia. *Journal of Asian Earth Sciences*, **29**, 751–777.

VECCHIOTTI, F. 2008. Calculating geomorphic indices in SE Asia using a SRTM derived DEM: a worked example from West Sulawesi, Indonesia. *In*: MICHEL, U., CIVCO, D.L., EHLERS, M. & KAUFMAN, H.J. (eds) *Remote Sensing for Environmental Monitoring*. GIS Applications and Geology VIII, SPIE Proceedings, **7110**, https://doi.org/10.1117/12.800240

VISSER, W.A. & HERMES, J.J. 1962. *Geological Results of the Exploration for Oil in Netherlands New Guinea*. Verhandelingen Koninklijk Nederlands Geologisch en Mijnbouwkundig Genootschap, Geologische Serie 20. Staatsdrukkerij-en Uitgeverijbedrijf, The Hague.

WALD, D.J., KANAMORI, H., HELMBERGER, D.V. & HEATON, T.H. 1993. Source study of the 1906 San Francisco earthquake. *Bulletin of the Seismological Society of America*, **83**, 981–1019.

WALKER, F. & ALLEN, M.B. 2012. Offset rivers, drainage spacing and the record of strike-slip faulting: the Kuh Banan Fault, Iran. *Tectonophysics*, **530–531**, 251–263.

WALLACE, R.E. 1968. Notes on stream channels offset by the San Andreas fault, southern Coast Ranges, California. *In*: DICKINSON, W. & GRANTZ, A. (eds) *Conference on Geologic Problems of San Andreas Fault System. Proceedings*. Stanford University Publications in the Geological Sciences, **11**, 6–21.

WALLACE, R.E. 1986. *Active Tectonics: Impacts on Society*. Studies in Geophysics. National Academy Press, Washington, DC.

WALLACE, R.E. (ed.) 1990. *The San Andreas Fault System, California*. US Geological Survey, Professional Papers, **1515**.

WALPERSDORF, A., VIGNY, C., SUBARYA, C. & MANURUNG, P. 1998. Monitoring of the Palu-Koro Fault (Sulawesi) by GPS. *Geophysical Research Letters*, **25**, 2313–2316.

WANG, Y., SIEH, K., SOE THURA TUN, LAI, K.-Y. & THAN MYINT, 2014. Active tectonics and earthquake potential of the Myanmar region. *Journal of Geophysical Research: Solid Earth*, **119**, 3767–3822, https://doi.org/10.1002/2013JB010762

WATKINSON, I.M. 2011. Ductile flow in the metamorphic rocks of central Sulawesi. *In*: HALL, R., COTTAM, M.A. & WILSON, M.E.J. (eds) *The SE Asian Gateway: History and Tectonics of the Australia–Asia Collision*. Geological Society, London, Special Publications, **355**, 157–176, https://doi.org/10.1144/SP355.8

WATKINSON, I.M., HALL, R. & FERDIAN, F. 2011. Tectonic re-interpretation of the Banggai-Sula–Molucca Sea margin, Indonesia. *In*: HALL, R., COTTAM, M.A. & WILSON, M.E.J. (eds) *The SE Asian Gateway: History and Tectonics of the Australia–Asia Collision*. Geological Society, London, Special Publications, **355**, 203–224, https://doi.org/10.1144/SP355.10

WELLS, D.L. & COPPERSMITH, K.J. 1994. New empirical relationships among magnitude, rupture length, rupture width, rupture area, and surface displacement. *Bulletin of the Seismological Society of America*, **84**, 974–1002.

WELLS, S.G., BULLARD, T.F. *ET AL.* 1988. Regional variations in tectonic geomorphology along a segmented convergent plate boundary, Pacific Coast of Costa Rica. *Geomorphology*, **1**, 239–265.

WESNOUSKY, S.G. 1988. Seismological and structural evolution of strike-slip faults. *Nature*, **335**, 340–342.

WESNOUSKY, S.G. 2006. Predicting the endpoints of earthquake ruptures. *Nature*, **444**, 358–360.

WHITE, L.T., HALL, R. & ARMSTRONG, R.A. 2014. The age of undeformed dacite intrusions within the Kolaka fault zone, SE Sulawesi, Indonesia. *Journal of Asian Earth Sciences*, **94**, 105–112.

WICHMANN, A. 1918. Die Erdbeben des Indischen Archipels Bis Zum Jahre 1857. *Verhandelingen der Koninklijke Akademie van Wetenschappen te Amsterdam*, **20**, 193 [in Dutch].

WU, J.E., MCCLAY, K., WHITEHOUSE, P. & DOOLEY, T. 2009. 4D analogue modelling of transtensional pull-apart basins. *Marine and Petroleum Geology*, **26**, 1608–1623.

WYSS, M. 2005. Human losses expected in Himalayan earthquakes. *Natural Hazards*, **34**, 305–314.

XU, X., YEATS, R.S. & YU, G. 2010. Five short historical earthquake surface ruptures near the silk road, Gansu Province, China. *Bulletin of the Seismological Society of America*, **100**, 541–561.

YEATS, R. 2010. *Active Faults of the World*. Cambridge University Press, Cambridge.

YILDIRIM, C. 2014. Relative tectonic activity assessment of the Tuz Gölü Fault Zone; Central Anatolia, Turkey. *Tectonophysics*, **630**, 183–192.

Sensitivity analysis for probabilistic seismic hazard analysis (PSHA) in the Aceh Fault Segment, Indonesia

AMALFI OMANG[1,2], PHIL CUMMINS[2,3]*, DAVID ROBINSON[3] & SRI HIDAYATI[1]

[1]*Badan Geologi, Jalan Diponegoro No. 57, Jawa Barat 40122, Indonesia*

[2]*Australian National University, Acton, ACT 2601, Australia*

[3]*Geoscience Australia, Cnr Jerrabomberra Avenue and Hindmarsh Drive, Symonston, ACT 2609, Australia*

**Correspondence: phil.cummins@anu.edu.au*

Abstract: Slip rate, locking width and ground-motion prediction equation (GMPE) selection are important in seismic hazard analysis because they are used to estimate earthquake recurrence, to limit the maximum magnitude in an earthquake source and to estimate earthquake ground shaking. In this study, we examine the sensitivity of probabilistic seismic hazard analysis (PSHA) to fault slip rates, fault locking width and the selection of GMPEs for the Aceh Fault Segment, Indonesia. The hazard level differences vary considerably owing to changes in these three parameters. Therefore, careful consideration is needed in applying PSHA in areas of high fault parameter uncertainty.

The Sumatran Fault System (SFS) in Indonesia is a major right-lateral system consisting of 19 segments. It runs parallel to the Sumatra Subduction Zone, where the Australian Plate subducts beneath the Burma Microplate and Sunda Block (Fig. 1). Sumatra is a classical example of a partitioned subduction zone, where much of the trench-normal plate motion is accommodated by the megathrust plate boundary, but much of the trench-parallel motion is accommodated by the SFS. According to historical and instrumental data, most of the 19 segments have ruptured within the last century. The largest recorded magnitudes for events on the segments vary between M_w 6.5 and 7.7 (Natawidjaja & Triyoso 2007).

Knowledge of the slip rate and locking depth along the SFS is sparse, and has changed little since the geological study of Sieh & Natawidjaja (2000) and the geodetic study of Genrich *et al.* (2000), and has been summarized more recently by Natawidjaja & Triyoso (2007). Geological studies generally suggest an increase in slip rate from 11 mm a^{-1} below the equator to 27 mm a^{-1} above it, while geodetic estimates of slip vary between 19 and 26 mm a^{-1}, with no apparent trend along strike. Locking depths determined from GPS measurements (Genrich *et al.* 2000) vary between 9 and 56 km. Knowledge of slip rate and locking depth is particularly sparse for the northernmost tip of the SFS, where it splits into two fault segments: the Aceh and the Seulimeum faults (Fig. 1). Slip rates for this segment vary from as low as 5 mm a^{-1} (Genrich *et al.* 2000) to 38 mm a^{-1} (Bennett *et al.* 1981). A more recent study by Ito *et al.* (2012) suggests a locking width of between 1 and 30 km, with a slip rate of 20 ± 6 mm a^{-1}.

This uncertainty in fault parameters is potentially important in seismic hazard analyses, both probabilistic and deterministic, which requires an assessment of the future earthquake potential in a region. In this paper, we explore the sensitivity of probabilistic seismic hazard analysis (PSHA) on the Aceh segment to three key parameters: slip rate; maximum magnitude as determined by locking width; and the selection of ground-motion prediction equations (GMPEs: also known as attenuation functions). Slip rate is important in determining seismic hazard because it is used to estimate earthquake recurrence (how often a fault is expected to host an earthquake). Earthquake hazard estimates are sensitive to slip rate because slip rate controls the overall rate of earthquake activity on a fault (e.g. Youngs & Coppersmith 1985; Kramer 1996).

Estimating the maximum expected earthquake on a specific fault or earthquake source is also important in PSHA (e.g. Beauval & Scotti 2004). It is rare, however, that the largest possible earthquakes along individual faults have occurred during the historical recording period. Therefore, maximum magnitude is often evaluated from estimates of fault dimensions (e.g. Wells & Coppersmith 1994; Leonard *et al.* 2007), and these dimensions are usually determined from a combination of geological and geodetic observations.

The latest Indonesian seismic hazard map (Irsyam *et al.* 2010) assumed a constant locking

From: CUMMINS, P. R. & MEILANO, I. (eds) 2017. *Geohazards in Indonesia: Earth Science for Disaster Risk Reduction*. Geological Society, London, Special Publications, **441**, 121–131.
First published online June 9, 2016, https://doi.org/10.1144/SP441.5

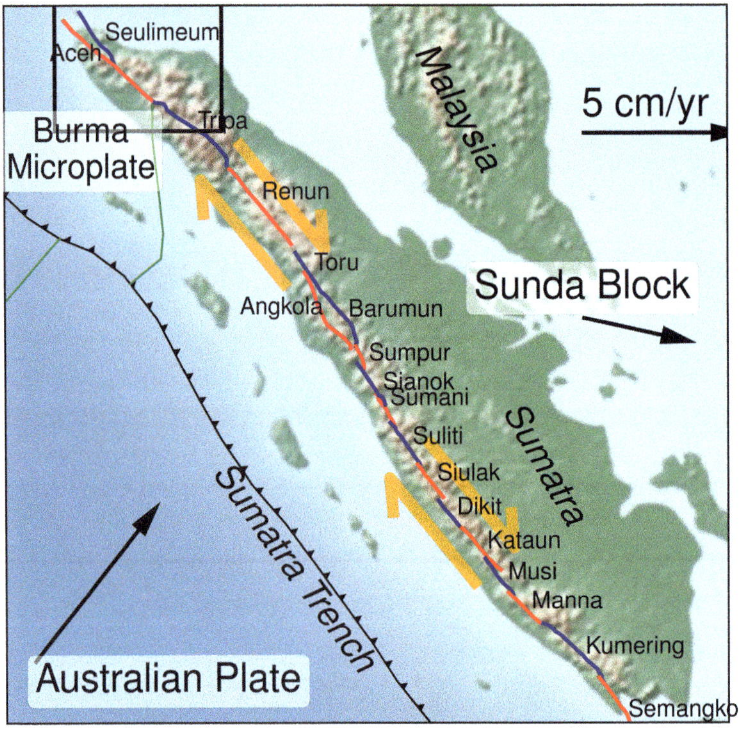

Fig. 1. The Sumatran Fault System (SFS) is a major right-lateral system that runs parallel to the Sumatra Subduction Zone, and is here indicated by its major segments plotted in alternating red and blue traces. The velocity vectors of the Australian Plate and Sunda Block (in the ITRF2002 reference frame) are indicated, along with the Burma Microlate, the velocity of which is omitted. The black rectangle indicates the area illustrated in Figure 2.

width of 17 km (depth of 3–20 km) for all SFS segments. However, Ito *et al.* (2012) used GPS to study the locking width of the Aceh Fault Segment and found uncertainty in its size, which ranged from a few kilometres up to 35 km, twice that used by Irsyam *et al.* (2010). Since earthquake magnitude depends on the area of rupture, which is the product of fault length and locking width, the maximum magnitude on a fault is expected to vary with locking width. However, the uncertainty in fault locking width is rarely considered in PSHA.

Another important aspect to consider in PSHA is the selection of the GMPEs. There are no GMPEs derived for the Indonesian region. Consequently, Indonesia's seismic hazard analyses adopt GMPEs from other regions. For example, Irsyam *et al.* (2010) used Boore & Atkinson (2008), Campbell & Bozorgnia (2003) and Chiou & Youngs (2008) in equal weights to estimate ground motion from crustal fault sources. Sabetta *et al.* (2005) demonstrated that the selection of GMPEs has a greater impact than the expert judgement applied in assigning relative weights to the GMPEs.

Sources of uncertainty in seismic hazard analysis

Seismic hazard may be analysed deterministically (e.g. when a particular earthquake scenario is assumed) or probabilistically, in which all possible earthquakes of various magnitudes, locations and probabilities of occurrence are explicitly considered (Kramer 1996). We focus here on PSHA, which can be summarized in four steps (Reiter 1990):

(1) Identification and characterization of earthquake sources. These include:
   (a) area source zones: in which an earthquake can occur anywhere in a certain area with a probability that is typically determined from an analysis of earthquake catalogue data (i.e. a recurrence relationship);
   (b) fault sources: where earthquakes can occur on particular faults with a specified probability typically determined from either catalogue data, or from slip rate and locking width estimated using either geological or geodetic observations.

(2) Application of GMPEs to forecast the level of shaking or propagation of ground motion. GMPEs are determined from analyses of strong-motion data.
(3) Integration (or summation) over the uncertainties in earthquake location, earthquake size and ground-motion parameters to obtain the probability that the ground-motion parameter will be exceeded during a particular time period.

Steps (1a), (b) and (2) all introduce uncertainty into the PSHA analysis. The GMPEs of step (2) are each associated with an aleatory uncertainty that reflects the observed variability in the ground-motion measurements and is formally included in the PSHA. In addition, the epistemic uncertainty associated with the lack of knowledge about which GMPE best describes the actual ground motion is usually accounted for via a logic tree.

However, none of the parameters describing the recurrence relationships of step (1a) and the fault behaviour in step (1b) are known perfectly, but their uncertainly is rarely considered (except for maximum magnitude, for which alternative values are sometimes considered in a logic tree). In this study, we explore the sensitivity of the PSHA results to the uncertainty in fault parameters, namely fault slip rate and maximum magnitude (as determined by locking width). Although PSHA results are also affected by uncertainty in recurrence parameters for aerial source zones (step 1a), an assessment of this sensitivity is beyond the scope of this paper.

In order to account for uncertainty in fault parameters in PSHA, we need to consider whether this uncertainty is best described as aleatory or epistemic. Aleatory is often defined as the natural variability of a quantity that cannot be reduced by making additional measurements (McGuire 2004). The uncertainty in GMPEs is normally regarded as an aleatory uncertainty. Epistemic uncertainty, however, is usually regarded as uncertainty due to a lack of knowledge and can, in principle, be reduced if additional data become available. Uncertainty in maximum magnitude, and that represented by choice of GMPE, are often regarded as epistemic uncertainties that are accounted for via the use of logic trees. We prefer the more practical definitions of Bommer (2003), in which aleatory uncertainty is defined as uncertainty that can be measured, while epistemic uncertainty is that which must be judged.

The estimations of the fault parameters slip rate and locking width used in this paper are based on measurements of offsets in dated geological formations and geodetic strain rates, respectively. Both types of observations are associated with measurable uncertainties. In the case of fault slip, the uncertainties are estimated from the measurement error in offset length and in the dating, and the uncertainty in locking width is estimated from the analysis of GPS velocities. We therefore consider the uncertainty in fault parameters as aleatory, and attempt to formally include this as part of the PSHA in a way analogous to that used for GMPEs.

Methodology for sensitivity analysis

The modelling in this study utilizes the Earthquake Risk Model (EQRM), which is a computer package for estimating earthquake hazard and earthquake risk (Robinson *et al.* 2005, 2006). The EQRM is developed by Geoscience Australia and can be freely downloaded from http:/code.google.com/p/eqrm/source/checkout

The EQRM uses an event-based approach. This approach differs from the traditional 'site-based' approach to PSHA integration over the magnitude and distance distributions on a site-by-site basis. The traditional approach was introduced by Cornell (1968) and was summarized by McGuire & Arabasz (1990). In contrast, an event-based analysis with the EQRM begins with the generation of a single simulated event catalogue, which is, in turn, used to generate ground-motion fields at all sites. Both the 'site-based' and 'event-based' approaches solve the same equations – they differ only in numerical mechanics. The generation of the synthetic event catalogue relies on a model for the seismicity in the region. Typically, the model of seismicity comes from an interpretation of historical earthquakes, geology and neotectonics (Robinson *et al.* 2005). In this paper, the seismicity model consists of the geometry and slip rate of the Aceh Fault Segment.

We focus on the Aceh Fault Segment, located in northern Sumatra, which typically hosts events with a strike-slip mechanism. Its length is approximately 200 km (Fig. 2). In the Irsyam *et al.* (2010) seismic hazard map, the slip of the Aceh segment is 2 mm a^{-1}, the locking width is 17 km and the hazard is derived using three GMPEs, all with equal weights. The GMPEs used are Boore & Atkinson (2008), Campbell & Bozorgnia (2003) and Chiou & Youngs (2008) (Irsyam *et al.* 2010).

We study the sensitivity of fault slip rates, uncertainty in locking width and different GMPEs for PSHA by comparing and analysing hazard curves at two different sites. Both sites are located near the middle of the fault trace, with Site 1 located on top of the fault line (0 km) and Site 2 located 20 km perpendicularly from the segment mid-point (see Fig. 2). We also compare the hazard maps for Aceh using different input parameters.

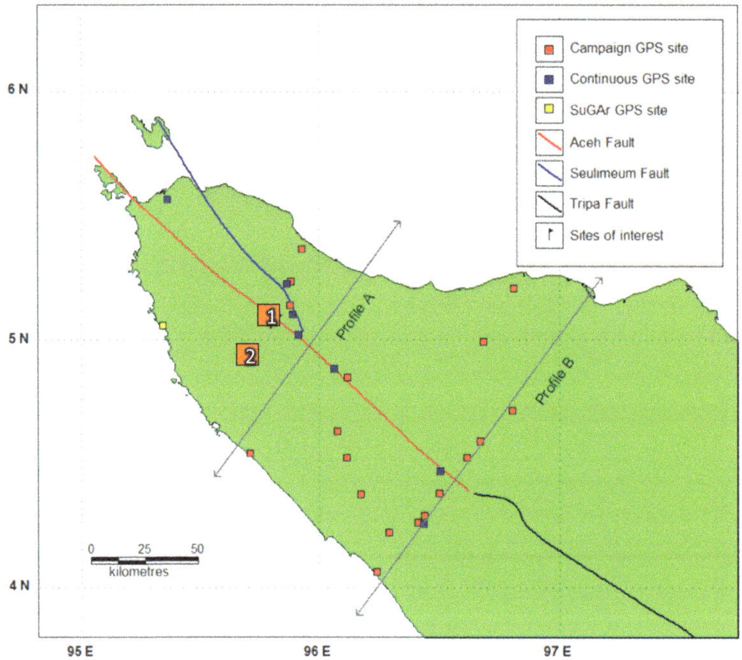

Fig. 2. The location of GPS measurements made by Ito *et al.* (2012), along with surface traces of the Aceh, Seulimeum and Tripa segments of the SFS. The two sites used for PSHA calculation are denoted by orange boxes marked 1 and 2.

Slip-rate sensitivity analysis

Reported slip rates for the Sumatran Fault System (SFS) are based on geological and geodetic studies. The latest seismic hazard map of Indonesia (Irsyam *et al.* 2010) assigned slip of 2 mm a^{-1} to the Aceh Fault. However, recent work by Ito *et al.* (2012) estimated the slip rate in the middle of the Aceh segment (Profile A in Fig. 2) to be about 20 ± 6 mm a^{-1}. Since slip-rate estimates are used to determine the recurrence interval, it is important to study how hazard varies with slip rate. Our study focuses on the sensitivity of PSHA to fault slip rate. We do not attempt to determine which slip-rate estimates are most suitable for the Aceh segment.

Figure 3a, b illustrates a comparison between the hazard curves for two different slip rates at sites 1 and 2. The green and red dashed lines are the hazard curves for 2 and 20 mm a^{-1} slip, respectively. Figure 3a shows the hazard curves comparison at a site located in the middle of the fault (Site 1). The hazard level is significantly higher for larger slip rate at all return periods. At a 500 year return period, which is commonly used in PSHA, the hazard discrepancy is significant: 0.83*g* for 20 mm a^{-1} slip and 0.18*g* for 2 mm a^{-1} slip. This represents a variation in hazard of approximately 400%. The comparison at the second site, located 20 km from the fault, also indicates a similar trend (Fig. 3b). At a 500 year return period, the hazard level for a higher slip rate reaches 0.31*g*, while the hazard level for a lower slip reaches 0.09*g*. This represents a 300% variation. As expected, the hazard level at the second site illustrates that hazard decreases as we move away from the fault. These hazard curve comparisons show that seismic hazard analysis is sensitive to fault slip rate (in fact, it is linear in slip rate) and demonstrate that the distance from an earthquake source plays an important role in the hazard level.

We also compare the hazard maps derived using these two slip rates for the Aceh Fault Segment. Figure 4 shows a comparison for the Aceh hazard maps at a 500 year return period. Figure 4a is the hazard map using 2 mm a^{-1}, while Figure 4b is the hazard map using 20 mm a^{-1}. Both maps demonstrate that the highest hazard level is located along the fault trace, and that hazard decreases as we move further from the fault. The main differences between these two maps are the maximum hazard level and the area impacted. A higher slip rate clearly leads to a greater maximum of ground shaking and a larger area impacted by future earthquakes. Consequently, the recent work by Ito *et al.* (2012) suggests that the hazard level in the Aceh area (Irsyam *et al.* 2010) is underestimated.

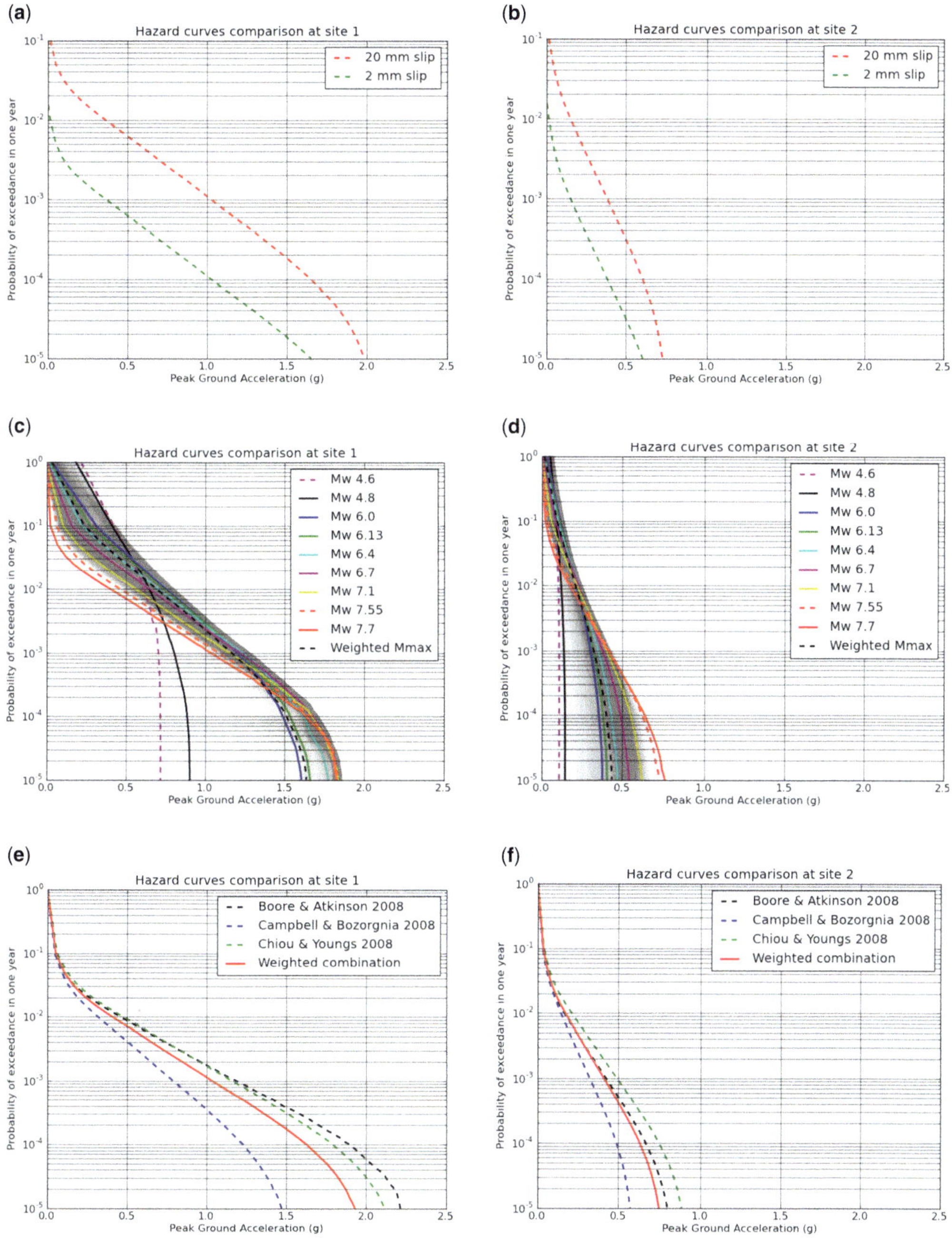

Fig. 3. Hazard curves comparison for different parameters in this study. (**a**) shows the comparison between hazard curves for slip rates of 2 and 20 mm a^{-1} at Site 1, while (**b**) shows the comparison at Site 2. (**c**) & (**d**) show the comparison of hazard curves comparison for different maximum magnitudes. The grey shaded hazard curves on (c) & (d) show the hazard curves derived from 1000 samples of the magnitude PDF at sites 1 and 2. (**e**) & (**f**) show the comparison of hazard curves using different GMPEs at sites 1 and 2.

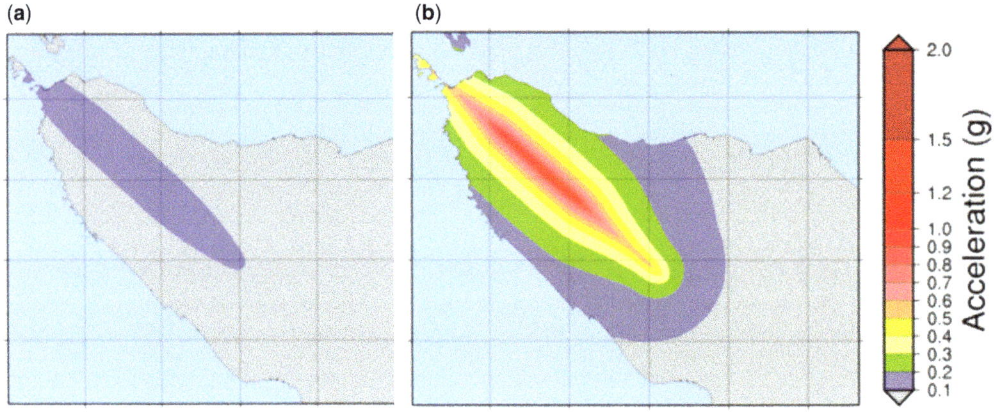

Fig. 4. Comparison of hazard maps for the Aceh Fault Segment at a return period of 500 years. (**a**) shows the hazard map for a slip rate of 2 mm a^{-1}, while (**b**) shows the hazard map for the 20 mm a^{-1} slip rate.

Locking width (maximum magnitude) sensitivity analysis

Another key parameter in PSHA is the maximum magnitude, which can be related to the locking width. Ito *et al.* (2012) studied GPS constraints on locking width for two cross-sections on the Aceh Fault Segment (Fig. 2). Their results are presented as locking width probability density functions (PDFs) for two profiles in Figure 5. We use the locking width data from Ito *et al.* (2012) to estimate a PDF for the maximum magnitude using the formula from Wells & Coppersmith (1994) (Table 1) that relate rupture width to magnitude. For the purpose of this exercise, we assume a mapped length of 200 km for the Aceh Fault Segment.

To study how broadly the hazard level might vary due to aleatory uncertainty in the locking width, we undertook a detailed sensitivity study by sampling the maximum magnitude PDF tightly (1000 samples) and then plotted all the hazard curves (the grey shaded hazard curves in Fig. 3c, d). The hazard curve from the largest maximum magnitude in the PDF is presented as the dashed red line, while the smallest maximum magnitude is presented as the magenta dashed line. These hazard curves indicate a large variation in hazard at all return periods. However, the largest discrepancies occur at longer return periods.

The maximum magnitudes based on the Ito *et al.* (2012) locking widths are also plotted as a maximum magnitude PDF (Fig. 6). The mean value for

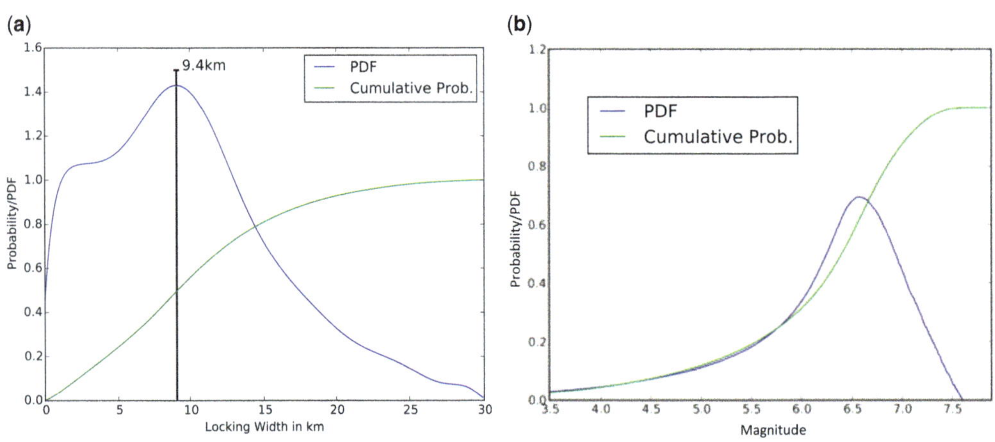

Fig. 5. Probability density functions for (**a**) the locking width, from the study of Ito *et al.* (2012), and their conversion to (**b**) the magnitude PDF (blue) and cumulative PDF (green curve).

Table 1. *Regressions of rupture (RW) width and moment magnitude (M) (Wells & Coppersmith 1994)*

Equation	Slip type	Coefficients *a*	Coefficients *b*	Magnitude range
$M = a + b^* \log(RW)$	Strike-slip	3.80	2.59	4.8–8.1
	Reverse	4.37	1.95	4.8–7.6
	Normal	4.04	2.11	5.2–7.3
$\log(RW) = a + b^* M$	Strike-slip	−0.76	0.27	4.8–8.1
	Reverse	−1.61	0.41	4.8–7.6
	Normal	−1.14	0.35	5.2–7.3

The regressions are derived using a worldwide database for 421 historical earthquakes.

the maximum magnitude from this PDF is M_W 6.13, which is considerably smaller than the M_W 7.7 used by Team-9 (Irsyam *et al.* 2010).

Standard PSHA tools, such as the EQRM we use here, do not provide for the incorporation of aleatory uncertainty in fault parameters, but allow only for epistemic uncertainty to be included using logic trees. Therefore, in order to incorporate the maximum magnitude aleatory uncertainty into our PSHA, we had to borrow the tools of epistemic uncertainty and then apply them to an aleatory uncertainty: that is, we used a logic tree approach with parameters chosen in such a way as to account for the aleatory uncertainty in the maximum magnitude. To achieve this, we sampled the maximum magnitude PDF by dividing it into five different quantiles with an equal one-fifth area below the PDF (Fig. 6). The median values for each of the five quantiles can be seen in Table 2.

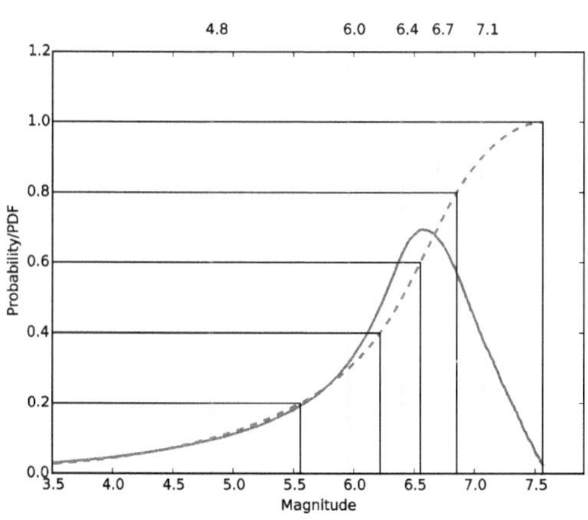

Fig. 6. Magnitude PDF derived from Ito *et al.* (2012) data. The solid curve is the probability density, while the dashed curve is the cumulative density. The PDF is divided into five quantiles, each of which has a probability of one-fifth. The dashed line indicates the median value for each quantile.

As described above, the five median (i.e. from each quantile) maximum magnitudes were used separately in the hazard calculations and used to derive the weighted hazard curve. The weighted hazard curve is the weighted sum of the five hazard curves using equal weights of one-fifth (see Table 2). We also compared the difference between the weighted hazard curves using three, five and seven quantiles. As the latter two results were similar, we reasoned that five quantiles are sufficient to represent the aleatory uncertainty.

To assess the differences between hazard curves utilizing different maximum magnitude estimates and the hazard curve that accounts for aleatory uncertainty by combining the five quantiles, we compared the hazard curves for two different locations in the Aceh Fault Segment (Fig. 1). The comparison between the hazard curves for the five median maximum magnitudes and also the corresponding weighted hazard curve derived from these five median hazard curves at sites 1 and 2 can be seen in Figure 3c, d.

Figure 3c shows a comparison for Site 1, located in the middle of the Aceh Fault. The hazard curves for different maximum magnitudes differ widely at most return periods. At shorter return periods, the hazard level for lower maximum magnitude is higher than the hazard curve with the largest maximum magnitude. For instance, a comparison between the M_W 4.8 and M_W 7.1 hazard curves at a 100 year return period show that the peak ground

Table 2. *The median and weight values used to sample the maximum magnitude PDF*

Division	Median	Weight
Five quantiles	4.8	1/5
	6.0	1/5
	6.4	1/5
	6.7	1/5
	7.1	1/5

acceleration (PGA) for the lower magnitude is 0.65*g*, and 0.56*g* for the higher magnitude. However, at a longer return period (100 kyr), the hazard level for the lower magnitude reaches 0.9*g*, while the hazard level for a higher magnitude is 1.84*g*. This result is consistent with that of Youngs & Coppersmith (1985), who studied a similar variation in hazard curve changes with maximum magnitude given a constant moment rate. They found that lowering the maximum magnitude with constant moment rate forces the activity rates of smaller earthquakes to increase, so that hazard levels increase for short return periods. The approach that we used does not assume a constant moment rate because, for each different maximum magnitude, we also change the moment rate based on the locking width for each magnitude. Nevertheless, we observe a similar effect.

The hazard curve at Site 2, located 20 km from the fault, is shown in Figure 3d. The hazard curves show a similar trend to those of the previous comparison. The main difference is that the hazard level decreases significantly at each maximum magnitude. This is simply due to the increased distance from the fault.

We are also interested in understanding the implication of uncertainty in the maximum magnitude on our understanding of hazard, as illustrated in the latest Indonesian seismic hazard map. To explore this, we considered the hazard contribution from the Aceh Fault Segment using the maximum magnitudes of the Aceh Fault corresponding to the M_w 7.7 used by Irsyam *et al.* (2010), the mean of the maximum magnitude PDF of M_w 6.13 of Ito *et al.* (2012), the weighted hazard curve using the five quantile maximum magnitudes (as given in Table 2) and results from 1000 maximum magnitudes sampled from the maximum magnitude PDF of Ito *et al.* (2012). These results for sites 1 and 2 (Fig. 2) can be seen in Figure 3c, d.

The first hazard curve comparison for Site 1 (Fig. 3c) shows that the hazard level for higher maximum magnitudes is smaller at shorter return periods. At a 500 year return period, the hazard level for M_w 7.7 is 0.85*g*, while the weighted magnitude and the M_w 6.13 are 0.99*g* and 1.08*g*. The difference is significant, reaching up to 0.23*g*. At a 1000 year return period, the difference in hazard level between these three different magnitudes becomes smaller, about 0.03*g*–0.17*g*. At a 2500 year return period, the hazard differences are much smaller, at between 0.01*g* and 0.07*g*. However, at very long return periods (100 kyr), the hazard level from the largest maximum magnitude (M_w 7.7) reaches 1.83*g*, which is larger than the weighted magnitude and M_w 6.13, which have hazard levels of 1.59*g* and 1.66*g*, respectively. The hazed grey lines, which represent the sampled maximum magnitudes from the maximum magnitude PDF derived from the results of Ito *et al.* (2012), show how wide the uncertainty in the hazard level might be due to the uncertainty in the locking width from the geodetic observations in northern Sumatra. The maximum hazard level from these data varies between 0.7*g* and 1.8*g*. These hazed grey plots also show a similar trend. The hazard level for a lower maximum magnitude at shorter return period would be higher than the hazard level for higher magnitude at a shorter return period.

The second comparison at Site 2 (Fig. 3d) shows a similar trend with the comparison at Site 1 (Fig. 3c). However, the probability for exceeding a certain ground-motion level (e.g. 0.5*g*) occurred at a longer return period. The difference varies between 0.04*g* and 0.06*g* at a return period of 500 years. At a return period of 1000 years, the difference is about 0.1*g*: while at a 2500 year return period, it reaches 0.16*g*. The overall hazard level has decreased, which is simply owing to the increased distance from the fault.

Since the hazard map consists of many sites at a single return period, it is important to note that the hazard level for lower maximum magnitude is higher than the hazard level for higher maximum magnitude at shorter period. Based on this result, a careful consideration is needed before we assign a certain maximum magnitude for PSHA work.

Ground-motion prediction equation (GMPE) sensitivity analysis

The other aspect that we study is the impact from the selection of GMPEs. Many studies mention the importance of the GMPE selection for PSHA (e.g. Lombardi *et al.* 2005; Sabetta *et al.* 2005; Bommer & Abrahamson 2006). The problem of GMPE selection is further complicated in Indonesia as there are no GMPEs derived from Indonesian earthquakes. The GMPEs used in the most recent hazard map of Irsyam *et al.* (2010) are Boore & Atkinson (2008), Campbell & Bozorgnia (2003) and Chiou & Youngs (2008). The seismic hazard map of Irsyam *et al.* (2010) used an equal weight for each of these GMPEs. We examine the contribution from these three GMPEs at sites 1 and 2, and compare the results with the weighted results. We also compare the hazard curves and hazard maps using each of the GMPEs used in the Irsyam *et al.* (2010) seismic hazard map. As with Irsyam *et al.* (2010), we use equal weights when combining the GMPEs.

Figure 3e, f show the hazard curve comparisons at sites 1 and 2. The first comparison at Site 1 (0 km) shows that Boore & Atkinson (2008) (BA08) and Chiou & Youngs (2008) (CY08) lead to larger hazard estimates than Campbell & Bozorgnia (2003) (CB08). The result for the weighted combination of GMPEs sits in the middle. The hazard level for the weighted combination is about 23% larger than CB08. However, compared to BA08 and CY08, the weighted combination hazard level is about 13 and 11% smaller, respectively. The hazard level for BA08 is 5% smaller than CY08 for shorter return periods (<600 years), and larger by approximately 10% for longer return period (>800 years).

The second comparison at Site 2 (20 km) shows a different result (Fig. 3f). The hazard level for CY08 is larger than BA08 and CB08 for all return periods. This shows that the hazard level dependence on GMPE depends on distance, highlighting the important role of GMPE selection on seismic hazard analysis. The weighted combination GMPEs in this case also sits in the middle between CB08, BA08 and CY08. The main difference compared to the first site is that the overall hazard level decreases owing to the increased distance from the fault.

Discussion

The sensitivity analysis for probabilistic seismic hazard analysis (PSHA) for two different sites near the Aceh Fault Segment, Indonesia, shows a significant sensitivity in hazard level to the parameters that we analysed: slip rates, locking widths (M_{max}) and GMPEs.

The hazard curves comparison for the Aceh segment using two different slip rates show significant differences in hazard level. The previous slip rate used for this fault segment (Irsyam *et al.* 2010) is only 10% of that suggested by more recent work (Ito *et al.* 2012). We clearly show that seismic hazard analysis is sensitive to such a large difference in fault slip rates, with a higher slip rate leading to a higher maximum hazard, and a larger area subject to high hazard levels. In general, geodetic studies in the northern part of Sumatra suffer from limited measurements in the far field because of the narrow land area available for measurements. Therefore, it seems important to account for uncertainty in geodetic measurements of slip rate there.

Fault geometry, particularly the locking width, influences the maximum potential magnitude of a fault and is, hence, important for hazard. It is common to use geological data to estimate maximum magnitude: however, recent developments in geodesy (Ito *et al.* 2012) can provide researchers with a PDF for locking width and, hence, a new option for determining the maximum magnitude on faults. The analysis for different maximum magnitudes show that the hazard level for lower maximum magnitude (a shorter return period) is higher than the hazard level for higher magnitude at shorter return periods. While this has been found to be the case when moment rate is held constant (Youngs & Coppersmith 1985), we find that it is still true for the case of the Aceh Fault, even when the moment rate increases as the locking width increases.

GMPE selection, however, shows a very significant difference in the hazard level, and we found that the difference may vary at difference distances. A GMPE that results in the highest hazard level at a certain distance might have smaller hazard level at a different distance. However, given the fact that there is no single GMPE derived for Indonesia, a further study is needed in order to select the most reliable GMPE(s) for use in seismic hazard analysis in Indonesia.

Conclusions

In this study, we present a sensitivity analysis for three parameters used in probabilistic seismic hazard analysis. We analysed hazard curves and maps for two different sites in the Aceh Fault Segment, Indonesia. The parameters that we studied were slip rates, locking widths and GMPEs.

The comparison between different slip rates shows a significant hazard difference between smaller (2 mm a^{-1}) and larger (20 mm a^{-1}) slip rates. The study at Site 1 shows that at a 500 year return period, the hazard level from smaller slip is 0.06*g*, while the larger slip hazard level is 0.55*g* (Fig. 3a), resulting in a difference of 0.49*g*, which is significant. Another comparison at a 2500 year return period shows that the hazard level for smaller slip reaches 0.45*g*, while the larger slip produces 1.13*g*, which still gives a difference of approximately 0.68*g*. A similar result also appears at Site 2 (Fig. 3b): however, owing to the increased distance from the fault, the hazard level is smaller.

Conversely, the maximum magnitude (as determined from the fault locking width) also plays an important role in determining the hazard level. The comparison between M_w 4.6 (magenta dashed line) and M_w 7.7 (red line) at Site 1 (Fig. 3c) shows two important points. The first point is that the hazard difference is 0.08*g* at a return period of 500 years. The second point is that the hazard level from a smaller maximum magnitude results in a higher hazard level at a return period of <700 years. At longer return periods, the hazard level from the higher maximum magnitude shows a significant difference. For example, at a 2500 year return period, a M_w 7.7 maximum magnitude results in a 1.06*g* exceedance in ground shaking, at M_w 4.6, however, the maximum magnitude results in only 0.68*g*. The comparison at Site 2 (Fig. 3d) also shows a similar result, but the hazard level is smaller owing to the increased distance.

The study of different GMPEs also shows the importance of the correct selection of the GMPE. At smaller return periods, the difference is relatively small, while at longer return periods the difference is larger, as expected. At a 500 year return period (Fig. 3e), CY08 resulted in a similar hazard level to BA08: for CB08, however, the hazard level is smaller by 0.21*g*. At a 2500 year return period, the hazard level difference with CY08 and BA08 was still small, but the difference with CB08 increased to about 0.41*g*. However, at a return period of >2500 years, the hazard difference between CY08 and BA08 is larger. The comparison at Site 2 (Fig. 3f) for the GMPE selection shows a different trend. The hazard level between the three GMPEs can be seen clearly at both short and long periods. The main difference compared to Site 1 is that the hazard level is smaller owing to the increased distance from the fault.

The results show that each of the parameters studied are equally important and affect the hazard level significantly. Therefore, careful consideration is needed as we work on seismic hazard analysis in areas where there is significant uncertainty in fault parameters and GMPEs. However, it is important to note that PSHA also depends on other factors, such as earthquake catalogue data and source zonation, that have not been considered in this study.

This work was supported by the Australian Department of Foreign Affairs Australian Aid Program and Australian Research Council Linkage Project LP11010525.

References

BEAUVAL, C. & O.& SCOTTI, 2004. Quantifying sensitivities of PSHA for France to earthquake catalog uncertainties, truncation of ground-motion variability, and magnitude limits. *Bulletin of the Seismological Society of America*, **94**, 1579–1594.

BENNETT, J.D., BRIDGE, D.McC. ET AL. 1981. *Geologic map of the Banda Aceh Quadrangle, North Sumatra.* 1:250,000. Geological Research and Development Centre, Bandung, Indonesia.

BOMMER, J.J. 2003. Uncertainty about the uncertainty in seismic hazard analysis. *Engineering Geology*, **70**, 165–168.

BOMMER, J.J. & ABRAHAMSON, N.A. 2006. Why do modern probabilistic seismic-hazard analyses often lead to increased hazard estimates? *Bulletin of the Seismological Society of America*, **96**, 1967–1977.

BOORE, D.M. & ATKINSON, G.M. 2008. Ground-motion prediction equations for the average horizontal component of PGA, PGV, and 5%-damped PSA at spectral periods between 0.01 s and 10.0 s. *Earthquake Spectra*, **24**, 99–138.

CAMPBELL, K.W. & BOZORGNIA, Y. 2003. Updated near-source ground-motion (attenuation) relations for the horizontal and vertical components of peak ground acceleration and acceleration response spectra. *Bulletin of the Seismological Society of America*, **93**, 314–331.

CHIOU, B.S.-J. & YOUNGS, R.R. 2008. An NGA model for the average horizontal component of peak ground motion and response spectra. *Earthquake Spectra*, **24**, 173–215.

CORNELL, C. 1968. Engineering seismic risk analysis. *Bulletin of the Seismological Society of America*, **58**, 1583–1606.

GENRICH, J.F., BOCK, Y. ET AL. 2000. Slip distribution at the northern Sumatra Fault System. *Journal of Geophysical Research*, **105**, 28327–28341.

IRSYAM, M., SENGARA, W. ET AL. 2010. *Development of Seismic Hazard Maps of Indonesia for Revision of Seimic Hazard Map in SNI 03-1726-2002*. Reseach report submitted to the Ministry of Public Works by the Team for Revision of Seismic Hazard Maps of Indonesia.

ITO, T., GUNAWAN, E. ET AL. 2012. Isolating along-strike variations in the depth extent of shallow creep and fault locking on the northern Great Sumatran Fault. *Journal of Geophysical Research*, **117**, B06409, https://doi.org/10.1029/2011JB008940

KRAMER, S. 1996. *Geotechnical Earthquake Engineering*. Prentice-Hall, Englewood Cliffs, NJ.

LEONARD, M., ROBINSON, D., ALLEN, T., SCHNEIDER, J., CLARK, D., DHU, T. & BURBIDGE, D. 2007. *Toward a Better Model of Earthquake Hazard in Australia.* Geological Society of America, Special Papers, **425-17**.

LOMBARDI, A.M., AKINCI, A., MALAGNINI, L. & MUELLER, C.S. 2005. Uncertainty analysis for seismic hazard in Northern and Central Italy. *Annals of Geophysics*, **48**, 853–865.

MCGUIRE, R.K. 2004. *Seismic Hazard and Risk Analysis*. Earthquake Engineering Research Institute, Oakland, CA.

MCGUIRE, R.K. & ARABASZ, W.J. 1990. An introduction to probabilistic seismic hazard analysis. *In*: WARD, S.H. (ed.) *Geotechnical and Environmental Geophysics, Volume III*. Society of Exploration Geophysics, Tulsa, OK, 333–353.

NATAWIDJAJA, D.H. & TRIYOSO, W. 2007. The Sumatran fault zone – from source to hazard. *Journal of Earthquake and Tsunami*, **1**, 21–47.

REITER, L. 1990. *Earthquake Hazard Analysis – Issues and Insights*. Columbia University Press, New York.

ROBINSON, D., FULFORD, G. & DHU, T. 2005. *EQRM: Geoscience Australia's Earthquake Risk Model.* Geoscience Australia, Record, **2005/0 1**.

ROBINSON, D., DHU, T. & SCHNEIDER, J. 2006. Practical probabilistic seismic risk analysis: a demonstration of capability. *Seismological Research Letters*, **77**, 453–459.

SABETTA, F., LUCANTONI, A., BUNGUM, H. & BOMMER, J.J. 2005. Sensitivity of PSHA results to ground motion prediction relations and logic-tree weights. *Soil Dynamics and Earthquake Engineering*, **25**, 317–329.

SIEH, K. & NATAWIDJAJA, D.H. 2000. Neotectonics of the Sumatran fault, Indonesia. *Journal of Geophysical Research*, **105**, 28295–28326.

WELLS, D.L. & COPPERSMITH, K.J. 1994. New empirical relationships among magnitude, rupture length, rupture width, rupture area and surface displacement. *Bulletin of the Seismological Society of America*, **84**, 974–1002.

YOUNGS, R.R. & COPPERSMITH, K.J. 1985. Implications on fault slip rates and earthquake recurrence models to probabilistic seismic hazard estimates. *Bulletin of the Seismological Society of America*, **75**, 939–964.

A probabilistic seismic hazard assessment for Sulawesi, Indonesia

A. CIPTA[1,2]*, R. ROBIANA[1], J. D. GRIFFIN[3], N. HORSPOOL[3], S. HIDAYATI[1] & P. CUMMINS[2,3]

[1]*Geological Agency of Indonesia, Jalan Diponegoro No. 57, Bandung 40122, Indonesia*

[2]*Australian National University, Research School of Earth Sciences, Building 142, Mills Road, Canberra, ACT 0200, Australia*

[3]*Geosicence Australia, Cnr Jerrabomberra Avenue and Hindmarsh Drive, Symonston, ACT 2609, Australia*

**Correspondence: wamditanka_santee@yahoo.com*

Abstract: A probabilistic seismic hazard assessment that includes the effect of site amplification is undertaken for the island of Sulawesi, Indonesia. High seismic activity rates, both along fast-slipping crustal faults including the major Palu-Koro–Matano Fault System and in regions of distributed deformation, contribute to moderate–high earthquake hazard over all but the SW part of the island. Of particular concern in terms of seismic risk are the numerous cities sited on soft sedimentary basins that have formed due to movement along presently active structures and that can be expected to amplify earthquake ground motions, including the provincial capitals of Palu and Gorontalo.

The island of Sulawesi is actively deforming and experiences frequent seismicity in all but the most SW part of the island. Although mass casualty (1000+ fatalities) earthquake events have not previously occurred, more-frequent smaller disasters regularly cause smaller numbers of fatalities and economic losses. The most severe historical event in terms of fatalities is the Toli-toli earthquake and tsunami (M_s 7.4, 1968, as in Pelinovsky *et al.* 1997) that killed 200 people (BNPB 2010). Steep topography both onshore and offshore means that landslide is a major secondary hazard, with the extension of faults offshore meaning there is potential for earthquake- and mass-movement-generated tsunami associated with earthquakes. Characterizing earthquake sources and the seismic hazard due to ground shaking is therefore an important step in understanding better the threat from earthquakes faced by the more than 17 million people (BPS 2010) who inhabit the island of Sulawesi.

Hazard maps resulting from this assessment were published by the Geology Agency of Indonesia (Badan Geologi) for the six provinces comprising the island of Sulawesi. The purpose of these hazard maps is to inform resilient building design, local spatial planning and the development of local disaster management plans. In order to be appropriate for these purposes, it is important that the effect of local site conditions on seismic hazard is considered. Site amplification is correlated with local surface geology, and therefore site amplification factors are estimated using proxy methods based on surface geological and morphological data. Furthermore, results are expressed both in terms of spectral acceleration for different periods of ground motion and as felt intensities – with the former being more appropriate for informing building practices, and the latter providing a simpler description of hazard to inform spatial planning and local disaster management planning.

This paper first provides a brief review of the geology and seismotectonics of the island of Sulawesi, and the implications for seismic hazard assessment. Secondly, details of the hazard assessment methodology, the Earthquake Risk Model (EQRM) software package and input parameters used in the model are described. A key aspect is a comparison of suitability of the site class proxy methods of Matsuoka *et al.* (2006) and Wald & Allen (2007) for estimating amplification due to local site effects. In the final part of the paper, the hazard assessment results are presented and discussed.

Geological and seismotectonic setting

The island of Sulawesi consists of a number of fragments of lithosphere that display a complex geological history of subduction, collision and local extension (Fig. 1). West Sulawesi, including the southern arm, is underlain by relatively strong, thick and cool lithosphere that was rifted from Australia and accreted to Sundaland in the Cretaceous (Hall 2011). The SE and east arms are derived

From: CUMMINS, P. R. & MEILANO, I. (eds) 2017. *Geohazards in Indonesia: Earth Science for Disaster Risk Reduction*. Geological Society, London, Special Publications, **441**, 133–152.
First published online April 26, 2016, https://doi.org/10.1144/SP441.6

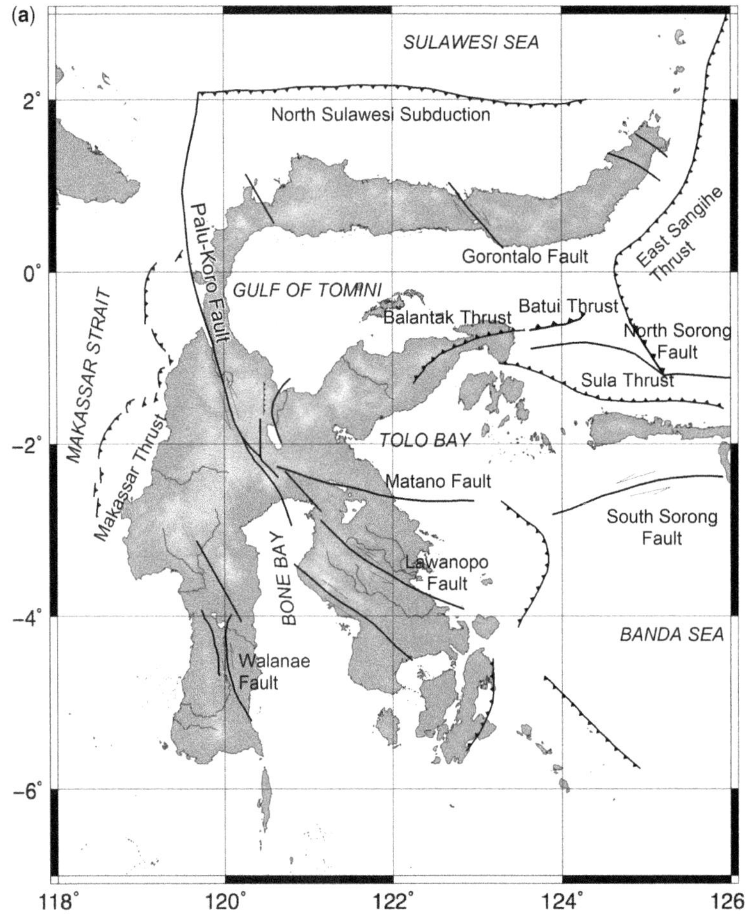

Fig. 1. Map of (**a**) the main active fault structures of the Sulawesi region.

from fragments of continental Australia (Sula Spur) that were accreted during the Miocene; the north arm represents a volcanic arc formed by north-dipping subduction in the early Miocene and associated thrusting of ophiolites onto the Sula spur during collision (Silver *et al.* 1983*a*; Hall 2011). Subduction polarity of the north arm has reversed and now the Celebes Basin subducts southwards beneath the north arm at the North Sulawesi Trench. Seismicity defines a north-dipping slab east of 122.3° E (McCaffrey & Sutardjo 1982 report 122.6 °E) and the south-dipping slab extends deeper west of here.

East of the North Sulawesi Trench there is double-subduction of the Moluccas Sea Plate (Silver & Moore 1978). Active thrust faulting above the Molluccas Sea Plate verges away from the collision zone as frontal arc sediments are overthrust onto the volcanic arc. This area is associated with intense seismic activity both within the slab and on the thrust faults. Deep earthquakes associated with the subducted slabs occur beneath North Sulawesi; there is also a cluster of deep earthquakes that appear to be associated with the enigmatic Una-Una Volcano in the middle of the Gorontalo Basin (Hall 2011). The Gorontalo Basin may represent a zone of extension resulting in the sharp topography changes between the north arm, Gorontalo Basin and Central Sulawesi (Hall 2011). Furthermore, low-angle detachment zones have been identified in the Tokorondo and Pompangeo mountains of Central Sulawesi (Spencer 2011). Thrusting also occurs in the SE along the Tolo Thrust and to the west along the Makassar Thrust. Seismic reflection profiles (Puspita *et al.* 2005) show that the Makassar Thrust is a significant feature that probably extends to meet the Palu-Koro Fault, NW of Palu. This is consistent with horizontal GPS velocities from Sulawesi that,

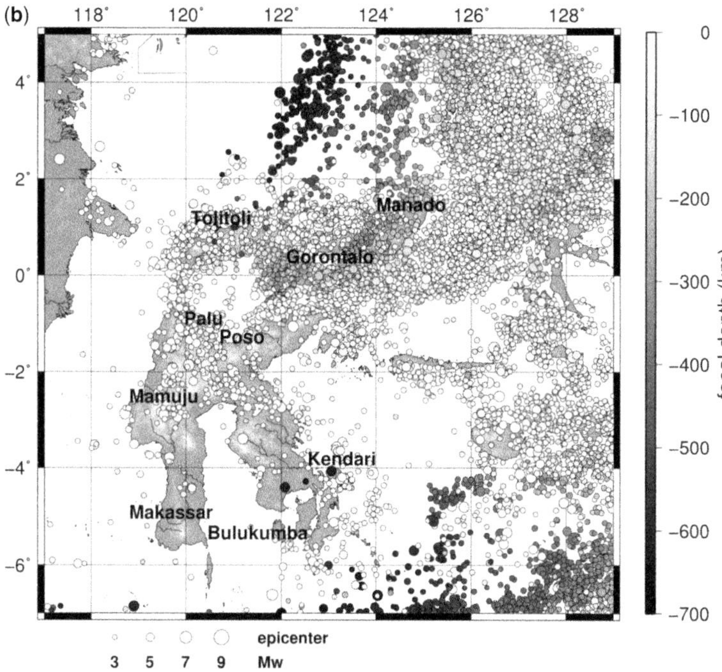

Fig. 1. (**b**) the distribution of earthquake epicentres.

in general, are directed to the NW quadrant with respect to a stable Sunda Block reference frame (Sarsito *et al.* 2011).

Various attempts have been made to interpret Sulawesi into a number of blocks or microcontinents from GPS and gravity anomaly studies (Socquet *et al.* 2006; Sarsito *et al.* 2011). It is unclear how many blocks are needed: however, there are probably more blocks than can be modelled with the available data (D.A. Sarsito pers. comm. 2011). Block boundaries interpreted from GPS measurements match the location of some major mapped faults (e.g. the Palu-Koro Fault and the North Sulawesi Trench), while others are more speculative (e.g. extending from the Poso Fault through the Gorontalo Basin towards the West Molluccas Sea Thrust). Furthermore, Socquet *et al.* (2006) suggested there are four strands of the Palu-Koro Fault that cover an area 50 km wide near Palu. Geological maps from the Geological Agency of Indonesia (Badan Geologi) show a highly faulted landscape, with numerous faults considered active. However, most structures have no information on activity rates or correlation with the GPS block models of Sarsito *et al.* (2011). This makes specific inclusion of individual faults in our hazard model difficult. Hall (2011) suggested that the region is better interpreted as a continuum of weak lithosphere overlying a heterogeneous basement that includes areas of old, relatively strong crust, such as that beneath West Sulawesi. This interpretation poses two particular challenges for how to approach development of an earthquake hazard model for the region:

- Developing a source model that balances inclusion of specific faults with inclusion of background source zones – including only fault sources that accommodate the GPS velocities completely is expected to cause an underestimate of hazard in areas of continuous and distributed deformation, and may overestimate hazard along the mapped faults if unmapped structures are accommodating some of the relative motion. The model presented here includes active fault structures where sufficient information is available, along with distributed source zones in less-well-defined regions, and attempts to balance the seismic moment between fault and background sources based on recorded seismicity.
- Application of ground-motion prediction equations (GMPEs) for different source regions – the heterogeneous basement structure, including both old Sundaland and Australian continental crust, active or recently active island arc settings and obducted oceanic crust, means that large differences in ground-motion propagation can be expected between different geological regions.

A lack of data with which to constrain ground-motion models means that largely default models are chosen.

Methods

The Earthquake Risk Model (EQRM)

The EQRM is an open-source, event-based earthquake hazard and risk calculator developed at Geoscience Australia (Robinson *et al.* 2006; source code freely available from: http://sourceforge.net/projects/eqrm/). The EQRM generated a synthetic earthquake catalogue based on input parameters defining recurrence and geometry properties for earthquake sources. Earthquake recurrence may be defined using either the bounded Gutenberg–Richter model (Youngs & Coppersmith 1985; Kramer 1996) or the characteristic earthquake model (Schwartz & Coppersmith 1984). Sources may be defined using three different geometric representations, being zones, faults or intraslab sources:

- Zone sources are defined by a geographical polygon, and a minimum and maximum depth: synthetic ruptures occur randomly within this zone, with strike and dip randomly sampled from a uniform distribution within a range of values specified by the user.
- Fault sources are rectangular planes defined by the updip surface projection of the fault trace. Synthetic rupture centroids are randomly distributed on the plane, and strike and dip are controlled by the geometry of plane.
- Intraslab sources are defined using the same functionality as fault sources, except now the plane defines the geometry of the dipping slab and individual synthetic events are allowed to rupture at some angle, or range of angles, out of the plane (out-of-dip rupture). This allows for realistic simulation of the variety of earthquake focal mechanisms that occur within subducting slabs. Out-of-dip rupture can also be used to simulate uncertainty in the dip for fault sources.

Inputs

Catalogue and background crustal source zones. The ISC catalogue was downloaded and declustered using the Seisan software package (Ottemöller *et al.* 2011). The catalogue was separated into shallow crustal (depth ≤ 35 km), instraslab (depth > 35 km) and megathrust events. The catalogue is considered complete for events magnitudes great than 4.8 and 5.0 for shallow crustal and intraslab earthquakes, respectively. Gutenberg–Richter *b* values were calculated for crustal and instraslab events using the maximum likelihood method (Aki 1965) and used for all crustal and intraslab source zones, respectively. Activity rates (λm_0) above a minimum magnitude of catalogue completeness (m_0) were calculated based on the average annual number of events above m_0 within each source zone. Crustal earthquakes within 10 km of mapped active faults were assumed to occur on these faults, allowing for hypocentre location errors (Husen & Hardebeck 2010), and therefore were excluded from the background zone analysis. Major crustal-scale faults were used as boundaries for the source zones, including the Palu-Koro and Matano faults. M_{max} for background crustal source zones was set to 7.5, as it is likely that there are large structures capable of earthquakes of this magnitude that are not included in the fault source model. Focal mechanisms from the Global Centroid Moment Tensor (GCMT) catalogue were used to determine faulting type and infer the geometry of fault structures with background zones.

Four background zones were defined. The Northern Sulawesi Zone, bounded to the south and west by the Palu-Koro–Matano Fault System, to the north by the North Sulawesi Trench and to the east by the West Molluccas Sea Thrust. It is characterized by crust that is undergoing high shear strain resulting from clockwise rotation of the Sula Block along the Palu-Koro–Matano Fault System. A number of strike-slip faults are present within the zone, including the Poso and Gorontalo faults, and also eastern strands of the Palu-Koro Fault (Socquet *et al.* 2006). Extension is probably occurring within the Gorontalo Bay (Hall 2011) and this structure may connect the Poso Fault to the Gorontalo Fault and the West Molluccas Thrust. This structure is also used to define a boundary in block model of Sarsito *et al.* (2011). There is also thrusting to the east along the Batui Thrust. The whole area is seismically active, making further delineation of specific structures difficult without more detailed fault studies. The high background activity rate means that a lack of resolution of specific faults should not adversely affect the results of the seismic hazard assessment (Fig. 2).

The West Sulawesi Zone encompasses the seismically active region west of the Palu-Koro Fault and east of the Makassar Thrust. Again, many faults are delineated on geological maps, but there is no information about activity rates.

The SW Sulawesi Zone encompasses the relatively seismically quiet zone along the SW arm of Sulawesi. This area consists of continental crust and is not actively deforming. The Walanae Fault crosses this structure but has low reported slip rates (2 mm a^{-1}: Irsyam *et al.* 2010). This fault zone is bounded on the east by the Bone Gulf.

The SE Sulawesi Zone is bounded to the north by the Matano Fault, to the east by the Tolo Thrust and

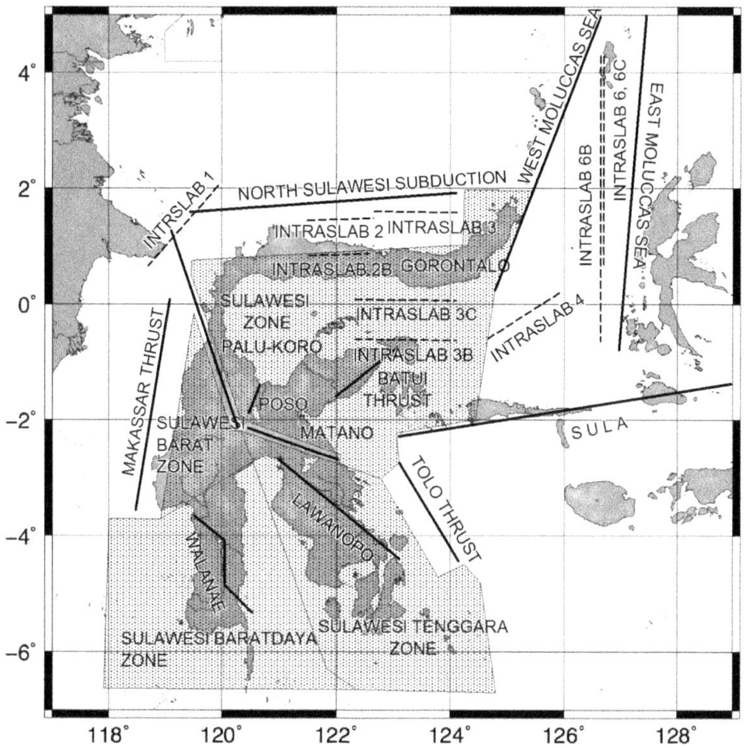

Fig. 2. Fault sources, intraslab sources (updip surface projection of slab trace) and background source zones.

to the west by the Bone Gulf. It is crossed by the Lawanopo Fault. Activity rates here are lower than in the North and West Sulawesi zones.

Fault sources. The initial reference fault model was that used for the 2010 revision of Indonesia's national seismic hazard map (Irsyam *et al.* 2010). However, several of the fault sources were modified to take into account new data regarding fault locations, geometry and earthquake recurrence. Furthermore, where fault sources were located within background source zones (rather than defining the zone boundary), fault-source slip rates derived from geological or geodetic methods (S_g) were reduced by the equivalent slip rate from the activity rate of the background source model. Slip rate (S) can be related to seismic moment rate (M^T) by (Youngs & Coppersmith 1985):

$$M^T = \mu A_f S$$

where A_f is the fault area and μ is the shear modulus. Therefore, the Gutenberg–Richter magnitude distribution, which describes the rate of seismicity and, hence, the seismic moment rate, can be used to relate the slip rate to the seismic activity rate. The background seismicity rate is used to calculate an equivalent slip rate for the fault that lies within it by rearranging from Youngs & Coppersmith (1985):

$$S_\lambda = \frac{b\lambda^f_{m_0} M_0^{\max - \beta(m_{\max} - m_0)}}{\mu A_f (c - b)(1 - e^{-\beta(m_{\max} - m_0)})}$$

where S_λ is the estimated slip rate for the fault derived from background seismicity only, $M_0^{\max}$ is the moment of the maximum magnitude earthquake ($m_{\max}$) for the background zone, $c = 1.5$ from the moment magnitude definition of Hanks & Kanamori (1979) and $\beta = 2.303$, and $\lambda^f_{m_0}$ is the activity rate of the source zone (μ_{m_0}) scaled by the volume of the 10 km buffer around the fault:

$$\lambda^f_{m_0} = \frac{\lambda_{m_0} V_{\text{buffer}}}{V_{\text{zone}} - V_{\text{buffer}}}.$$

Therefore the final input slip rate for the fault is

$$S = S_g - S_\lambda.$$

Reported slip rates for some faults were lower than that calculated from seismicity, giving a negative value for S and demonstrating the difficulty in characterizing individual faults for seismic hazard assessment within a zone of high shear strain and seismicity. However, as these faults clearly exist, are active and pose a threat, they are still included with a nominal slip rate of 2 mm a^{-1}, acknowledging that this probably results in an overestimate of total seismic moment rate compared with that derived purely from seismicity (although we cannot be sure owing to the incompleteness of the instrumental catalogue).

Fault input parameters. The fault parameters used in this study are shown in Table 1. A brief summary of the key differences between this paper and Irsyam *et al.* (2010) is given below.

The Minahasa (North Sulawesi) Trench. This fault is mapped at the trench axis, as evident in bathymetry (Silver *et al.* 1983*b*; Irsyam *et al.* 2010). However, although a few thrust events have occurred here, the majority of thrust events are located 50–100 km to the south. Focal mechanisms (Fig. 3) show steeper thrusting in the east (average dip 24°) than in the west (average dip 18°), consistent with Silver *et al.* (1983*b*), who observed a steeper frontal slope in the east in seismic reflection profiles. These data show a shallow-dipping main thrust plane with steeper splay faults in the overriding wedge, including north-dipping

Table 1. *Summary of faults parameters used in this paper and Irsyam* et al. *(2010)*

Fault	Slip rate (mm)		M_{max} (M_w)		Fault type	*A*, *b* values
	Irsyam *et al.* (2010)	This paper	Irsyam *et al.* (2010)	This paper		
North Sulawesi			8.2	8.2	M	4.82, 0.914
Palu-Koro	30 (0.25) 35 (0.50) 45 (0.25)	35	7.94	7.9	SS	–, 0.95
Poso	2	2	6.93	6.9	SS	–, 0.95
Matano	37 (0.50) 44 (0.50)	41	7.9	7.9	SS	–, 0.95 –, 0.95
Lawanopo	25	20.3	7.59	7	SS	–, 0.95
Walanae	2	–	7.53	–	SS	–, 0.95
Walanae N	–	1.7	–	6.6	SS	–, 0.95
Walanae M	–	1.7	–	6.6	SS	–, 0.95
Walanae S	–	1.7	–	6.6	SS	–, 0.95
Gorontalo	11	5	7.06	7.6	SS	–, 0.95
Batui	2	2	7.06	7.3	R	–, 0.95
Tolo	9 (0.50) 19 (0.50)	14	7.94	7.5	R	–, 0.95 –, 0.95
Makassar	4 (0.50) 13 (0.50)	9	7.46	7.5	R	–, 0.95 –, 0.95
Sula	10	14	7.19	7.7	R	–, 0.95
West Mollucca Sea	13	13	8.47	7.9	R	–, 0.95
East Mollucca Sea	29	29	8.47	8.1	R	–, 0.95
Intraslab 1	–	–	–	–	I	0.690, 1.21
Intraslab 2A (shallow)	–		–	8	I	1.100, 1.21
Intraslab 2B (deep)	–	–	–	8	I	5.200 1.21
Intraslab 3	–	–	–	8	I	1.900, 1.21
Intraslab 3A (shallow)	–		–	8	I	6.080, 1.21
Intraslab 3B (deep)	–	–	–	8	I	6.675, 1.21
Intraslab 4A (shallow)	–		–	8	I	8.125, 1.21
Intraslab 6	–	–	–	8	I	15.575, 1.21
Intraslab 6A (west dipping)	–		–	8	I	12.6, 1.21
Intraslab 6B (east dipping)	–		–	8	I	23.625, 1.21
Background seismicity zone						
Sulawesi	–	–	–	7.5	C	10.825, 0.95
Sulawesi Timur (East Sulawesi)	–	–	–	7.5	C	2.925, 0.95
Sulawesi Barat (West Sulawesi)	–	–	–	7.5	C	1.100, 0.95
Sulawesi Baratdaya (SW Sulawesi)		–	–	7.5	C	0.175, 0.95

Note: M, megathrust; SS, strike-slip fault; R, reverse fault; N, normal fault; I, instraslab, variable mechanism; C, crustal, variable mechanism. In the 'Slip rate' column, the number in the brackets indicates the weight value. A-value, event rate for certain region, describes seismic activity; b-value, tectonic parameter, determined by slope of the frequency-magnitude distribution curve.

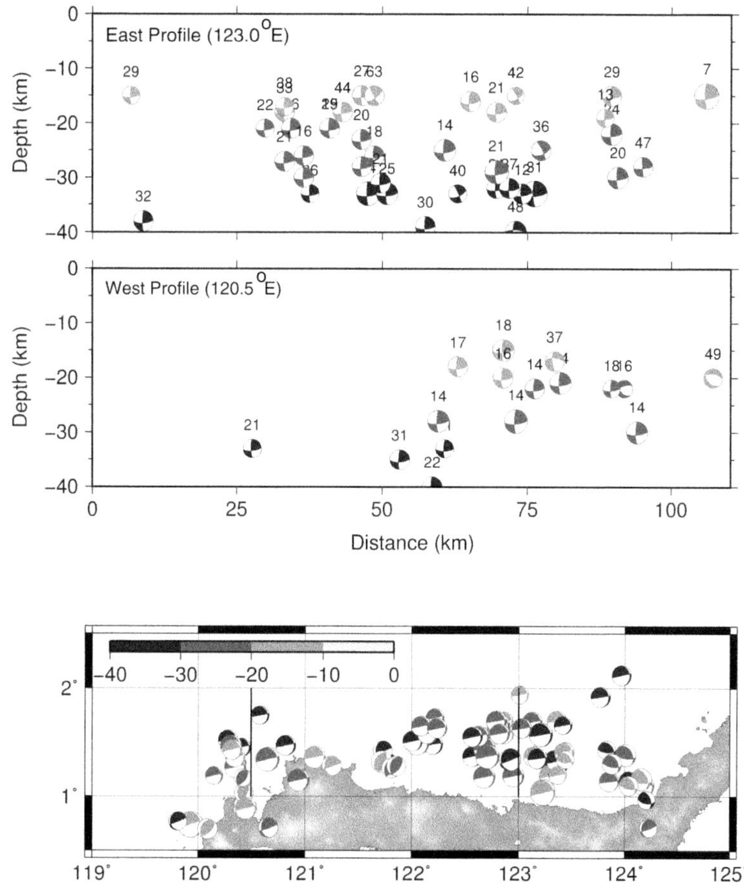

Fig. 3. Focal mechanisms for the Minahasa Trench along north–south profiles showing steeper average dips in the west than in the east. Numbers over each beachball in the upper and middle panels indicate the dip. The area in this figure is indicated by black rectangle in Figure 1a.

structures in the eastern section of the fault (Silver *et al.* 1983*b*).

It is unclear whether the lack of earthquakes near the trench is because strain is not being accumulated within weak sediments (i.e. the theory of Wang & Hu 2006) or if, in fact, these sediments are slowly accumulating strain and could host a large earthquake, as has been observed in the 2004 Great Sumatra–Andaman Earthquake (where there is a thick sedimentary wedge: Gulick *et al.* 2011) and in source models for the 2011 Tohoku, Japan earthquake (Ammon *et al.* 2011; Koketsu *et al.* 2011). If this is the case, then the fault could contain potential to generate a significant tsunami. However, as short-period ground motions generated by subduction interface earthquakes decrease rapidly with distance (e.g. Youngs *et al.* 1997), we place the source fault landwards of the trench axis in order to reproduce, as accurately as possible, the observed seismicity on the main section of the subduction interface. The fault plane is taken to dip at an angle of 21° (mean dip from focal mechanisms): however, synthetic events are allowed to thrust out of this plane by randomly sampling a uniform distribution of dips between 11° and 43°. Events rupturing out of the fault plane are further constrained to extend no more than maximum of 5 km from the fault plane. Fault width is 100 km and the maximum depth is 35 km.

The Batui Thrust. Recent work using high-resolution multi-beam bathymetric and seismic data allowed Watkinson *et al.* (2011) to reassess the tectonics of this region. They found no evidence for the Batui Thrust, as previously mapped to the north of Poh Head (Silver *et al.* 1983*b*; Irsyam *et al.* 2010). Instead, this fault is reinterpreted as occurring as a thrust zone to the SW of the Poh Head peninsula.

We follow this new interpretation of the location of the Batui Thrust. Slip rates from Irsyam *et al.* (2010) of 2 mm a^{-1} are poorly constrained: however, in the absence of further information, we apply this value to the new location of the fault. This rate is lower than that calculated from background seismicity in the surrounding source zone.

The Balantak Fault. This fault is a right-lateral strike-slip structure that cuts Poh Head to the north of the Batui Thrust. It has a clear geomorphic expression and has been associated with recent seismicity (Watkinson *et al.* 2011). This fault is not included in Irsyam *et al.* (2010) and there are no available data on slip rates. Therefore, we do not explicitly include this fault in our assessment.

The Makassar Thrust. Irsyam *et al.* (2010) include this fault offshore of the western bulge of Sulawesi near the city of Majene. We extend this thrust fault to the north to meet the Palu-Koro Fault NW of Palu as seismic reflection profiles (Puspita *et al.* 2005) show clear evidence of thrusting offshore of much of West Sulawesi. It is unclear whether this structure is one continuous structure, as we include it, or a number of segmented thrust faults and fold zones.

Irsyam *et al.* (2010) also identified a number of other active fault structures throughout Sulawesi that were not included in their hazard model owing to a lack of information on slip rates. In addition, the active fault database of the Geological Agency of Indonesia also contains numerous mapped structures, many of which are discernable as topographical lineaments using the SRTM digital elevation model. A lack of geological data and geodetic data to constrain slip rates, and even to confirm the faults are active structures, means that slip rates for individual faults cannot be defined. Therefore, zone sources covering areas of distributed deformation are used instead.

Intraslab sources. Deep seismicity is conspicuous throughout northern Sulawesi, and is the result of a complex system of active and inactive subduction systems, as described above. Although much of the deep seismicity can be explained through these systems, the catalogue also contains some more enigmatic events that are difficult to assign to any structures and suggest a complex pattern of Cenozoic collision that is not fully understood. Identifying structures is further complicated by formal catalogue hypocentre depth uncertainties of up to 25 km (Husen & Hardebeck 2010). We define a number of intraslab sources that attempt to explain the seismicity and fit with geological interpretations of the area.

Table 2. *Ground-motion prediction equations used for each source region and associated logic tree weight*

Region	Ground-motion model	Weight
Crustal	Boore & Atkinson (2008)	0.33
	Campbell & Bozorgnia (2008)	0.34
	Chiou & Youngs (2008)	0.33
Subduction interface	Youngs *et al.* (1997)	0.25
	Atkinson & Boore (2003)	0.25
	Zhao *et al.* (2006)	0.5
Intraslab	Atkinson & Boore (2003) – Cascadia	0.33
	Youngs *et al.* (1997)	0.34
	Atkinson & Boore (2003) – worldwide	0.33

Ground-motion models. At present, ground-motion prediction equations (GMPEs) based on strong-motion data from Indonesia do not exist. At present, work is being started to develop GMPEs from a new strong-motion accelerometer network being deployed by the Indonesian Meteorology, Climatology and Geophysics Agency (BMKG): however, no results are yet available either for Indonesia in general or Sulawesi in particular (Rudyanto 2013). Therefore, we use the logic tree of Irsyam *et al.* (2010), as shown in Table 2.

Site amplification

Amplification of seismic waves in shallow soil near the surface can contribute significantly to variations in seismic hazard between nearby areas exposed to the same seismic sources. To apply site amplification to a regional-scale hazard assessment in Sulawesi where field measurements are limited, proxy methods are used to estimate the average shear-wave velocity in the upper 30 m of the Earth (V_{s30}). Estimated V_{s30} values are then either incorporated directly into those GMPEs that include this term in their functional form or, for those that do not, classified into the NEHRP site classes, and associated amplification factors (Borcherdt 1994) are used to scale ground-motion estimates from the GMPEs at each site.

Two proxy methods for estimating V_{s30} are compared against field measurements of the response spectral ratio of horizontal and vertical components (H/V) of ambient noise measurements. The first proxy method is that of Wald & Allen (2007). This method estimates V_{s30} based purely on topographical slope using empirical relationships derived from data from California and Taiwan. Global SRTM topography has been used to generate a global grid of V_{s30} using this method, which is freely available online from the United States Geological Survey's Global V_{s30} Server.

The second proxy method follows Matsuoka *et al.* (2006) and estimates V_{s30} as an empirical function of geomorphology, elevation, slope, and the distance from hills and mountains. Empirical factors are derived from analysis of field measurements from Japan. The classification of geomorphology for this method is shown in Table 3. Geomorphological and geological maps held by the Geological Agency of Indonesia (1:250 000 scale, 1973–98) were used along with SRTM topography data to classify Sulawesi's geomorphology using the scheme (ABC) of Matsuoka *et al.* (2006) and then to calculate the V_{s30} estimates.

Sulawesi can be divided into 14 geomorphic units using the Matsuoka *et al.* (2006) scheme. Of these, the classes of Pre-Tertiary mountain, Tertiary mountain, hill, mountain footslope, volcanic footslope and volcano cover more than 75% of the island's area. These classes of geomorphic units are characterized as being in steep and elevated areas subject to erosion, where the ground surface is assumed to be composed of hard and compacted material with high average V_{s30} values. V_{s30} is lower for volcanic than non-volcanic mountains, a key distinction that is useful for application in Indonesia. The remaining 25% of Sulawesi Island is characterized by more undulating or flat morphology where sediments are accumulating.

Figure 4 shows the geomorphology of Sulawesi using the classification of Matsuoka *et al.* (2006) and V_{s30} values derived from empirical relationships. Mountainous regions are characterized by high V_{s30} values (NEHRP site class A or B), footslopes and volcanic areas have moderate V_{s30} values (site class C), while lowland areas (valley, delta and coastal) have lower V_{s30} values and are classified as site class D or E.

Estimated V_{s30} values taken from both the Wald & Allen (2007) and the Matsuoka *et al.* (2006) methods are compared with V_{s30} calculated from ambient noise field measurements in the cities of Gorontalo, Palu and Manado. Measurement locations were classified as volcanic footslope, mountain footslope, delta and coastal lowland, and valley bottom lowland in the Matsuoka *et al.* (2006) scheme.

Field measurements were performed using single-station three-component 1 Hz L4–3D seismometers. Noise was recorded for approximately 20 min and the spacing between measurements was 1–2 km. The fundamental frequency of the ratio of horizontal to vertical components (H/V)

Table 3. *Geomorphological unit classification*

Slope (°)	Elevation (m)	Lithology	Geomorphological unit (Matsuoka *et al.* 2006)	Notes
>15	>700	Tertiary rock	Tertiary Mountain	
		Pre-Tertiary rock	Pre-Tertiary Mountain	
	<700	Tertiary products	Hill	
		Volcanic (Q)	Volcanic Hill	Volcanic product
5–15			Mountain footslope	
		Volcanic (Q)	Volcanic footslope	Volcanic product
≤5		Alluvium, colluviums, fluvial etc.	• Valley bottom lowland • Alluvial fan • Back marsh • Abandoned river channel • Delta and coastal lowland • Marine sand and gravel bars	Colluviums Fluvial
Active volcano			Volcano	Eruption centres
Terrace			• Rocky strath terrace • Gravelly terrace	• Brecciated, geological structure • Alluvium, geological structure
		Volcanic ash	Terrace covered by volcanic ash soil	Volcanic ash
5–15		Sand (Q)	Sand dune • Reclaimed land • Filled land • Natural levee	Sand • Engineered • Engineered • Natural barrier parallel to river/coast, sand bar, flood plain

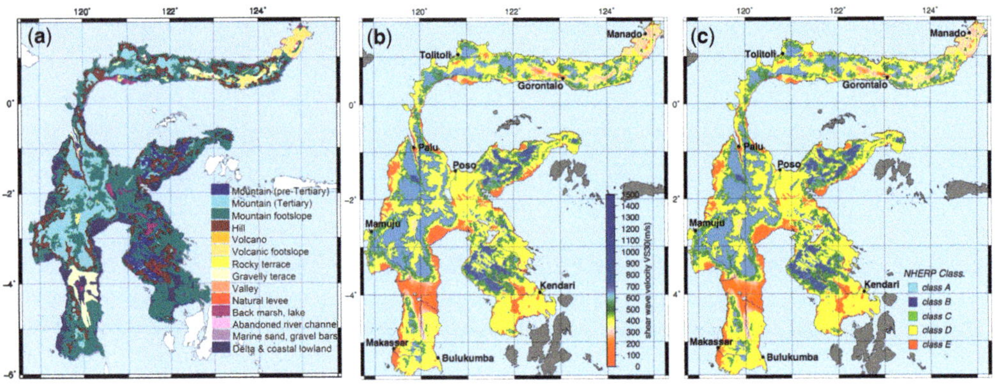

Fig. 4. (**a**) Geomorphology of Sulawesi using the Matsuoka *et al.* (2006) classification, (**b**) V_{s30} values for Sulawesi derived from the Matsuoka *et al.* (2006) method (higher V_{s30} values are indicated in blue and lower V_{s30} values are indicated in red (right)) and (**c**) NEHRP's soil classification.

was determined using Geopsy processing software and then used to estimate V_{s30} following Zhao (2011). This method uses the relationships $T_{V_{s30}} = 120/V_{s30}$, where $T_{V_{s30}}$ is the site period assuming bedrock is reached at a depth of 30 m. In using this method, we assume that site period and V_{s30} are correlated, and that the fundamental period of the H/V ratio is equivalent to $T_{V_{s30}}$. V_{s30} values are then classified into NEHRP site classes (Borcherdt 1994). We note that there is a large degree of uncertainty inherent both in the assumption of the correlation between fundamental period and V_{s30}, and in the application of V_{s30} itself as a predictor of site effects. Overall, the method is more robust for shorter site periods (Zhao 2011).

Figure 5 shows histograms of the difference between V_{s30} values estimated using proxy methods and those based on field measurements. Both proxy methods estimate higher V_{s30} values that those derived from H/V measurements, although the United States Geological Survey (Wald & Allen 2007) method overestimates (222 ± 212 m s^{-1}) by about twice that of the Matsuoka *et al.* (2006) method (102 ± 194 m s^{-1}).

In terms of site classification, higher estimates of V_{s30} using the Wald & Allen (2007) method means that this method is more likely to correctly assign site class C, while the Matsuoka *et al.* (2006) method is more likely to correctly assign site class D (Fig. 6). The Matsuoka *et al.* (2006) method gives the same site class as that derived from H/V measurements (Zhao 2011) at approximately 25% of the measured locations compared with the Wald & Allen (2007) method, which gives the same site class at 15% of the measured sites.

In general, both the Matsuoka *et al.* (2006) and Wald & Allen (2007) methods result in higher estimates of V_{s30} compared with H/V measurements. Weathering is one possible explanation for this difference as the degree of weathering is not taken into account in the proxy methods. For a tropical country like Indonesia, high rainfall levels mean that weathering rates can be much higher than the regions where these proxy methods where derived (Japan and California). This can create deep weathering profiles. It can be said that a wetter climate is prone to faster weathering. Physical weathering breaks the original rock down into soil-like material and, consequently, highly weathered rock loses its engineering properties significantly. Furthermore, clay minerals introduced by weathering reduce effective shear strength and, hence, the shear-wave velocity (Arikan & Aydin 2012).

Hazard modelling results

Probabilistic seismic hazard results are shown for annual probabilities of exceedance of 0.002 and 0.0004 (equivalent to return periods of 500 and 2500 years), and were produced for response spectral acceleration (RSA) of 0.2 and 1.0 s, and peak ground acceleration (PGA). These results are shown in Figure 7, including site amplification. The hazard maps show high hazard in all but the south arm of Sulawesi. Hazard is highest along the Palu-Koro, Matano and Lawanopo faults, where slip rates are greater than 30 mm a^{-1}.

Hazard results by region

North arm. Earthquake hazard in the northern arm of Sulawesi is controlled by the North Sulawesi Subduction Zone and associated intraslab sources to the north, with an additional hazard from intraslab sources to the east of the arm. At longer

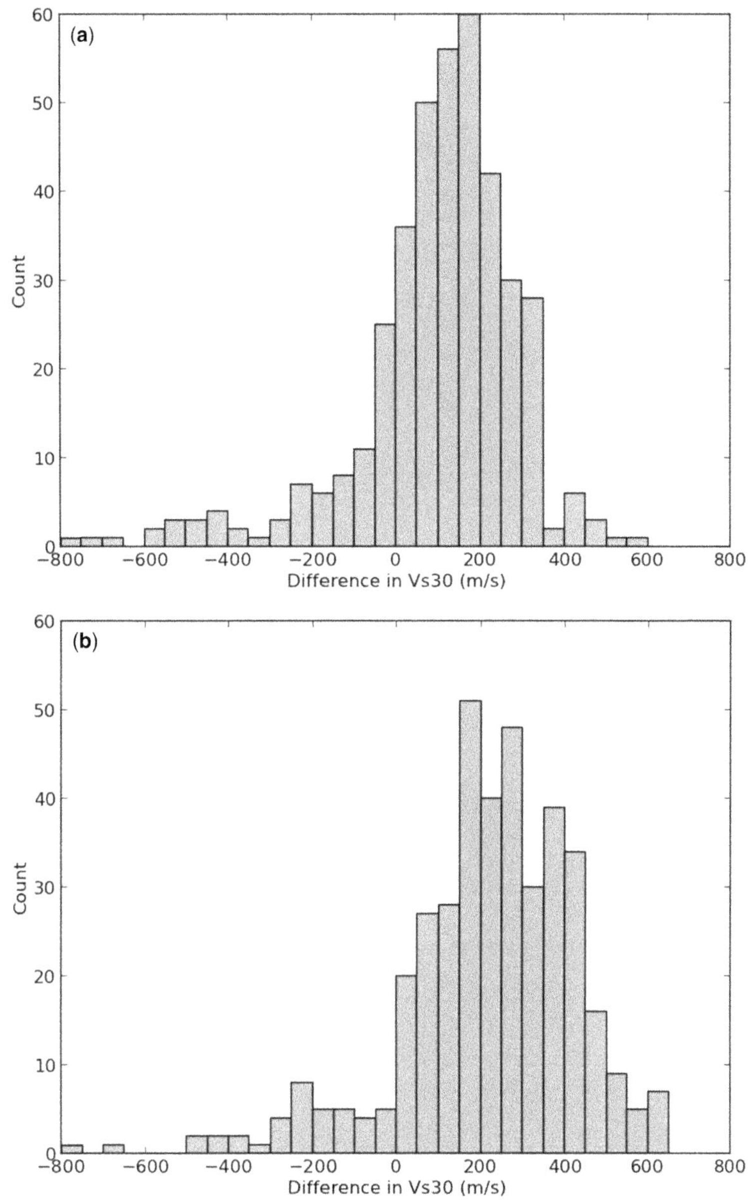

Fig. 5. Histogram of differences in V_{s30} values for (**a**) the Matsuoka *et al.* (2006) method minus H/V-derived measurements, and (**b**) the United States Geological Survey (Wald & Allen 2007) method minus H/V-derived measurements.

return periods (1000 and 2500 years), the Gorontalo Fault contributes significantly to the hazard level. Although the frequency of large earthquakes occurring on the Gorontalo Fault is lower than for other source regions, when these events do occur they can cause high levels of ground shaking as the source is shallow and on land. Earthquake hazard in this area is amplified by soft sediments along the depression created by the Gorontalo Fault, most notably surrounding Lake Limboto between the cities of Limboto (population 50 000) and Gorontalo (population 200 000). Earthquake hazard in the northern arm of Sulawesi is highest in the Buol and Toli-toli districts.

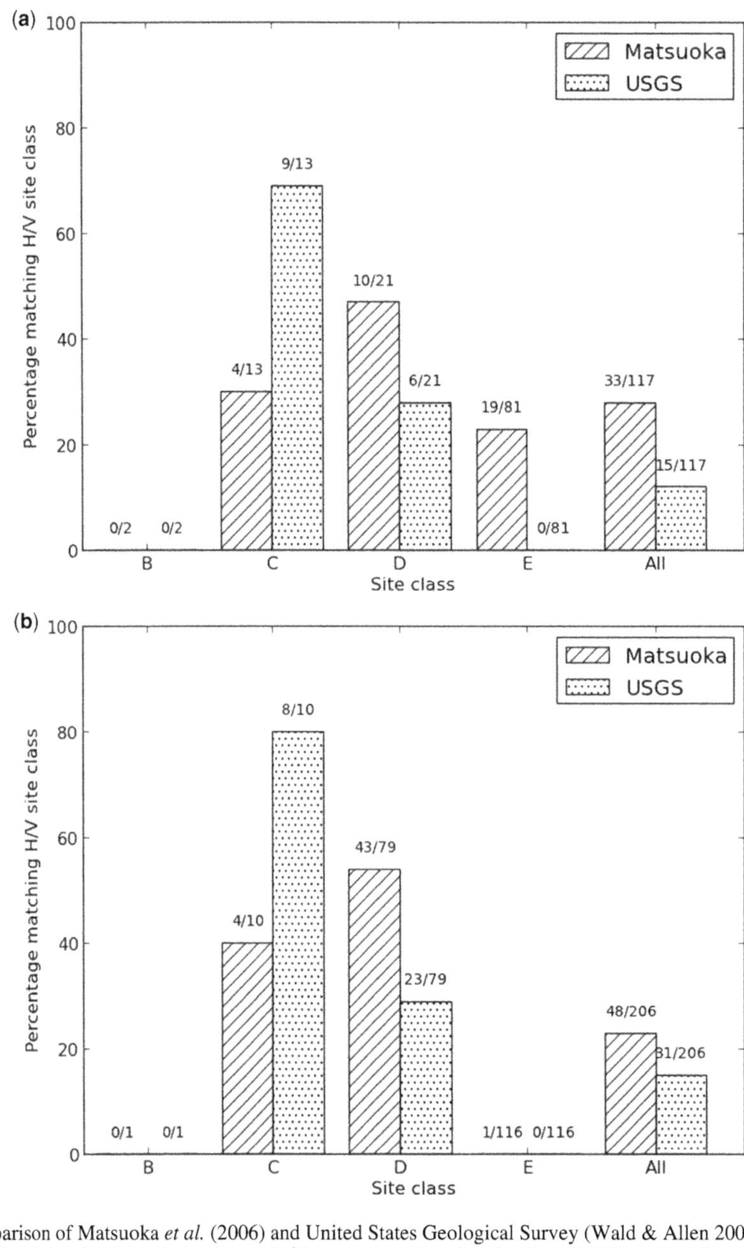

Fig. 6. Comparison of Matsuoka *et al.* (2006) and United States Geological Survey (Wald & Allen 2007) proxy estimates of site class with those derived from H/V measurements using the Zhao (2011) method for (**a**) Palu, (**b**) Gorontalo.

Central and East Sulawesi. Earthquake hazard in Central and East Sulawesi is very high. Hazard is extremely high along fast-moving onshore faults, in particular the Palu-Koro–Matano Fault System and the Balantak–Batui faults. High background seismicity rates drive a high hazard in the area far away from known faults. This pattern is consistent for all return periods and spectral periods simulated. Extremely high hazard (PGA $> 0.8g$ at a 500 year return period) occurs in the city of Palu (population 335 000), located on a pull-apart basin created by strands of the Palu-Koro Fault.

West Sulawesi. Earthquake hazard is generally high in West Sulawesi owing to a high background seismicity rate. Hazard is higher in the east, near the

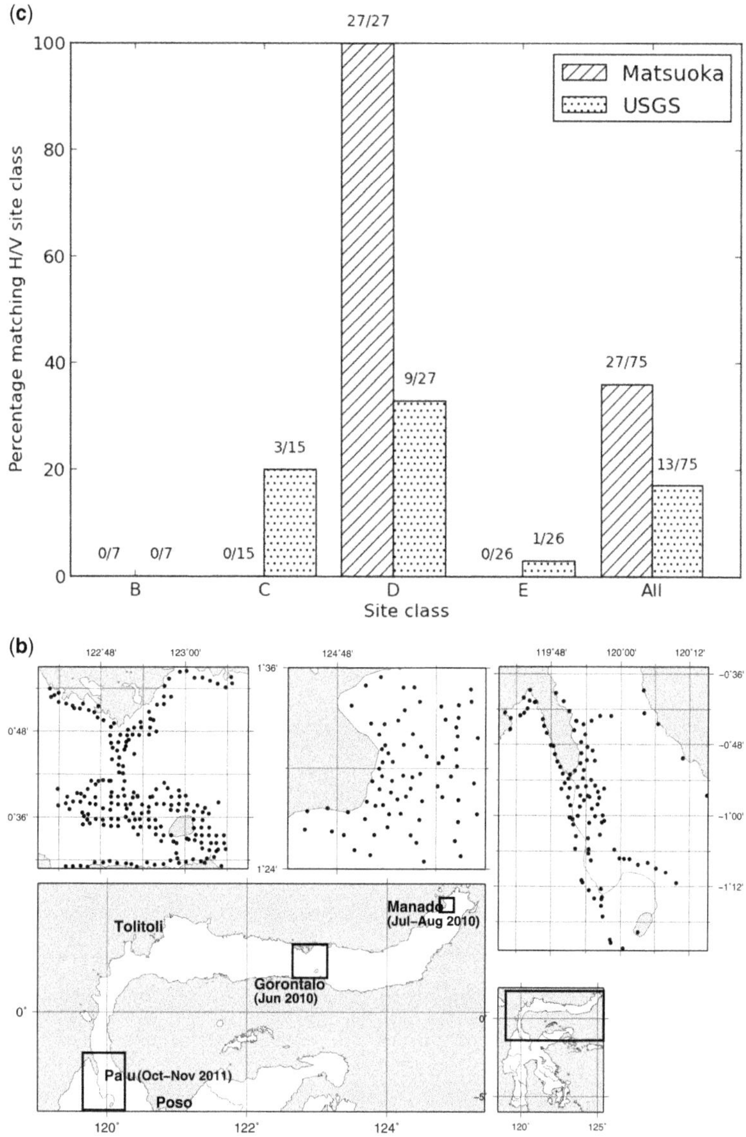

Fig. 6. (**c**) Manado. (**d**) shows measurement points, the year that the measurements were made is written in brackets.

Palu-Koro Fault, and to the west, where the offshore Makassar Thrust is significant.

SE Sulawesi. Earthquake hazard in SE Sulawesi is extremely high along the Matano and Lawanopo faults. The Lawanopo Fault runs just to the north of the city of Kendari (population 200 000), which is sited on a delta and surrounding coastal lowlands (site class E) near Kendari Bay. Background seismicity rates are lower to the south of the Matano Fault. Therefore, to the south of the Matano and Lawanopo faults, earthquake hazard, while still being high, is lower relative to Central and East Sulawesi.

South Sulawesi. Earthquake hazard is high along the Walanae Fault, with lowlands along the depression created by the fault leading to high amplification (site classes D and E) of ground motions. Background seismicity is low, with only a few historical earthquakes recorded in the background zone, and therefore away from the Walanae Fault hazard is

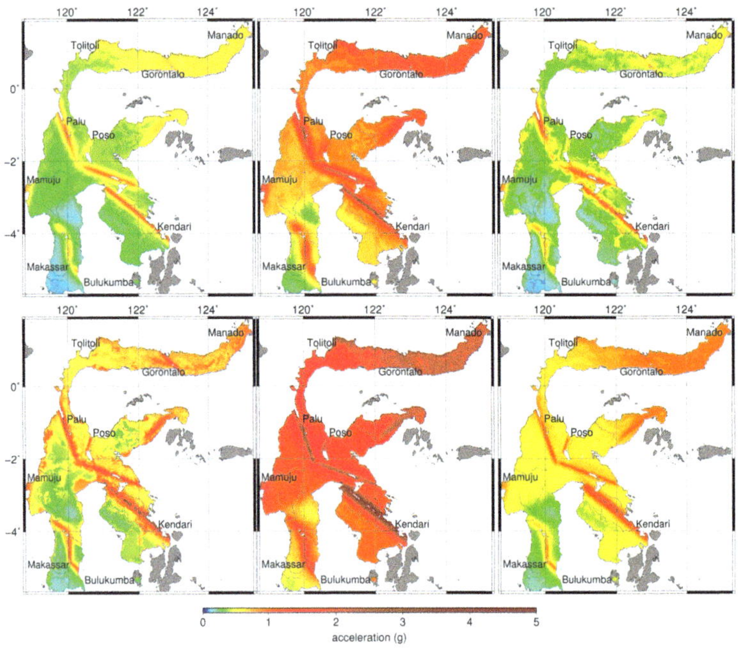

Fig. 7. Peak ground acceleration (PGA) and response spectra (RSA) of 0.2 and 1.0 s for (top) a 500 year return period and (bottom) a 2500 year return period.

much lower than for the rest of Sulawesi. The largest population centre in Sulawesi, Makassar (population 1 340 000), is located in a region of lower hazard relative to the rest of the island.

Using the EQRM, we obtained ground acceleration at each point on a 1 × 1 km grid for all return periods and spectral periods. These results are informative in seismic-resistant building and infrastructure design. For spatial planning purposes at the local scale and for construction of non-engineered buildings, the expected felt seismic intensity is a simpler and more easily communicated measure of hazard. Therefore, in the final step, we convert acceleration to Modified Mercalli Intensity (MMI) using the equation formulated by Atkinson & Kaka (2007) for acceleration at 1.0 s period, including site effects. MMI is then classified into four classes: high hazard zone (MMI $\geq$ VIII); medium hazard zone (MMI VII–VIII); low hazard zone (MMI V–VII); and very low hazard zone (MMI $<$ V) based on the 500 year return period. This is similar to, but slightly different from, the classification of Arya *et al.* (2014).

Comparison against historical data

The conversion of a 500 year recurrence interval 1.0 s hazard to intensity (MMI) is shown in Figure 8 and is overlain with the location of historical damaging earthquakes (Supartoyo & Surono 2008). All historical damaging events occur in regions classified as moderate (four events) or high (30 events) hazard, with the exception of the 1828 Bulukumba earthquake, which occurs in a low hazard region.

Discussion

A probabilistic seismic hazard assessment for the island of Sulawesi, Indonesia, including the effects of site amplification through proxy methods, was undertaken. Much of the island has high hazard, particularly along major, fast-slipping crustal faults, in the highly sheared region north of the Palu-Koro–Matano Fault System and along the north arm of Sulawesi. The SW arm of Sulawesi, including the city of Makassar, is lower hazard compared with the rest of the island, although all regions may experience damaging events at longer return periods. The high hazard regions correlate well with locations of historical earthquake damage, with the exception of the damaging 1828 earthquake in Bulukumba (MMI VIII–IX: Supartoyo & Surono 2008), on the south coast of South Sulawesi, which we speculate may have been caused by either the Walanae Fault or an unmapped offshore extension of this structure to the SE.

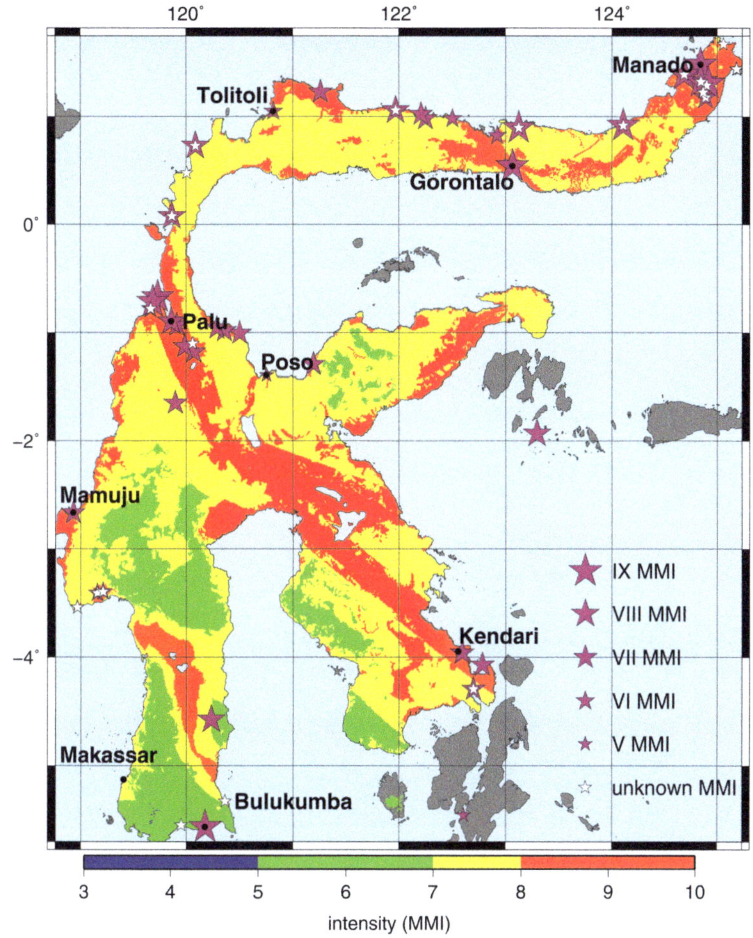

Fig. 8. Intensity map of Sulawesi (500 year recurrence interval 1.0 s) and historical damage location (the size of the star indicates the MMI scale).

A key challenge in developing this hazard map has been reconciling abundant recent seismicity with geological evidence of active structures. With the exception of the Palu-Koro Fault, the geological and geodetic evidence for activity rates is generally weak, while formal uncertainties in earthquake location (Husen & Hardebeck 2010) make attributing specific earthquakes to specific structures problematic. In order to address these issues, large crustal-scale faults that form the boundaries of different domains, such as the Palu-Koro–Matano Fault System, are included as fault sources with their geodetic slip rates and do not overlap background zone sources. Smaller faults that fall within zone sources have their slip rate adjusted to avoid double counting of seismicity. This is done by assuming a volume around the fault and using the seismicity of this volume (based on data in the surrounding source zone scaled to this volume) to calculate the equivalent slip rate for the fault. This is then balanced against the reported slip rate for the fault. In some cases, such as the Poso Fault in north Central Sulawesi, the activity rate estimated from seismicity in the surrounding zone is higher than the estimated slip rate from Irsyam *et al.* (2010). This suggests that the estimated slip rates in Irsyam *et al.* (2010) may be underestimates: however, more detailed geodetic studies are needed to confirm this.

As a result of this approach, hazard results in the northern half of Sulawesi (north of the Palu-Koro–Matano Fault System) are generally high and the role of individual crustal faults in driving the hazard is relatively minor. This is even more so along the northern arm, where subduction interface and intraslab sources contribute significantly to the hazard, with the effect of the onshore Gorontalo Fault only becoming significant at long return periods (>1000 years). South of this region, and most noticeably in South Sulawesi, background seismicity is lower and seismic activity is concentrated along individual faults, such as the Walanae Fault. Seismic hazard in the central part of Sulawesi is concentrated along the fast-moving Palu-Koro–Matano Fault System and Lawanopo Fault. Several pull-apart basins are present along these fault systems, where sediments have accumulated and resulting in high site amplifications. This is particularly evident near the city of Palu, which sits on a basin of alluvial and coastal sediments bounded to the east and west by two strands of the Palu-Koro Fault.

Palu and Gorontalo are the two most significant population centres located in coastal basins formed by active faulting, although other centres are also in similar situations, including Kendari. These centres are therefore exposed to a combination of high earthquake activity from near-field sources, amplified ground motions, liquefaction and landslide potential, as well as the threat of localized tsunami if earthquake deformation extends offshore or triggers submarine landslides (e.g. tsunami near Palu in 1927 and 1968: Pelinovsky *et al.* 1997). These population centres should be prioritized for more detailed site-amplification studies, hazard assessment for secondary earthquake hazards and risk-reduction activities.

A comparison of our bedrock hazard results with those of Irsyam *et al.* (2010) shows high hazard in the northern part of the island, particular along the north arm, in our assessment (Fig. 9). This may be due to siting the North Sulawesi Subduction Zone source close to the island near the region of most

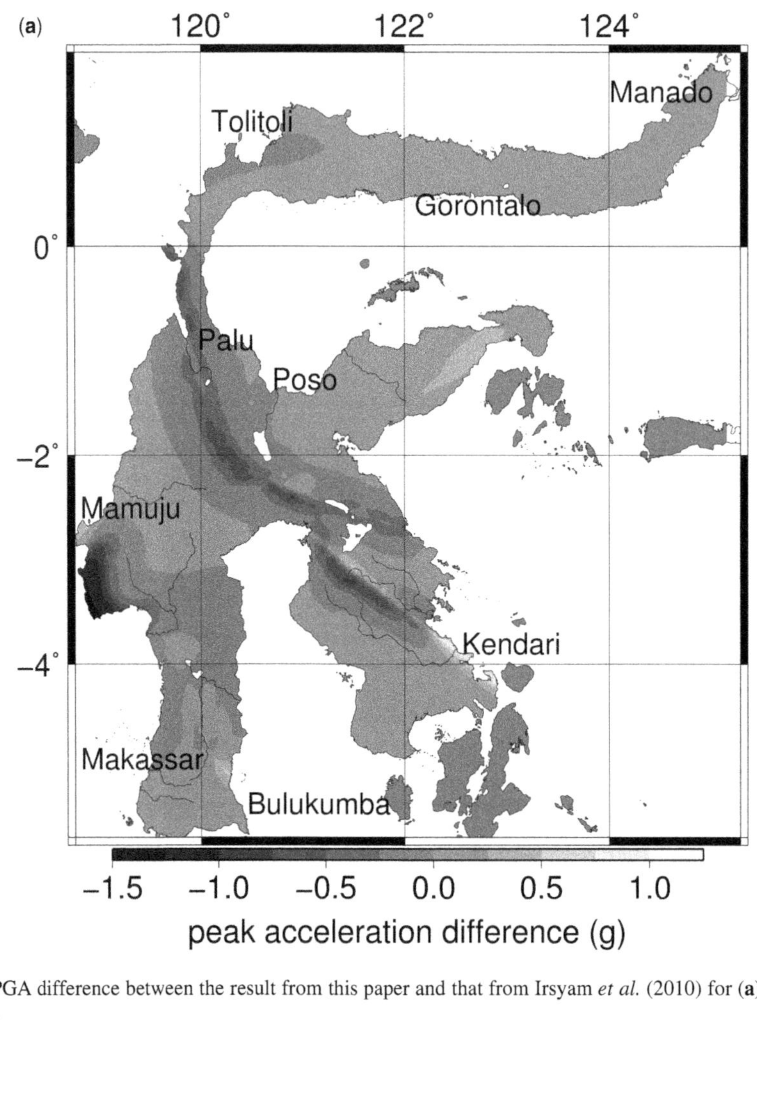

Fig. 9. The PGA difference between the result from this paper and that from Irsyam *et al.* (2010) for (**a**) a 500 year return period.

frequent seismicity, rather than at the trench. We also explicitly include intraslab sources using fault planes with out-of-plane ruptures, allowing us to more realistically capture the geometry of intraslab sources, while Irsyam *et al.* (2010) use smoothed seismicity within stepped rectangular volumes to represent the slab. The Irsyam *et al.* (2010) results are higher along the major crustal faults, including the Palu-Koro–Matano Fault System (Fig. 9).

The assessment presented here uses the same ground-motion prediction equations, and logic tree weightings, as Irsyam *et al.* (2010). None of these GMPEs have been developed using strong-motion data from Indonesia, and therefore there is considerable uncertainty in their application. Analysis of initial data from the newly established Indonesian strong-motion network will begin to reduce uncertainty, although research to date has focused on Java and Sumatra (Rudyanto 2013). Furthermore, the geological heterogeneity of Sulawesi – including thick continenal crust derived from both the Sundaland and Australian continents, overthrust ophiolites and mélange complexes, actively deforming crustal regions, and active and inactive arc systems – means that a high degree of heterogeneity in ground motions is to be expected within Sulawesi.

Proxy methods used to estimate site amplification are subject to considerable uncertainty. The Matsuoka *et al.* (2006) method based on geomorphology is slightly more accurate, correctly classifying site class at approximately 25% of the measured locations compared with the topographical slope method of Wald & Allen (2007), which is correct for approximately 15% of the measured sites. The Wald & Allen (2007) method more accurately assigns site class C, while the Matsuoka *et al.* (2006) method more accurately predicts site class D. Noting the considerable uncertainty in the use of

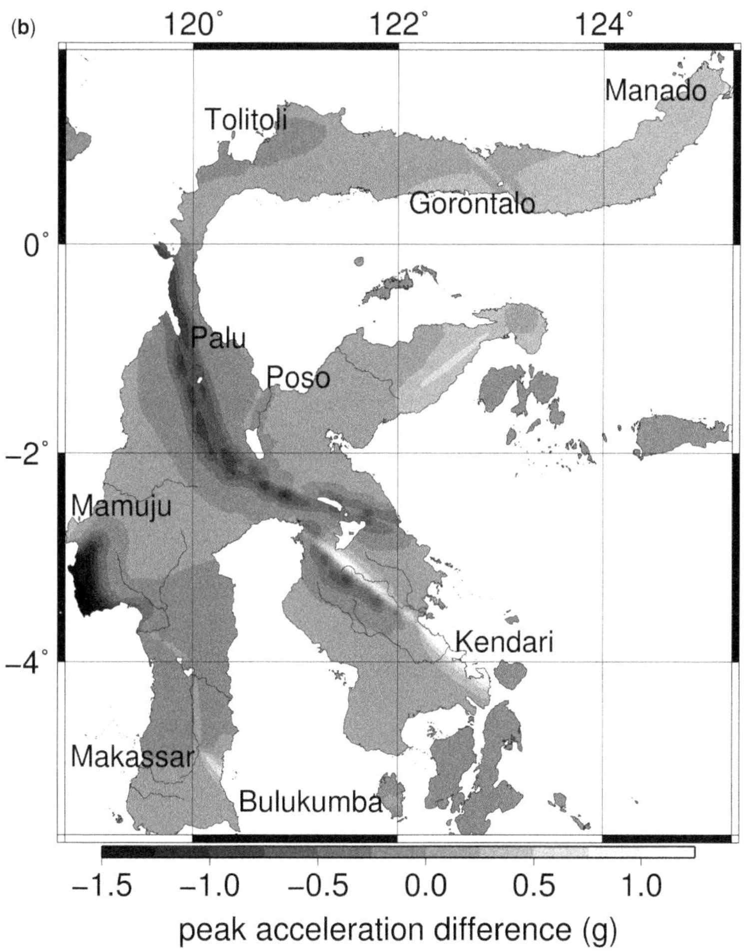

Fig. 9. (**b**) a 2500 year return period.

H/V measurements as a basis for estimating site effects (Ghasemi *et al.* 2009; Zhao 2011), the main conclusion that can be drawn is that the Matsuoka *et al.* (2006) proxy method is probably more suitable for Sulawesi, but that site amplification remains a major source of uncertainty in the hazard results.

By discriminating between volcanic and non-volcanic mountains, the method of Matsuoka *et al.* (2006) is expected to have a distinct advantage over slope-based approaches as there are both active volcanoes and much older non-volcanic mountain regions in Sulawesi. Volcanic product deposited on the footslopes of volcanoes may have low shear-wave velocities (V_{s30} *c.* 200 m s^{-1}), while a terrace covered by volcanic materials may have even lower values (V_{s30} *c.* 160 m s^{-1}) (Matsuoka *et al.* 2006). These geomorphic features, such as footslopes and terraces, are suitable for the deposition of thick loose pyroclastic flow deposits such as ash, pumice and scoria. These deposits have low shear-wave velocities. Low estimates of V_{s30} for areas of active volcanic deposition is consistent with Nunziata *et al.* (1999), who measured shear-wave velocities on recent pyroclastic deposits (younger than 12 000 years) in the Campi-Flegrei caldera (Italy). Although shear-wave velocities were highly variable, they were generally low, varying from less than 200 up to 650 m s^{-1}. Variability in shear-wave velocity of pyroclastic deposits is due to the variability in the materials of particular deposits and the degree of welding that has occurred during diagenesis.

Unfortunately, at present, we do not have any field data with which to test these regions against the Matsuoka *et al.* (2006) classification. Furthermore, many of the major cities are situated in valleys and coastal lowlands and delta. In many cases, these geomorphic environments are a direct result of active faulting in the area, and therefore accurate characterization of site amplification at the local scale will be important for designing critical infrastructure and spatial planning in order to reduce exposure in the areas of greatest amplification. Future field measurements to better understand site amplification should use more robust techniques for calculating site amplification, such as multi-channel analysis of surface waves (MASW), N-SPT (Standard Penetration Test), cone penetration tests (CPT) or borehole measurements. These measurements would be able to validate both the proxy methods and the V_{s30} estimates derived using H/V methods.

Conclusions

A first probabilistic seismic hazard assessment including the effects of site amplification has been undertaken for the island of Sulawesi, Indonesia. Most of the island, with the exception of South Sulawesi, is undergoing rapid deformation, leading to high hazard in most regions (PGA > 0.4*g* at a 500 year return period including site effects), with extremely high hazard (PGA > 0.8*g* at a 500 year return period) along fast-slipping crustal-scale faults, such as the Palu-Koro–Matano Fault System and the Lawanopo Fault. Active subduction and a complex array of active and inactive subducted slabs in the north drive very high hazard along the north arm of the island. Reported slip rates for active faults are balanced against background seismicity rates to avoid double-counting seismicity: in many cases, background rates exceed report fault slip rates, highlighting the need for further geodetic and geological studies to better constrain slip rates. Site amplification is included using proxy methods and, combined with conversion to intensity and classification into hazard zones, facilitates use of the hazard maps for spatial planning purposes. A high degree of uncertainty associated with these methods means there is a need for further local-scale studies to better characterize site effects. This is particularly important for the many significant population centres that are sited in sedimentary basins created by currently active faults, including provincial capitals Palu and Gorontalo, and therefore exposed to both high seismicity rates and high amplification of ground motions. Building design in these areas should combine bedrock hazard results with locally measured site effects to determine appropriate design criteria.

The next plan is to try to validate the amplification factor inferred from H/V using a geotechnical approach. Geotechnical investigation, such as N-SPT and CPT, may be a promising method to validate the H/V method to achieve a more reliable amplification factor and, hence, produce a better seismic hazard map.

This work was supported by the Australia Agency for International Development and the Indonesian National Disaster Management Agency through the Australia–Indonesia Facility for Disaster Reduction. The authors are grateful to Masyhur Irsyam (Institut Teknologi Bandung) for sharing input parameters and results from the 2010 revision of the national seismic hazard maps. The authors acknowledge the hard work and fruitful discussions with the EQRM development team (Duncan Gray, Vanessa Newey, David Robinson, Geoscience Australia). We would like to thank Dr Irwan Meilano and the anonymous reviewer for their very useful comments and suggestions.

References

Aki, K. 1965. Maximum likelihood estimate of *b* in the formula $\log N = a - bM$ and its confidence limits. *Bulletin of the Earthquake Research Institute*, **43**, 237–239.

AMMON, C.J., LAY, T., KANAMORI, H. & CLEVELAND, M. 2011. A rupture model of the 2011 off the Pacific coast of Tohoku earthquake. *Earth, Planets and Space*, **63**, 693–696.

ARIKAN, F. & AYDIN, N. 2012. Research article: influence of weathering on the engineering properties of dacites in Northeastern Turkey. *ISRN Soil Science*, **2012**, article ID 218527. https://doi.org/10.5402/2012/218527

ARYA, A.S., BOEN, T. & ISHIYAMA, Y. 2014. *Guidelines for Earthquake Resistant Non-engineered Construction*. UNESCO.

ATKINSON, G.M. & BOORE, D.M. 2003. Empirical ground-motion relations for subduction-zone earthquakes and their application to cascadia and other regions. *Bulletin of the Seismological Society of America*, **93**, 1703–1729.

ATKINSON, G.M. & KAKA, SL.I. 2007. Relationships between Felt Intensity and Instrumental Ground Motion in the Central United States and California. *Bulletin of the Seismological Society of America*, **97**, 497–510, https://doi.org/10.1785/0120060154

BNPB 2010. *Disaster Data and Information Indonesia*. Badan Nasional Penanggulangan Bencana (Indonesian National Disaster Management Agency), Jakata, http://dibi.bnpb.go.id

BOORE, D.M. & ATKINSON, G.M. 2008. Ground-motion prediction equations for the average horizontal component of PGA, PGV, and 5%-damped PSA at spectral periods between 0.01 sand 10.0 s. *Earthquake Spectra*, **24**, 99–138.

BORCHERDT, R.D. 1994. Estimates of site-dependent response spectra for design (methodology and justification). *Earthquake Spectra*, **10**, 617–653.

BPS 2010. *Population Census of Indonesia*. Badan Pusat-Statistik (Indonesian Central Agency for Statistics), Jakata, http://sp2010.bps.go.id

CAMPBELL, K.W. & BOZORGNIA, Y. 2008. NGA ground motion model for the geometric mean horizontal-component of PGA, PGV, PGD and 5% damped linear elastic response spectra for periods ranging from 0.01 to 10.0 s. *Earthquake Spectra*, **24**, 139–171.

CHIOU, B. & YOUNGS, R. 2008. A NGA model for the average horizontal component of peakground motion and response spectra. *Earthquake Spectra*, **24**, 173–215.

GHASEMI, H., ZARE, M., FUKUSHIMA, Y. & SINAEIAN, F. 2009. Applying empirical methods in site classification, using response spectral ratio (H/V): a case study on Iranian strong motion network (ISMN). *Soil Dynamics and Earthquake Engineering*, **29**, 121–132.

GULICK, S.P., AUSTIN, J.A. JR ET AL. 2011. Updip rupture of the 2004 Sumatra earthquake extended by thick indurated sediments. *Nature Geoscience*, **4**, 453–456.

HALL, R. 2011. Australia–SE Asia collision: plate tectonics and crustal flow. *In*: HALL, R., COTTAM, M.A. & WILSON, M.E.J. (eds) *The SE Asian Gateway: History and Tectonics of the Australia–Asia Collision*. Geological Society, London, Special Publications, **355**, 75–109, https://doi.org/10.1144/SP355.5

HANKS, T.C. & KANAMORI, H. 1979. A moment magnitude scale. *Journal of Geophysical Research: Solid Earth*, **84**, 2348–2350.

HUSEN, S. & HARDEBECK, J.L. 2010. Earthquake location accuracy. *Community Online Resource for Statistical Seismicity Analysis*, **10**, https://doi.org/10.5078/corssa-55815573. Available at http://www.corssa.org

IRSYAM, M., SENGARA, W. ET AL. 2010. *Development of Seismic Hazard Maps of Indonesia for Revision of Seimic Hazard Map in SNI 03-1726-2002*. Reseach report submitted to the Ministry of Public Works by the Team for Revision of Seismic Hazard Maps of Indonesia.

KOKETSU, K., YOKOTA, Y. ET AL. 2011. A unified source model for the 2011 Tohoku earthquake. *Earth and Planetary Science Letters*, **310**, 480–487.

KRAMER, S.L. 1996. *Geotechnical Earthquake Engineering*. Prentice–Hall International Series in Civil Engineering and Engineering Mechanics. Prentice-Hall, Upper Saddle River, NJ.

MATSUOKA, M., WAKAMATSU, K., FUJIMOTO, K. & MIDORIKAWA, S. 2006. Average shear-wave velocity mapping using Japan engineering geomorphologic classification map. *Journal of Structural Mechanics and Earthquake Engineering, JSCE*, **23**, 57–68.

MCCAFFREY, R. & SUTARDJO, R. 1982. Reconnaissance microearthquake survey of Sulawesi, Indonesia. *Geophysical Research Letters*, **9**, 793–796.

NUNZIATA, C., MELE, R. & NATALE, M. 1999. Shear wave velocities and primary influencing factors of Campi-Flegrei_Neapolitan deposits. *Engineering Geology*, **54**, 299–312.

OTTEMÖLLER, V., VOSS, P. & HAVSKOV, J. 2011. *SEISAN Earthquake Analysis Software for Windows, Solaris, Linux and Macosx*, http://www.geosig.com/files//GS_SEISAN_9_0_1.pdf

PELINOVSKY, E., YULIADI, D., PRASETYA, G. & HIDAYAT, R. 1997. The 1996 Sulawesi Tsunami. *Natural Hazards*, **16**, 29–38.

PUSPITA, S.D., HALL, R. & ELDERS, C.F. 2005. Structural styles of the offshore West Sulawesi fold belt, North Makassar Straits, Indonesia. *In*: *Proceedings of the Indonesian Petroleum Association, 30th Annual Convention Jakarta*, August 2005, 519–542.

ROBINSON, D., FULFORD, G. & DHU, T. 2006. *EQRM: Geoscience Australia's Earthquake Risk Model*. Geoscience Australia, Record, 2005/01.

RUDYANTO, A. 2013. *Development of strong-motion database for Sumatra–Java region*. Masters thesis, Australian National University.

SARSITO, D.A., ABIDIN, H.Z., SAPIIE, B., TRIYOSO, W., MEILANO, I. & SIMONS, W.J.F. 2011. Geometric and kinematic modelling of Sulawesi–East Kalimantan zone based on GNSS-GPS and global gravitation data. *Paper presented at the IUGG XXV General Assembly Earth on the Edge: Science for a Sustainable Planet*, Melbourne, Australia, 27 June–8 July 2011.

SCHWARTZ, D.P. & COPPERSMITH, K.J. 1984. Fault behaviour and characteristic earthquakes: examples from the Wasatch and San Andreas fault zones. *Journal of Geophysical Research: Solid Earth*, **89**, 5681–5698.

SILVER, E.A. & MOORE, J.C. 1978. The Molucca Sea Collision Zone, Indonesia. *Journal of Geophysical Research: Solid Earth*, **83**, 1681–1691.

SILVER, E.A., MCCAFFREY, R., JOYODIWIRYO, Y. & STEVENS, S. 1983*a*. Ophiolite emplacement by collision between the Sula Platform and the Sulawesi island arc, Indonesia. *Journal of Geophysical Research: Solid Earth*, **88**, 9419–9435.

SILVER, E.A., MCCAFFREY, R. & SMITH, R.B. 1983*b*. Collision, rotation, and the initiation of subduction in the evolution of Sulawesi, Indonesia. *Journal of Geophysical Research*, **88**, 9407–9418.

SOCQUET, A., SIMONS, W. ET AL. 2006. Microblock rotations and fault coupling in SE Asia triple junction (Sulawesi, Indonesia) from GPS and earthquake slip vector data. *Journal of Geophysical Research*, **111**, B08409.

SPENCER, J.E. 2011. Gently dipping normal faults identified with Space Shuttle radar topography data incentral Sulawesi, Indonesia, and some implications for fault mechanics. *Earth and Planetary Space Science*, **308**, 267–276.

SUPARTOYO & SURONO 2008. *Catalog of Indonesian Destructive Earthquake, Year 1629–2007*. Badan Geologi (Geological Agency of Indonesia), Jawa Barat.

WALD, D.J. & ALLEN, T.I. 2007. Topographic slope as a proxy for seismic site conditions and amplification. *Bulletin of the Seismological Society of America*, **97**, 1379–1395.

WANG, K. & HU, Y. 2006. Accretionary prisms in subduction earthquake cycles: the theory of dynamic Coulombwedge. *Journal of Geophysical Research*, **111**, B06410, https://doi.org/10.1029/2005JB004094

WATKINSON, I.M., HALL, R. & FERDIAN, F. 2011. Tectonic re-interpretation of the Banggai-Sula–Molucca Sea margin, Indonesia. *In*: HALL, R., COTTAM, M.A. & WILSON, M.E.J. (eds) *The SE Asian Gateway: History and Tectonics of the Australia–Asia Collision*. Geological Society, London, Special Publications, **355**, 203–224, https://doi.org/10.1144/SP355.10

YOUNGS, R.R. & COPPERSMITH, K.J. 1985. Implications of fault slip rates and earthquake recurrence models to probabilistic seismic hazard estimates. *Bulletin of the Seismological Society of America*, **75**, 939–964.

YOUNGS, R.R., CHIOU, S.J., SILVA, W.J. & HUMPHREY, J.R. 1997. Strong ground motion attenuation relationships for subduction zone earthquakes. *Seismological Research Letters*, **68**, 58–73.

ZHAO, J.X. 2011. Comparison between VS30 and site period as site parameters in ground-motion prediction equations for response spectra. *Paper presented at the 4th IASPEI/IAEE International Symposium: Effects of Surface Geology on Seismic Motion*, Santa Barbara, California, 23–26 August 2011.

ZHAO, J.X., ZHANG, J. ET AL. 2006. Attenuation relations of strong motion in Japan using site classification based on predominant period. *Bulletin of the Seismological Society of America*, **96**, 898.

Development of an engineering bedrock map beneath Jakarta based on microtremor array measurements

M. RIDWAN[1,2*], S. WIDIYANTORO[3,4], M. IRSYAM[4,5], AFNIMAR[3] & H. YAMANAKA[6]

[1]*Graduate Research on Earthquakes and Active Tectonics, Faculty of Earth Science and Technology, Institut Teknologi Bandung, Jalan Ganesa No. 10, Bandung 40132, Indonesia*

[2]*Research Institute for Human Settlements, Ministry of Public Works, Jalan Panyaungan, Cileunyi Wetan, Kabupaten Bandung 40393, Indonesia*

[3]*Faculty of Mining and Petroleum Engineering, Institut Teknologi Bandung, Jalan Ganesa No. 10, Bandung 40132, Indonesia*

[4]*Research Centre for Disaster Mitigation, Institut Teknologi Bandung, Jalan Ganesa No. 10, Bandung 40132, Indonesia*

[5]*Faculty of Civil and Environment, Institut Teknologi Bandung, Jalan Ganesa No. 10, Bandung 40132, Indonesia*

[6]*Department of Environmental Science and Technology, Tokyo Institute of Technology, 2-12-1 Ookayama, Meguro-ku, Tokyo 152-8550, Japan*

**Correspondence: m.ridwan@pu.go.id*

Abstract: Jakarta has been selected as a study area for seismic microzonation by considering its population and its infrastructure growth, seismicity and geological setting. One of the important factors in seismic hazard analysis is the characteristics of the ground overlying bedrock, and, as yet, this has not been studied for Jakarta. This study was intended to estimate the depth to bedrock by applying microtremor array exploration. The phase velocity of microtremors was estimated by the spatial autocorrelation (SPAC) method, whereas the S-wave velocity profile was estimated by inversion using a genetic algorithm. Microtremor array analysis has been conducted at 55 sites that covered the whole of Jakarta. The result of one-dimensional (1D) S-wave velocity profile estimation indicates that the engineering bedrock depth has pronounced differences between northern (>700 m) and southern Jakarta (about 300 m). A 3D S-wave velocity model was constructed from 1D profiles resulting from a second inversion in which the depth of constant-velocity layers, with velocities of 500, 700 and 900 m s^{-1}, was determined. The constructed 3D velocity structures show a bedrock morphology that has a depth range of 350–725 m, with depths increasing towards northern Jakarta. The implication of differences in bedrock depth is that estimated seismic amplification for the thicker sediment will be higher.

The city of Jakarta is the capital of Indonesia, with a very high rate of population and infrastructure growth. Although there are no known sources of seismicity in Jakarta itself, there are at least four fault zones that can be identified within a distance of 500 km that might host earthquakes large enough to cause significant ground motion in Jakarta: the Cimandiri, Lembang, Sunda and Semangko faults, and the Sumatra and Java megathrusts in the Indonesian Ocean. An Indonesian earthquake catalogue for the years 1900–2009, compiled from various sources (Irsyam *et al.* 2010), shows a clustering of earthquake hypocentres around these seismic source zones. Geologically, Jakarta is located in the Ciputat Sub-basin of the NW Java Basin, which is comprised of very thick sedimentary layers (Padmosukismo & Yahya 1974), so that there may be pronounced amplification of ground shaking when an earthquake occurs.

As a first step in developing seismic risk reduction strategies, it is important to undertake a seismic hazard and vulnerability assessment in Jakarta. For a seismic hazard analysis at city scale (seismic microzonation), the subsurface geology should be considered as one of the input parameters for site response analysis, including information on bedrock depth and soil properties. However, the available data for constraining such parameters in Jakarta

From: Cummins, P. R. & Meilano, I. (eds) 2017. *Geohazards in Indonesia: Earth Science for Disaster Risk Reduction*. Geological Society, London, Special Publications, **441**, 153–165.
First published online July 28, 2016, https://doi.org/10.1144/SP441.7

are still in very short supply: shallow borehole data have only been collected in the centre of Jakarta, while deep boreholes are very limited. More detailed studies and soil investigations are required.

In general, subsurface investigations can be conducted using active-source geophysical and geotechnical methods (e.g. seismic refraction, seismic downhole test, standard penetration test (SPT)), but these are difficult to undertake in urban areas, particularly for deep layers. The microtremor array method was applied in this study to obtain subsurface S-wave velocity profiles beneath Jakarta. This method has been used widely, not only for scientific but also for engineering purposes, owing to its simplicity of operation and the fact that it does not require an active source. The microtremor data analysis was carried out using the spatial autocorrelation (SPAC) method, first developed by Aki (1957), to obtain a Rayleigh-wave dispersion curve on a circular array. Further development was conducted by many researchers, such as, among others, Okada (1998, 2003) and Morikawa *et al.* (2004). Asten *et al.* (2004) also proposed the Multi Mode SPAC (MMSPAC) method for array processing, where observed coherency–frequency curves can be inverted in coherency space to identify multiple modes by using a semi-circular array.

Previous geological studies were conducted in the Jakarta area by Turkandi *et al.* (1992) for regional geological mapping, and by Fachri *et al.* (2002) who focused on hydrostratigraphy for groundwater basins. Ridwan *et al.* (2013, 2014) conducted microtremor measurements at two lines in a west–east direction and one line in a north–south direction to obtain a rough stratigraphic model based on S-wave velocity profiles.

The main objective of this study was to identify the depth to engineering bedrock in the Jakarta area by analysing microtremor array measurements at many sites throughout the study area. The results are presented as layer depth maps for each S-wave velocity, including an estimated engineering bedrock depth of Jakarta. A comparison of these 1D S-wave velocity profiles resulting from microtremor measurements with SPT and seismic downhole tests at the KMAL site is also discussed.

Geology of the Jakarta area

The NW Java Basin is located in the NW part of the island of Java, and is controlled by a block-faulting system aligned in a north–south direction. This region was formed as a result of tectonic processes acting over a long period. Block-faulting systems divide the NW Java Basin into six sub-basins: Tanggerang; Ciputat; Rengasdengklok; Pasir Putih; Pamanukan; and Jati Barang. The sedimentation process occurred in several stages that began in the Eocene. Bedrock in the NW Java Basin is defined as pre-Tertiary igneous or metamorphic rock (Padmosukismo & Yahya 1974).

In general, the sedimentary rock cover throughout Jakarta consists of Quaternary sediments: alluvial; beach ridge deposits; alluvial fan; and banten tuff (Fig. 1). Turkandi *et al.* (1992) proposed a stratigraphic sequence based on outcrops of Tertiary sediments around Jakarta, which, from oldest to youngest, is as follows: the Rengganis Formation (sandstones, conglomerate, claystone); the Bojongmanik Formation (sandstones, clay, limestone) in the western part, which interfingers with the Jatiluhur Formation (marl, claystone) in the eastern part; the Klapanunggal Formation (coral limestone, marl, sandstone); the Genteng Formation (tuff claystone and sandstone); and the Serpong Formation (conglomerate, sandstone, claystones). Based on Jakarta's geological setting in the Ciputat Sub-basin, this study defines seismic bedrock as pre-Tertiary igneous or metamorphic rock, and the Klapanunggal Formation was assumed to comprise the engineering bedrock.

Data and method

Array microtremor observation

Previous studies using microtremor array measurements in Jakarta were conducted by Ridwan *et al.* (2013, 2014) at 11 sites along two lines in an east–west direction and eight sites along a line in a north–south direction, to obtain stratigraphic models for the Jakarta area. In the present study, microtremor array measurements were conducted at an additional 36 sites, and these were combined with the 19 sites from the microtremor data of Ridwan *et al.* (2013, 2014), so that there was a total 55 sites available for further analysis.

The locations of microtremor array measurements for all sites were chosen to follow a grid with a 2–3 km interval that covered the entire area of Jakarta (Fig. 2). Turkandi *et al.* (1992) showed that the surface geology in the Jakarta area is comprised of alluvial, beach ridge deposits, alluvial fan and banten tuff, and also that there are very thick sedimentary rocks under these Quaternary sediments. The thickness of sediments is thought to gradually increase towards the north, while in a west–east direction there is relatively little variation in thickness.

The array configuration used for microtremor measurements was triangular in shape, utilizing four seismographs of high magnification, in which one station was placed in the centre and the other three stations were deployed on a circle. The installation of the devices was set-up on a flat area on a hard

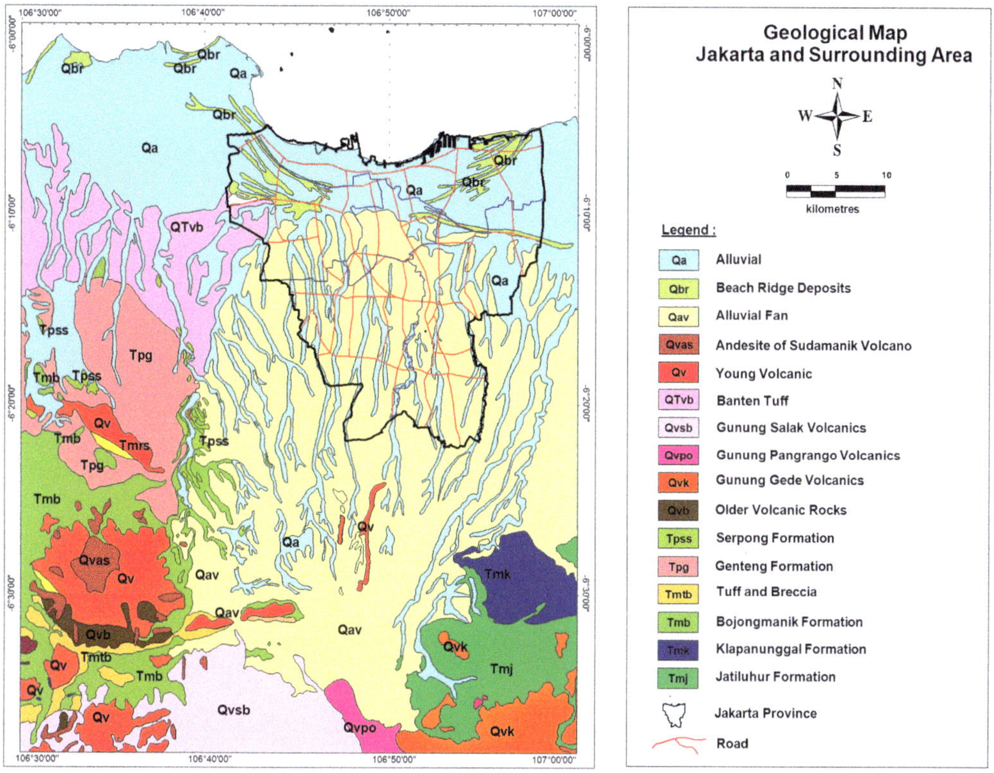

Fig. 1. Geological map of Jakarta and surrounding areas (modified from Turkandi *et al.* 1992; Effendi *et al.* 1998).

ground or concrete surface. Three triangular arrays, with sizes of 600, 300 and 150 m (see Fig. 3a), were considered for deep surveys, with the assumption that engineering bedrock can be reached using this array size. Since only four seismographs were used in the fieldwork, the microtremor measurements were conducted for each triangular array size in succession. Simultaneous recording at the four stations was conducted for approximately 1–1.5 h (Okada 1998, 2003) using a 32-bit A/D converter and a sampling frequency of 100 Hz. The same configuration of array with regard to shape and size was employed at all sites for microtremor data recording. The instruments used in the fieldwork were portable vertical-component seismometers manufactured by OYO Corporation, which have a characteristic frequency band of 0.1–200 Hz and use GPS for time synchronization.

The spatial auto-correlation (SPAC) method

The SPAC method is based on the theory developed by Aki (1957) to obtain the dispersion curves for the fundamental mode of Rayleigh waves. In a circular array that consists of many stations, the phase velocity can be calculated based on the average coherence of all pairs of stations. The optimal array configuration is a dense circular array of seismographs: however, in our fieldwork, it was only possible to deploy a few instruments. We used a modified triangular array shape, consisting of four sensors to reduce difficulties in the fieldwork, with which it was possible to make the calculations required for the SPAC method (Okada 2003).

As microtremors are dominated by surface waves consisting of Rayleigh and Love waves, the microtremor measurements were conducted using vertical-component seismometers to analyse the Rayleigh wave component of microtremor energy. The SPAC method was applied to estimate phase velocity after digital data were recorded at a 100 Hz sampling interval and divided into several time blocks. For a triangular array configuration with distance r between two stations, the SPAC function, $\varphi(\omega, r, \theta)\,d\theta$, was calculated for each pair of stations. Then, the SPAC coefficient, $\rho(\omega, r)$, was estimated by averaging over all azimuths, θ.

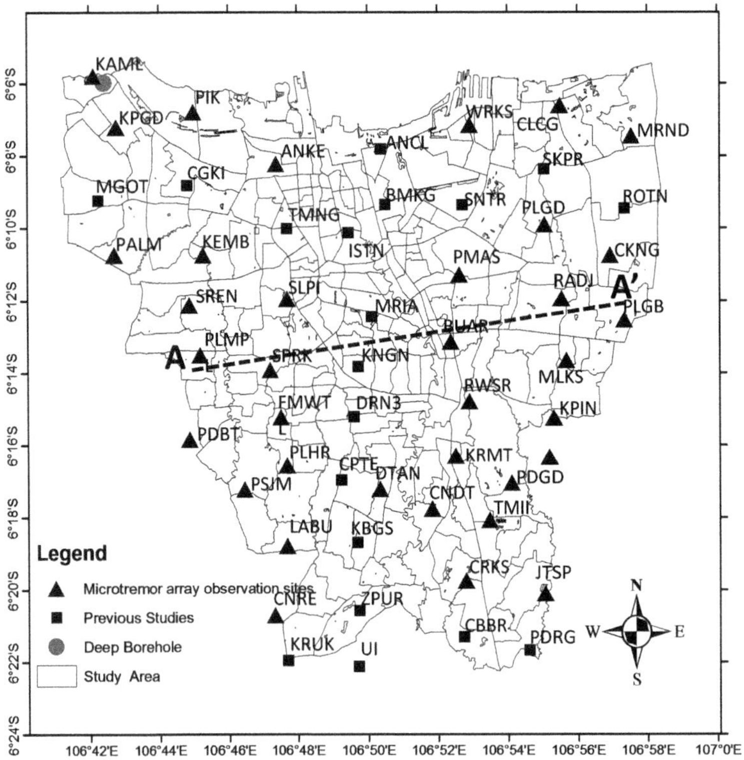

Fig. 2. Microtremor array observation sites in Jakarta city. A–A′ is a cross-section line in a west–east direction.

This can be expressed by the following equation (Aki 1957):

$$\rho(\omega, r) = \frac{1}{2\pi\varphi(\omega, 0, 0)} \int_{0}^{2\pi} \varphi(\omega, r, \theta)\,\mathrm{d}\theta \quad (1)$$

The SPAC coefficient for a circular array configuration can also be expressed as a Bessel function, as follows (Okada 2003):

$$\rho(\omega, r) = J_0\left(\frac{\omega r}{c(\omega)}\right) \quad (2)$$

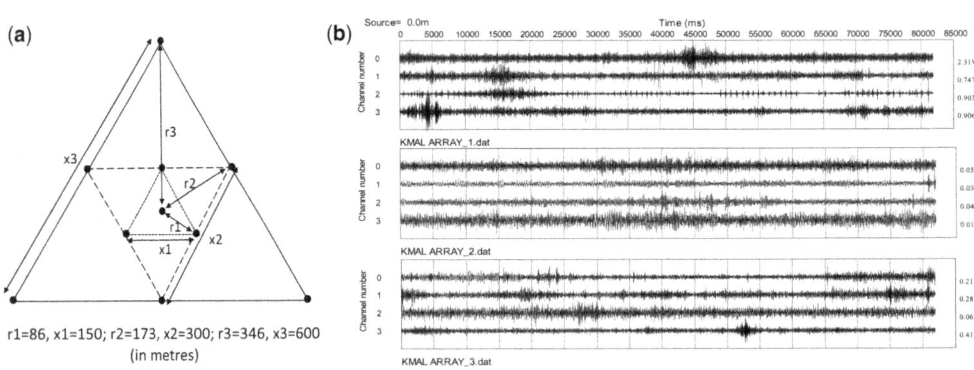

Fig. 3. (**a**) Triangular array configuration and (**b**) an example of a selected microtremor record at the KMAL site.

where J_0 is the Bessel function of the first kind of zero order, $c(\omega)$ is the phase velocity as a function of frequency ω, and r is the distance between the two stations. A more detailed explanation of the SPAC method can be found in Aki (1957), Okada (1998, 2003) and Morikawa *et al.* (2004).

Inversion by genetic algorithm

Estimation of S-wave velocity profiles was conducted by a non-linear inversion using a genetic algorithm applied to the dispersion curves for specified initial models. The genetic algorithm is a technique of searching for solutions to an optimization problem based on concepts from biological evolution. This technique can be applied to an initial population that consists of a set of individuals that is generated randomly, where each individual is a possible solution. Before the genetic operations are generated, each variable in the search area was encoded in the form of binary numbers (0 and 1), which consists of the smallest genetic components, namely genes. The quality of individuals in a population was expressed by the fitness function $f_j = 1/\varphi_j$, where φ_j is the misfit function for the j-th individual. The chromosomes that have greater fitness will have more chance of being selected in the new population. The next generation is formed by the evolution process of the selected chromosome through interactions with some genetic operations (i.e. cross-over, mutation and elite selection). The application of genetic algorithms in the field of seismology was introduced by Yamanaka & Ishida (1996) for inversion of surface-wave dispersion data.

The misfit function for the j-th individual (φ_j) was defined as the average of the root-mean square of the differences between observed phase velocities $U_o(T_i)$ and calculated ones $U_c(T_i)$, and can be expressed as follows (Yamanaka & Ishida 1996):

$$\varphi_j = \frac{1}{N}\sum_{i=1}^{N}\left(\frac{U_o(T_i) - U_c(T_i)}{\sigma(T_i)}\right)^2 \quad (3)$$

where N is the number of observed data and $\sigma(T_i)$ is the standard deviation of the observed phase velocity at period T_i.

We assumed that microtremors consist of the Rayleigh wave fundamental mode in the calculation of theoretical phase velocities in the inversion. The S-wave velocity and thickness of each layer were chosen as the model parameters to be estimated. The genetic algorithm was applied to an initial model of five layers, so the model parameters consisted of five velocities and four thicknesses, with each variable divided into 6-bit binary strings. The operation of the genetic algorithm was applied to the initial model, which had population size $(Q) = 20$. For parameters governing cross-over, mutation and elite selection, we refer to Yamanaka & Ishida (1996). The inversions were carried out for 20 iterations, after which it was assumed to have converged to an optimal solution.

Results and discussion

Estimation of the 1D S-wave velocity profile

Phase velocities were calculated at each site by applying the SPAC method to the vertical component of microtremor records. Each data record was divided into several time blocks, each with a duration of 163.84 s. The coherence of each pair of stations was selected, and only data with relatively good coherence were used to calculate the SPAC coefficients using equation (1). The phase velocity can be calculated from the SPAC coefficient through a Bessel function (equation 2). The variation in estimated phase velocity with respect to frequency defines the dispersion curve. The result of data processing in this study shows totally different dispersion curves at all sites for both phase velocity and frequency, as they are influenced by the varying local geology. In general, the phase velocity for deep layers is obtained in the frequency range of 0.2–5.0 Hz, but the phase velocities are different: in northern Jakarta, the phase velocity is lower than in southern Jakarta.

The estimation of 1D S-wave velocity profiles was conducted with an inversion technique using specified initial models that took into account geological constraints. In this study, a genetic algorithm was used on each dispersion curve at each site using the same parameters and constraints. By referring to geological information for the Jakarta area, the Klapanunggal Formation, which consists of limestones, marl and sandstones, was assumed to be the engineering bedrock. Furthermore, four sediment layers over the base rock were assumed, and a five-layer model was used for an initial model to be inverted.

An example of a selected microtremor data record is shown in Figure 3b for site KMAL, while Figure 4 shows the coherence curve for each distance between two stations, the fitting of the observed and calculated dispersion curves, and the estimated 1D S-wave velocity profile. The Indonesian National Standard SNI-03-1726-2012 (Badan Standardisasi Nasional 2012) defines engineering bedrock as a rock–soil layer that has a S-wave velocity (V_S) of >750 m s^{-1}. It can be identified from the 1D S-wave velocity profiles resulting from the inversion at each site. Hence, the engineering bedrock at site KMAL can be estimated to be the fifth layer at a depth of 649 m and with a S-wave velocity

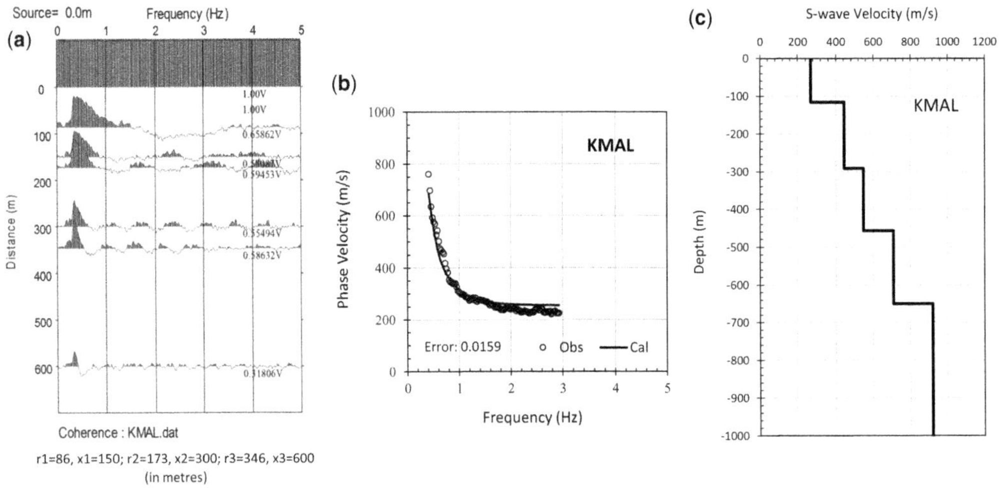

Fig. 4. An example of microtremor data processing at the KMAL site: (**a**) the coherence curve at each distance between two stations; (**b**) fitting of calculated and observed dispersion curves; and (**c**) the 1D S-wave velocity profile for the deep layer.

of 920 m s^{-1}. Examples of 1D S-wave velocity profiles at other sites are shown in Figure 5 for sites SLPI and CRKS.

One-dimensional S-wave velocity profiles at all observation sites were estimated using the same procedure, and the results for the depth of engineering bedrock throughout the Jakarta area are given in Figure 6. In general, the bedrock depth is in the range of 350–725 m, with the depth increasing towards northern Jakarta. The deepest part is in the Cilincing area of NE Jakarta and the shallowest part is in the south.

Construction of 2D and 3D S-wave velocity models

One-dimensional S-wave velocity profiles resulting from individual inversions describe layered models at each site, with pronounced differences between the different profiles for the S-wave velocities in

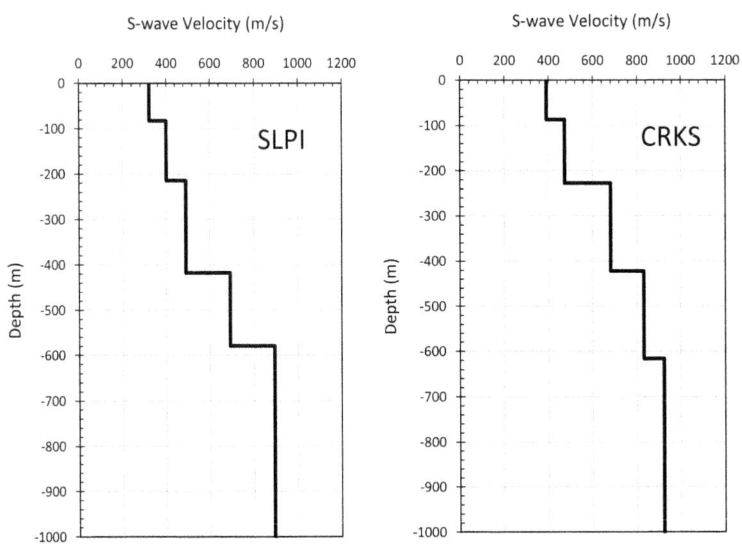

Fig. 5. One-dimensional S-wave velocity profile obtained from the first inversion at sites SLPI and CRKS.

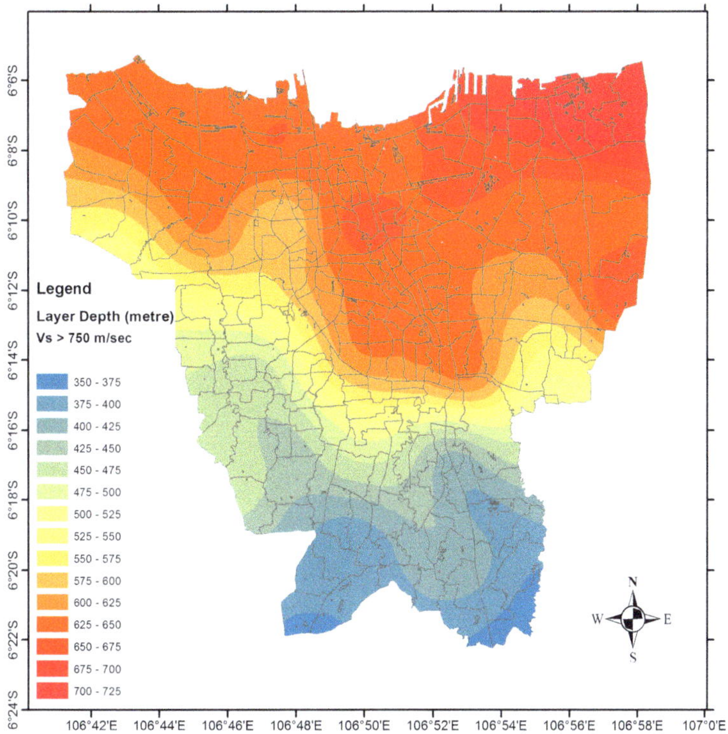

Fig. 6. Map of layer depth for $V_s > 750$ m s^{-1}.

each layer. Consequently, it is difficult to conduct stratigraphic correlation to construct 2D and 3D models. Therefore, a second inversion was conducted using initial models with constant velocity derived by averaging the S-wave velocities for each layer (Yamanaka & Yamada 2006). A sketch of stratigraphic correlation for all observation sites was required before determining the fixed S-wave

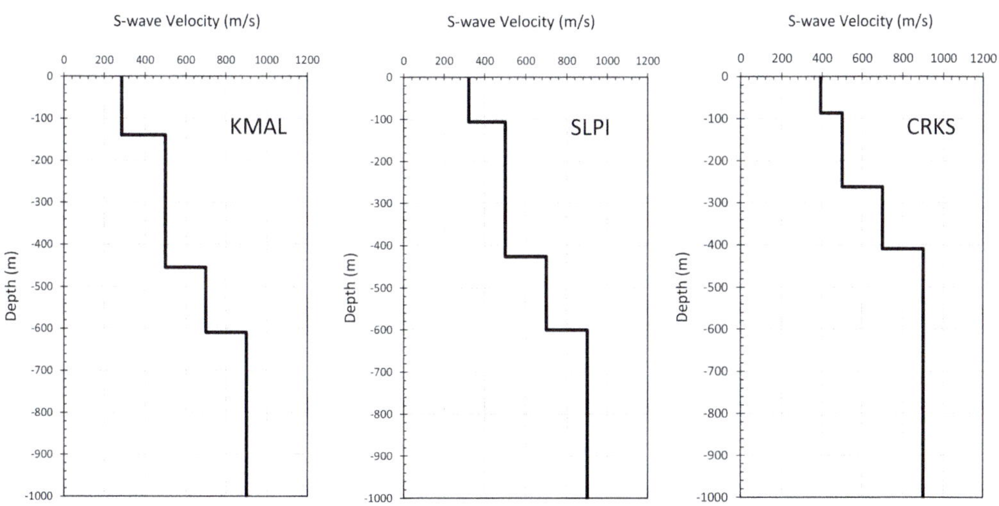

Fig. 7. One-dimensional S-wave velocity profile obtained from the second inversion at sites KMAL, SLPI and CRKS.

Table 1. *The location of microtremor array exploration and estimated layer depths*

No.	Site	Longitude (°)	Latitude (°)	Elevation (m)	Layer depths (m)			
					$V_s > 750$	$V_s = 500$	$V_s = 700$	$V_s = 900$
1	KMAL	106.702	−6.096	3	649	139	479	612
2	KPGD	106.713	−6.120	4	643	115	543	670
3	MGOT	106.704	−6.154	7	590	115	436	615
4	PALM	106.712	−6.179	7	535	102	414	565
5	PIKK	106.750	−6.113	4	631	137	531	668
6	CGKI	106.747	−6.147	5	644	116	481	636
7	KEMB	106.755	−6.179	7	614	96	429	615
8	SRNG	106.748	−6.202	11	535	82	363	530
9	PLMP	106.753	−6.225	14	490	105	315	546
10	PDBT	106.748	−6.264	29	460	102	312	549
11	PSJM	106.774	−6.287	34	469	109	307	493
12	ANKE	106.790	−6.137	5	645	136	491	648
13	TMNG	106.795	−6.167	7	581	126	404	563
14	SLPI	106.795	−6.199	14	580	106	450	600
15	SPRK	106.787	−6.232	17	481	86	295	568
16	FTMW	106.792	−6.254	27	461	85	256	513
17	PLHR	106.795	−6.276	34	437	106	283	534
18	LABU	106.795	−6.313	49	418	61	228	388
19	CNRE	106.789	−6.345	65	392	99	282	404
20	KRUK	106.795	−6.366	68	369	83	195	374
21	ANCL	106.840	−6.130	4	634	109	426	608
22	BMKG	106.842	−6.156	7	651	101	456	605
23	ISTN	106.824	−6.169	10	653	114	415	572
24	MRIA	106.835	−6.208	13	636	117	411	538
25	KNGN	106.829	−6.231	15	620	112	428	532
26	DRN3	106.827	−6.254	22	543	93	319	526
27	CPT	106.821	−6.283	36	462	101	277	468
28	DTAN	106.839	−6.287	37	465	122	273	513
29	KBGS	106.828	−6.312	49	394	72	201	384
30	ZPUR	106.829	−6.343	61	381	72	232	407
31	UI	106.828	−6.369	79	374	56	173	359
32	WRKS	106.882	−6.119	4	674	127	523	677
33	SNTR	106.878	−6.156	6	653	128	507	659
34	PMAS	106.877	−6.188	10	631	122	444	594
35	BUAR	106.873	−6.219	17	641	133	391	591
36	RWSR	106.882	−6.246	20	612	108	337	537
37	KRMT	106.875	−6.272	31	443	115	324	516
38	CNDT	106.864	−6.296	34	430	123	284	528
39	CRKS	106.880	−6.329	44	422	87	263	410
40	CBBR	106.879	−6.355	67	387	64	152	353
41	CLCG	106.925	−6.110	4	717	131	509	685
42	SKPR	106.918	−6.140	6	659	138	524	683
43	PLGD	106.918	−6.165	7	637	126	482	603
44	RADJ	106.926	−6.199	9	611	110	510	652
45	MLKS	106.928	−6.227	15	549	92	453	582
46	KPIN	106.922	−6.254	16	501	116	380	598
47	PDGD	106.902	−6.284	34	443	113	294	548
48	TMII	106.891	−6.301	41	395	101	267	495
49	JTSP	106.918	−6.335	44	373	71	201	370
50	PDRG	106.910	−6.361	70	365	64	165	399
51	MRND	106.959	−6.124	4	719	141	515	714
52	ROTN	106.956	−6.158	6	637	142	506	656
53	CKNG	106.949	−6.179	8	645	121	530	684
54	PLGB	106.956	−6.209	13	661	116	458	641
55	KLDL	106.920	−6.272	25	469	122	349	582

V_s, S-wave velocity in m s^{-1}.

layer velocity. Adjacent layers that have almost the same velocity were assumed to be one layer, and then the S-wave velocity of that layer was fixed, and a second inversion was undertaken to determine the depths of these constant-velocity layers. Based on the sketched stratigraphic correlation, the entirety of observation sites can be modelled with four layers that have fixed S-wave velocities of about 500, 700 and 900 m s^{-1} for the second, third and fourth layers, respectively.

The second inversion was conducted to determine the depths of the four layers with constant S-wave velocities for the second, third and fourth layers of 500, 700, and 900 m s^{-1}, respectively, while the depth was allowed to vary. Figure 7 shows examples of S-wave velocity profile resulting from second inversion at sites KMAL, SLPI and CRKS, where the engineering bedrock can be identified in the fourth layer ($V_s = 900$ ms^{-1}) at a depth of 612, 600 and 410 m respectively. The complete results of estimated layer depths for all sites are presented in Table 1.

Construction of 2D and 3D S-wave velocity models can be undertaken based on the 1D S-wave velocity profiles resulting from the second inversion for all observation sites: therefore, subsurface models now consist of four layers in which each layer has the same velocity. It is then easy to conduct stratigraphic correlation for the entire study area. A 2D S-wave velocity along the cross-section in a west–east direction is shown in Figure 8.

A–A′ depicts the subsurface model for central Jakarta in a west–east direction, where it can be seen that sedimentary thickness is increasing eastwards. A similar feature was indicated by Ridwan *et al.* (2013) in the west–east cross-section in northern Jakarta, however in the southern cross-section sediment thickness was relatively uniform. The cross-section in the north–south direction was shown in a previous study (Ridwan *et al.* 2014) to have sedimentary thickness increasing to the north. A constructed 2D S-wave velocity structure shows very clearly the position of the engineering bedrock: that is, the fourth layer, which has an S-wave velocity of 900 m s^{-1}.

The estimated 1D S-wave velocity profiles resulting from the second inversion for all the sites were also utilized to construct a 3D model to describe the morphology of the engineering bedrock beneath Jakarta city. Figure 9 shows the maps of layer depths for the second, third and fourth layers with S-wave velocities of 500, 700 and 900 m s^{-1}, respectively. The main trend is that the sedimentary layers in northern Jakarta are thicker than those in southern Jakarta, and that the bedrock depth is slanting to the NE.

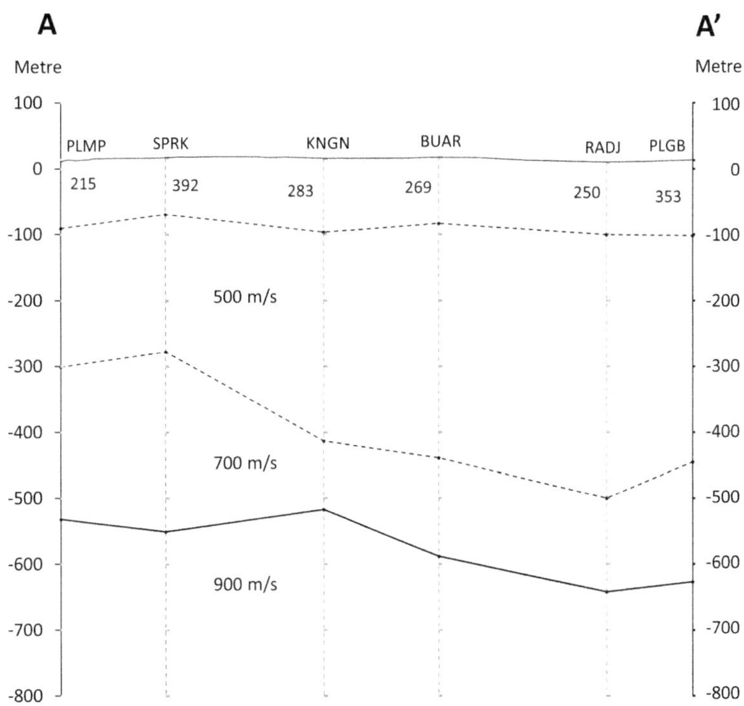

Fig. 8. A cross-section in a west–east direction from PLMP to PLGB (see Fig. 2).

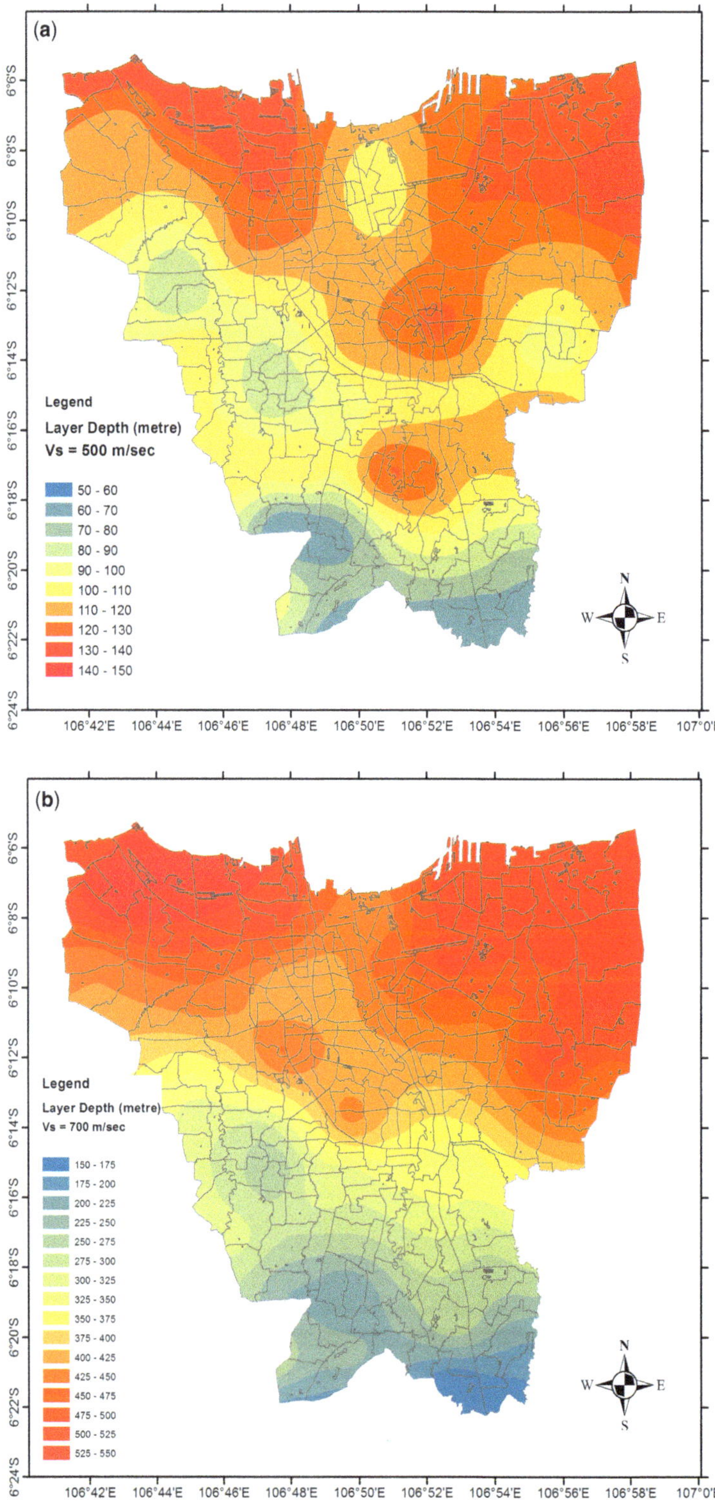

Fig. 9. Maps of layer depth for (**a**) $V_s = 500$ m s^{-1}. (**b**) $V_s = 700$ m s^{-1} and

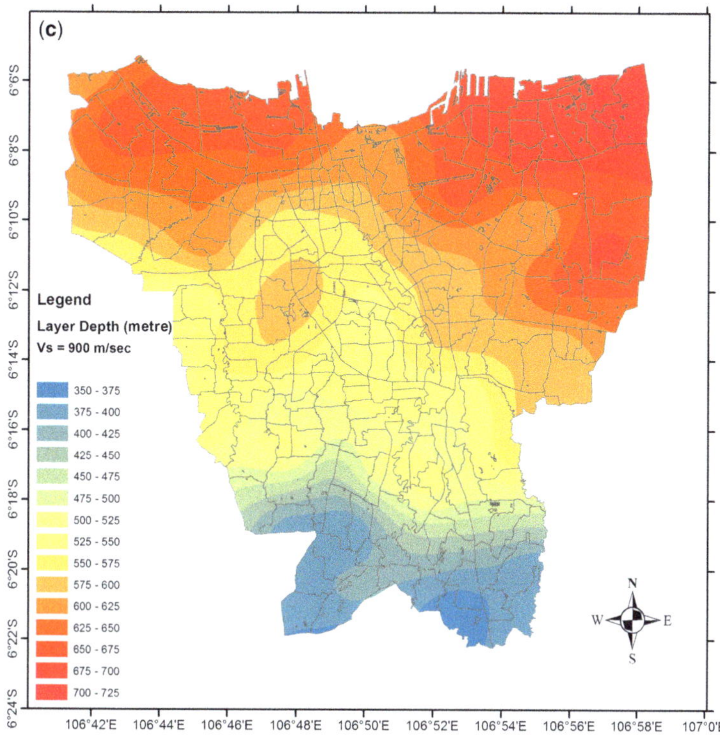

Fig. 9. (**c**) $V_s = 900\ \mathrm{m\ s^{-1}}$.

Verification by borehole data

The estimated S-wave velocity structures for deep layers have been obtained for all observation sites in this study. The result of the microtremor analysis shows that the sedimentary thickness over the bedrock exceeds 350–725 m, so it cannot be readily verified by geotechnical means. One deep

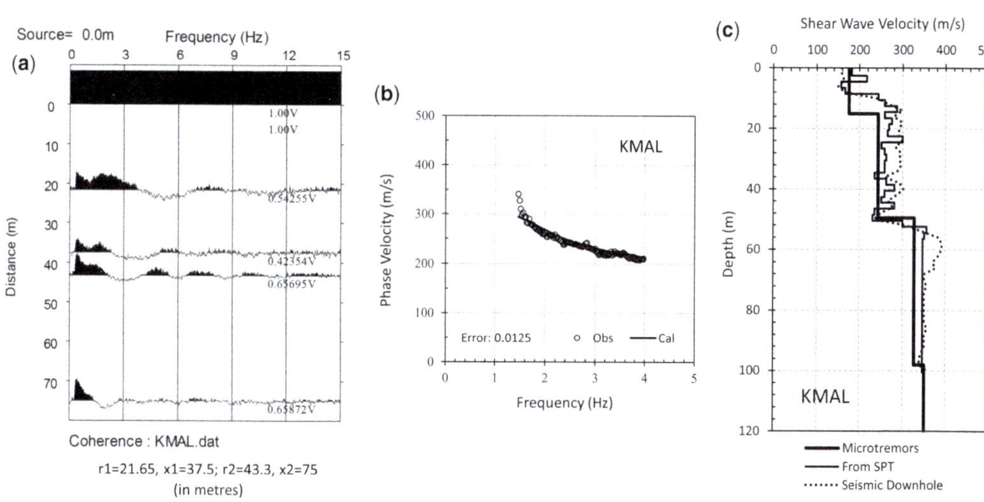

Fig. 10. (**a**) Coherence curve for a small array at the KMAL site; (**b**) fitting of calculated and observed dispersion curves; (**c**) comparison of S-wave velocity profiles obtained from microtremor arrays (thick line), SPT (thin line) and SDT (dot line) for the location at the KMAL site.

borehole was drilled at site KMAL, which involved standard penetration tests (SPTs) and seismic downhole tests (SDTs) to a depth of 100 m. The location of each borehole was less than 50 m from the corresponding microtremor observation site. In order to compare those results in terms of the S-wave velocity parameter, the SPT data were converted to the S-wave velocity by the averaging of some empirical formulae (Irsyam *et al.* 2008), i.e. from Imai & Tonouchi (1982), Imai (1977) and Ohta & Goto (1978).

As existing SPT and SDT data only reach a depth of 100 m, the microtremor data analysis down to a depth of 100 m was conducted by using dispersion curves derived from array sizes of 37.5 and 75 m, and inversion was applied to the four-layer initial model. Figure 10 shows a comparison of the S-wave velocity profiles obtained from microtremor array measurements, SPTs and SDTs. In general, a positive correlation is shown, especially for depths >25 m. A pronounced difference in velocity profiles appears at shallow depths (<25 m), presumably because the array size used for the microtremor survey was too large, so that only the low-frequency phase velocity of microtremors could be detected. Smaller array sizes are required in order to obtain high resolution for shallow layers.

Conclusions

Based on the results of a microtremor array analysis that was conducted at 55 surveyed sites, this study obtained a model for the subsurface geology beneath the city of Jakarta. The Indonesian National Standard SNI-03-1726-2012 (Badan Standardisasi Nasional 2012) defines engineering bedrock as having an S-wave velocity >750 m s^{-1}, and this can be identified directly on the 1D S-wave velocity profiles. Inversion results of the individual dispersion curves at all sites indicate that the resulting 1D S-wave velocity profiles have five layers with various depths and velocities. Engineering bedrock can be identified in the fourth and fifth layers. This result shows that the depth of engineering bedrock in northern Jakarta reaches 725 m, while in southern Jakarta it is at a depth of about 350 m. The morphology of engineering bedrock gradually increases northwards.

The 2D and 3D S-wave velocity models were obtained by correlation of the 1D S-wave velocity profiles resulting from the second inversion of the dispersion curves, in which the number and the velocities of the layers were set as constant, but depth was allowed to vary. The resulting 2D and 3D models consist of four layers, with S-wave velocities for the second, third and fourth layers being 500, 700, and 900 m s^{-1}, respectively. The thickness of sedimentary layers is shown to increase towards the north, particularly for the second layer. The fourth layer, which has an S-wave velocity of 900 m s^{-1}, was identified as engineering bedrock. The morphology of the bedrock is shown in Figure 9 and has a depth range of 350–725 m, slanting towards the north. The deepest part is in NE Jakarta (Cilincing and Marunda): this area is estimated to have higher seismic wave amplification than the southern part as a result of the sediment thickness.

Similar features are discernable in a statigraphic model for the Jakarta area derived from borehole data (Fachri *et al.* 2002). For example, although Fachri *et al.* (2002) estimated the depth to bedrock for the groundwater basin beneath Jakarta to be about 100 m in the south, in the north bedrock it could not be detected at depths less than 300 m. This trend in depth correlates positively with the result from the microtremor analysis.

The authors gratefully acknowledge the Research Institute for Human Settlement (RIHS), Ministry of Public Works, Republic of Indonesia, the seismic microzonation study team, and the Research Centre for Disaster Mitigation (PPMB), Institute of Technology of Bandung (ITB), for supporting data and assistance during this study. This research is part of the microzonation study for Jakarta supported by RIHS, ITB and local government.

References

AKI, K. 1957. Space and time spectra of stationary stochastic waves, with special reference to microtremors. *Bulletin of the Earthquake Research Institute*, **35**, 415–456.

ASTEN, M.W., DHU, T. & LAM, N. 2004. Optimised array design for microtremor array studies applied to site classification; comparison of results with SCPT logs. Paper 2903 presented at the 13th World Conference on Earthquake Engineering, 1–6 August 2004, Vancouver, BC, Canada.

BADAN STANDARDISASI NASIONAL 2012. *SNI-03-1726-2012. Guidelines of Earthquake Resistant Design for Building Structures*. Badan Standardisasi Nasional (National Standardization Agency of Indonesia), Jakata [in Indonesian].

EFFENDI,, A.C., KUSNAMA, & HERMANTO, B. 1998. *Geological Map of Bogor Quadrangles Jawa*. Geological Research and Development Centre, Ministry of Energy and Mineral Resources, Jakarta.

FACHRI, M., DJUHAENI,, HUTASOIT, L.M. & RAMDHAN, A.M. 2002. Stratigraphy and hydrostratigraphy of ground water basin for Jakarta. *Buletin Geologi*, **34**, 169–190 [in Indonesian].

IMAI, T. 1977. P- and S-wave velocities of the ground in Japan. *In*: *Proceedings of the 9th of International Conference Soil Mechanics and Foundation Engeneering*, Tokyo, 257–260.

IMAI, T. & TONOUCHI, K. 1982. Correlation of N-value with S-wave velocity and shear modulus. *In*: *Proceedings of the 2nd European Symposium on Penetrating*

Testing, 24–27 May 1982, Amsterdam, The Netherlands. A.A. Balkema, Rotterdam, 67–72.

Irsyam, M., Sahadewa, A. & Darjanto, H. 2008. *Soil Dynamics and Machine Foundation*. Institut Teknologi Bandung, Jawa Barat [in Indonesian].

Irsyam, M., Sengara, W. *et al*. 2010. *Summary: Development of Seismic Hazard Maps of Indonesia for Revision of Seismic Hazard Map in SNI-03-1726-2002*. Research Report Submitted to the Ministry of Public Works by Team for Revision of Seismic Hazard Maps of Indonesia.

Morikawa, H., Sawada, S. & Akamatsu, J.A. 2004. Method to estimate phase velocities of rayleigh wave using microseisms simultaneously observed at two sites. *Bulletin of the Seismological Society of America*, **94**, 961–976.

Ohta, Y. & Goto, N. 1978. Empirical shear wave velocity equation in terms of characteristic soil indexes. *Earthquake Engineering and Structural Dynamics*, **6**, 167–187.

Okada, H. 1998. *Microtremors as an Exploration Method. Geoexploration Handbook, Volume 2*. Society of the Exploration Geophysicists of Japan, Tokyo.

Okada, H. 2003. *The Microtremors Survey Method*. American Geophysical Union, Geophysical Monograph Series, **12**.

Padmosukismo, S. & Yahya, I. 1974. The basement configuration of the north west Java area. *In*: *Proceedings of the 3rd Annual Convention of the IPA, Volume 1*. Indonesian Petroleum Association, Jakarta, 129–152.

Ridwan, M., Widiyantoro, S., Irsyam, M. & Afnimar, 2013. Identification of engineering bedrock in jakarta by using array observations of microtremors. *In*: *3rd International Symposium on Earthquake and Disaster Mitigation 2013 (ISEDM 2013)*, Yogyakarta, Indonesia, 17–18 December 2013.

Ridwan, M., Afnimar,, Widiyantoro, S., Irsyam, M. & Yamanaka, H. 2014. Estimation of S-wave velocity structures by using microtremor array measurements for subsurface modeling in Jakarta. *Journal of Mathematical and Fundamental Sciences*, **46**, 313–326.

Turkandi, T., Sidarto,, Agustiyanto, D.A. & Hadiwodjoyo, M.M.P. 1992. *Geological Map of Jakarta and Kepulauan Seribu Quadrangles Jawa*. Geological Research and Development Centre, Ministry of Energy and Mineral Resources, Jakarta.

Yamanaka, H. & Ishida, H. 1996. Application of genetic algorithms to an inversion of surface wave dispersion data. *Bulletin of the Seismological Society of America*, **86**, 436–444.

Yamanaka, H. & Yamada, N. 2006. Estimation of 3D S-wave velocity model of Kanto basin, Japan, using Rayleigh wave phase velocity. *Bulletin of the Earthquake Research Institute*, **81**, 295–230.

Towards real-time earthquake impact alerting in Indonesia

SIGIT PRAMONO[1], TREVOR I. ALLEN[2]*, CRAIG BUGDEN[3], RAKHINDRO PANDHU[1], ISABELLA NINDYA[1], HADI GHASEMI[3] & MASTURYONO[1]

[1]*Geophysics Division, Badan Meteorologi, Klimatologi, dan Geofisika, Jakarta, Indonesia*

[2]*Pacific Geoscience Centre, Natural Resources Canada, Sidney, BC, Canada*

[3]*Community Safety and Earth Monitoring Division, Geoscience Australia, Symonston, ACT, Australia*

**Correspondence: trevor.allen@canada.ca*

Abstract: The Government of Indonesia has committed to deploying a network of 500 strong-motion sensors throughout the nation. The data from these sensors have the potential to provide critical near-real-time information on the level of ground shaking and potential impact from Indonesian earthquakes near communities. We describe the implementation of real-time 'ShakeMaps' within Indonesia's Agency of Meteorology, Climatology and Geophysics (BMKG). These ShakeMaps are intended to underpin real-time earthquake situational awareness tools.

The use of the new strong-motion network is demonstrated for two recent earthquakes in northern Sumatra: the 2 July 2013 M_w 6.1 Bener Meriah, Sumatra and the 10 October 2013 M_w 5.4 Aceh Besar earthquakes. The former earthquake resulted in 35 fatalities, with a further 2400 reported injuries. The recently integrated ShakeMap system automatically generated shaking estimates calibrated by BMKG's strong-motion network within 7 min of the Bener Meriah earthquake's origin, which assisted the emergency response efforts. Recorded ground motions are generally consistent with theoretical models. However, more analysis is required to fully characterize the attenuation of strong ground motion in Indonesia.

In the wake of large earthquakes around the world, it can take days to determine the full impact of the disaster. This is because communication and transport networks are often disabled, hampering the assessment and response to the emergency. For example, following the 27 May 2006 Yogyakarta, Indonesia earthquake, it took nearly to 3 days for the reported death toll to near the final estimate of almost 6000 from media reports (Earle *et al.* 2009). Relying solely on media reports to dispatch an appropriate level of emergency search and rescue would delay the humanitarian response, prolong suffering and potentially increase the death toll (Macintyre *et al.* 2006). Consequently, there is a critical need to rapidly assess the potential impacts of earthquakes in Indonesia, and elsewhere in the world. While these assessments can be, and often are, performed using theoretical models (Earle *et al.* 2009; Wald *et al.* 2011), near-real-time earthquake impact assessments can be significantly strengthened by a reliable network of real-time strong-motion sensors. These instruments can provide 'ground-truthed' measurements of observed ground shaking at locations near large earthquakes. Data from these instruments can subsequently be incorporated into impact models to help validate, and calibrate, real-time impact estimates to nearby communities.

Indonesia is the fourth most populous country in the world, with Java being the most densely populated island within the archipelago. The region is also one of the most seismically active, and suffers from frequent catastrophic earthquakes (Earthquake Engineering Research Institute 2006; Cosgrave 2007; So *et al.* 2008; Allen *et al.* 2009; Sengara *et al.* 2010). This, combined with a vulnerable building inventory (Pribadi *et al.* 2008), necessitates an Indonesian national earthquake response system that can provide rapid advice on the likely impacts of local and regional earthquakes.

Since the great 2004 M_w 9.1 Sumatra–Andaman Islands earthquake and tsunami, significant advances in safeguarding Indonesians from earthquake hazards have been realized. In particular, the development of the German–Indonesian Tsunami Early Warning System (GITEWS: Rudloff *et al.* 2009; Münch *et al.* 2011) has provided Indonesia's Agency of Meteorology, Climatology and Geophysics (Badan Meteorologi, Klimatologi, dan Geofisika (BMKG)) with the capability to rapidly identify potentially tsunamagenic earthquakes and to

From: CUMMINS, P. R. & MEILANO, I. (eds) 2017. *Geohazards in Indonesia: Earth Science for Disaster Risk Reduction*. Geological Society, London, Special Publications, **441**, 167–178.
First published online May 18, 2016, https://doi.org/10.1144/SP441.4

disseminate warnings to vulnerable coastal communities. The focus of GITEWS (also known as the Indonesian Tsunami Early Warning System (InaTEWS)) has focused on tsunami protection where the prompt dissemination of earthquake and tsunami information can save many lives in coastal communities. InaTEWS, while providing rapid earthquake source information, was not designed to assess the shaking-related impacts from Indonesian earthquakes (i.e. potential casualties and economic impacts through building failure).

Following the success of the InaTEWS programme, the Government of Indonesia has invested significant resources to develop a national strong-motion network of some 500 stations (Prakoso & Sukanta 2012). The data from these stations, coupled with data from additional donor-funded networks, have the potential to provide critical near-real-time information on the level of ground shaking and potential impact from Indonesian earthquakes near communities. The objective of this activity is to use these new data at BMKG and to develop a real-time 'ShakeMap' system.

ShakeMap is a product of the United States Geological Survey (USGS) Earthquake Hazards Program that provides near-real-time maps of ground motion and shaking intensity following significant earthquakes (Wald *et al.* 1999). The software ingests near-real-time peak ground motions from spatially distributed strong-motion instruments and interpolates these data with theoretical predictions to provide a grid of ground shaking of different intensity metrics (e.g. peak ground acceleration (PGA) and velocity (PGV), pseudo-spectral acceleration at 0.3, 1.0 and 3.0 s and instrumental intensities) (Worden *et al.* 2010). The primary use of ShakeMap is for the provision near-real-time situational awareness in regions affected by earthquakes. Originally designed for California, the system is now commonly being used by regional networks worldwide (e.g. Clinton *et al.* 2006; Ionescu *et al.* 2007; Michelini *et al.* 2008; Cauzzi *et al.* 2015). In regions with dense accelerometer networks, ShakeMaps are strongly influenced by actual observations of ground shaking. For those regions with little to no seismic instrumentation, ShakeMaps are largely based on empirical predictions using basic hypocentre and magnitude information. With the densification of strong-motion recording equipment throughout the Indonesian archipelago, the utility of these new data in calibrating ShakeMaps will provide more robust estimates of earthquake ground shaking, which can be used to refine rapid estimates of potential impact. This, in turn, will lead to improved, and more timely, decision-making around emergency management and response.

While the primary objective of this activity was to build a real-time ShakeMap situational awareness capability within BMKG, there were several important intermediary steps required to realize this objective. The activity was underpinned by:

- the development of standard operating procedures for the maintenance and management of the strong-motion network in both Bahasa Indonesian and English;
- improved management of strong ground-motion and instrument metadata;
- the installation of real-time data acquisition and processing software;
- the development of systematic data archival systems to support ongoing earthquake hazard research in Indonesia.

This contribution outlines the processes undertaken and challenges that were overcome to deliver a real-time ShakeMap system that makes full utility of Indonesian strong-motion data. We show recent examples where BMKG strong-motion data were used to provide rapid ShakeMaps to improve situational awareness following local earthquakes.

Indonesian strong-motion networks

Since the great 2004 M_W 9.1 earthquake and tsunami, there has been significant capital investment in Indonesia's earthquake monitoring infrastructure, both from the Government of Indonesia and from international donors (Fig. 1). Strong-motion instrumentation from three international donors is currently operational in Indonesia. These are the German, Chinese and Japanese provided networks. In addition, the Government of Indonesia has funded the acquisition of some 500 new strong-motion instruments (Prakoso & Sukanta 2012), of which almost 270 stations are now installed. These sites are being dispersed across the Indonesian archipelago, with 160 sensors co-located with InaTEWS broadband sensors using the existing Nanometrics Libra II systems for satellite-based communications. The remaining sites will be stand-alone strong-motion sensors used for ground-motion and engineering purposes.

While there are obvious benefits for the augmentation to Indonesia's earthquake monitoring capability, it also presented some challenges for the management of the strong-motion network and the archiving of the recorded data. The key challenges included:

- different systems used to capture and archive the strong-motion data from the different networks;
- strong-motion data not being routinely archived in some cases;
- data on dial-up communication systems with downloads having to be triggered manually (often only during office hours);

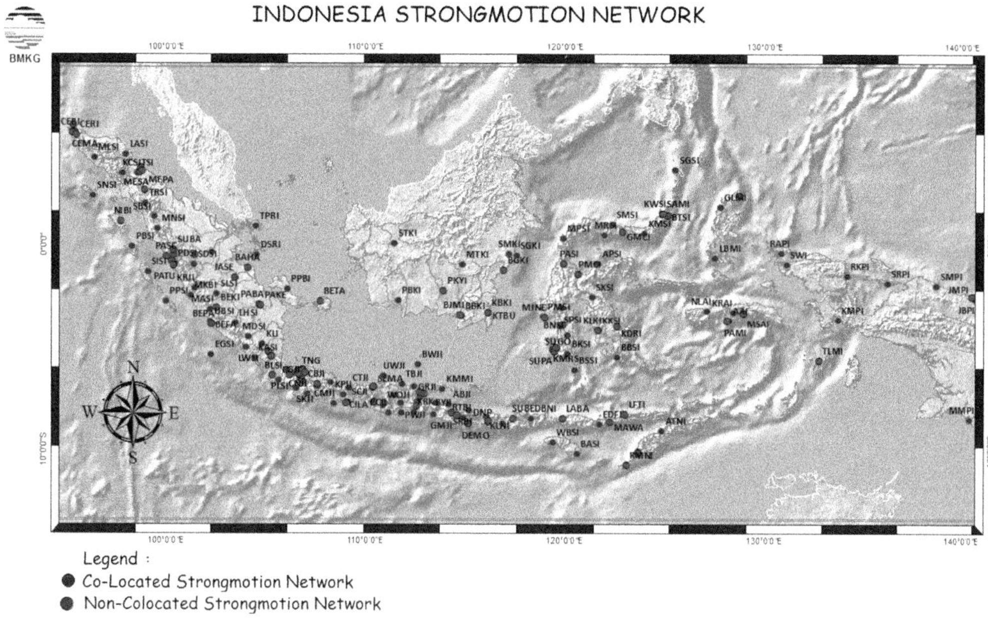

Fig. 1. Location of strong-motion instruments in Indonesia. The co-located strong-motion network are located together with InaTEWS broadband stations.

- limited information on equipment configuration of each site;
- little or no documentation on the donor-provided systems;
- no means for BMKG staff to easily monitor the operation and the state of health of the stations in real time.

Figure 2 shows a diagram of the data flow from the strong-motion networks prior to the work conducted herein. Through discussions between the authors of this manuscript, consensus decisions were made to redesign the data acquisition and storage architecture within BMKG. This effort will be discussed in more detail in the next section.

System architecture and strong-motion data consolidation

Prior to any work being undertaken, the major decision on the data-collection system to be used had to be made. Because InaTEWS is built on the SeisComP3 data-collection and processing system, it was decided to preserve the use of this system for the strong-motion networks and take advantage of BMKG's in-house expertise. SeisComP3 is a powerful and free software package for seismological data acquisition and real-time data exchange (Weber *et al.* 2007). It provides excellent graphical user interface capabilities, which have utility for

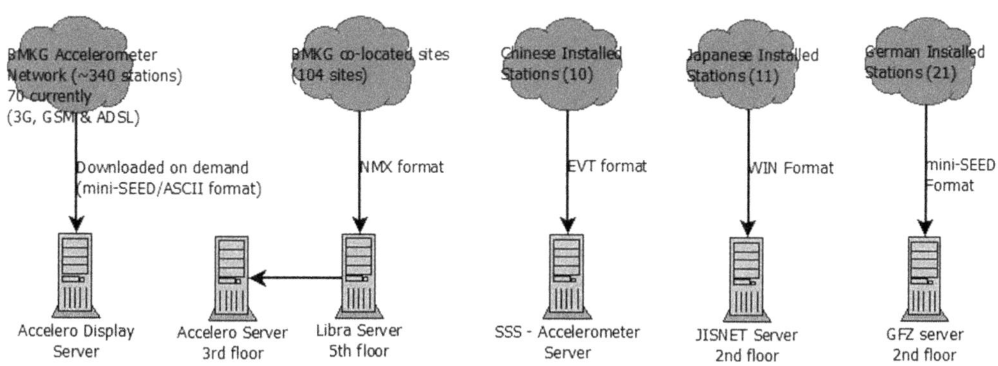

Fig. 2. BMKG strong-motion network servers prior to the redesign of the system architecture.

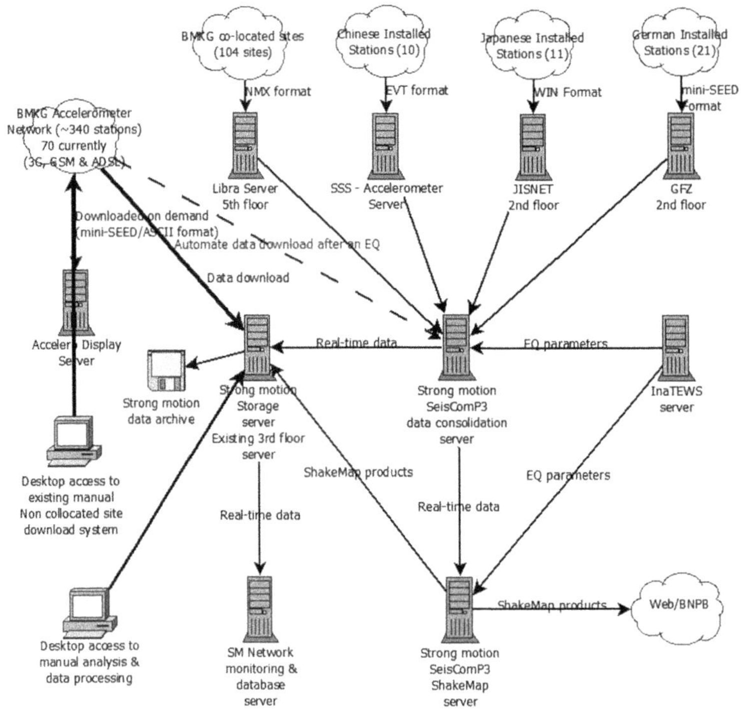

Fig. 3. Proposed BMKG strong-motion network server design. New servers are the data-consolidation server, the ShakeMap server and the database server.

real-time waveform inspection and quality assessment. The ability to monitor real-time data streams was a desirable asset for this activity.

Once the decision to use SeisComP3 was made, the system architecture for the data flow was designed. Figure 3 shows the proposed system configuration, based on consensus decisions by the authors. This design consolidates all of the strong-motion data into a single server. Data are then pushed (instantaneously) into a ShakeMap server, as well as the storage server and a network-monitoring server. This system design enables BMKG to better monitor the stations because all data are being streamed to one place, and systems to automatically monitor stations can be developed. The design also includes an archive system to prevent the loss of data as the independent network disks reach capacity. New servers were purchased by BMKG with Government of Indonesia funds and delivered in December 2011. Once the servers arrived, work commenced on installing software and configuring the two main servers: the data consolidation server and the ShakeMap server.

With the cost of modern strong-motion instrumentation and data storage decreasing significantly in the last decade, seismic network operators now have the potential to capture vast amounts of data. The challenge for BMKG, however, was how to telemeter these data in real time. Many of BMKG's stations necessitated the use of dial-up communications because of limitations in bandwidth available on traditional telecommunications networks. The limitations in communications bandwidth also limited the data sample rate that could be telemetered, which, in many cases, were set at 40 samples per second (samples/s). However, the rapidly increasing penetration of Indonesia's mobile communication infrastructure is now allowing for better real-time monitoring solutions. Throughout the duration of this project, BMKG recognized the importance of high sample-rate strong-motion data telemetry in real time. Consequently, communications on the BMKG-installed network (the non-co-located stations) were all migrated to real-time GSM communications with three-channel data being telemetered in real time at 100 samples/s. Strong-motion stations in the pre-existing co-located network using Nanometrics Libra II systems were also migrated to 100 samples/s.

An important requirement for incorporating recorded strong-motion data into ShakeMaps is to ensure data from each instrument is calibrated appropriately. It was found in some cases that the station calibration information for some networks was not documented. Through this project, missing station metadata were gathered and a robust station metadata management process was established using the Portable Data Collection Center (PDCC) software. The PDCC enables seismic network operators to describe, maintain and exchange their instrument metadata using the FDSN SEED format (Casey 2012). This software is easy to use and can output calibration files for use in SeisComP3.

Processing strong-motion data in real time

At a minimum, ShakeMap requires earthquake origin information (i.e. time, hypocentral location and magnitude) to produce maps of estimated peak ground shaking. These maps can be produced in the absence of any strong-motion data. While the strong-motion SeisComP3 server can be configured to perform automated locations to provide this information, it was determined that the ShakeMaps should be based on the official locations and magnitudes calculated by InaTEWS, which operates on a closed network independent of the strong-motion network. Consequently, the ShakeMap server was configured so that waveform data are only retrieved from the SeisComP3 server after an earthquake trigger has been received from InaTEWS. This data-retrieval process was initiated to reduce bandwidth across the BMKG network and to minimize data latency issues.

For use in ShakeMap, peak ground-motion values (PGA, PGV, 5% damped pseudo-spectral acceleration at 0.3, 1.0 and 3.0 s, and instrumental intensities) must be calculated from continuous data streams in near real-time. In collaboration with Eidgenössische Technische Hochschule Zürich (ETHZ) and Geological and Nuclear Sciences New Zealand, a waveform parameterization module (*scwfparam*) was developed for SesisComP3 to extract these peak ground motions from continuous real-time data (Cauzzi *et al.* 2013). Through this collaboration, ETHZ proposed to engage gempa (http://www.gempa.de), the developers of SeisComP3, to develop the module to ensure consistent real-time data-handling protocols. This module was delivered and testing has begun on its effectiveness at delivering near-real-time strong-motion parameters for ShakeMap. Examples of the utility of *scwfparam* are shown below.

ShakeMap configuration

While Indonesia is one of the most seismically active countries in the world, there have been few advances in developing attenuation relationships that estimate the level of ground shaking from

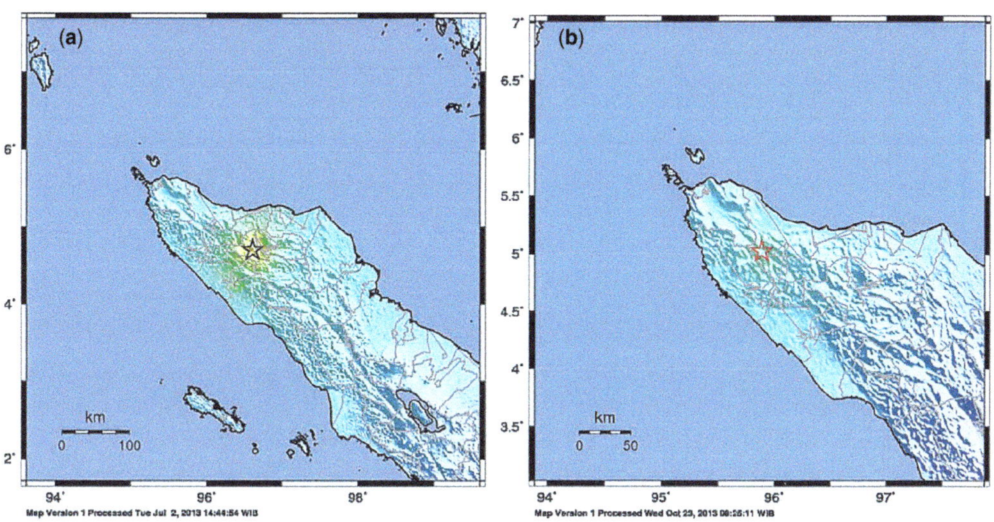

PERCEIVED SHAKING	Not felt	Weak	Light	Moderate	Strong	Very strong	Severe	Violent	Extreme
POTENTIAL DAMAGE	none	none	none	Very light	Light	Moderate	Mod./Heavy	Heavy	Very Heavy
PEAK ACC.(%g)	<0.05	0.3	2.8	6.2	12	22	40	75	>139
PEAK VEL.(cm/s)	<0.02	0.1	1.4	4.7	9.6	20	41	86	>178
INSTRUMENTAL INTENSITY	I	II-III	IV	V	VI	VII	VIII	IX	X+

Fig. 4. ShakeMap computed following (**a**) the 2 July 2013 M_w 6.1 Bener Meriah, Sumatra; and (**b**) the 10 October 2013 M_w 5.4 Aceh Besar, Sumatra earthquakes.

earthquakes. In addition, there has been very little work carried out in assessing the appropriateness of existing ground-motion prediction equations (GMPEs). In the past, this has been due to the limited strong-motion recording networks in Indonesia. Recent work by Rudyanto (2013) compiles existing Indonesian strong-motion data and attempts to provide an indication of the preferred GMPEs that could be used for hazard studies in Indonesia. However, the results of Rudyanto (2013) are presently inconclusive because the dataset used was comprised of too few data within the magnitude and distance range of engineering significance. As a consequence, we chose to use the Zhao *et al.* (2006) suite of GMPEs developed from strong-motion data from Japan. The Zhao *et al.* (2006) models are available for both shallow crustal and subduction zone earthquakes. This choice ensures continuity between ground-shaking levels of both shallow and deep events. That is, the GMPE used by ShakeMap is dependent on the hypocentral depth of the earthquake. Consequently, relatively minor changes in hypocentral depth can have dramatic effects on the estimated level of ground shaking if using GMPEs developed by different authors. As more data become available, it is expected that the attenuation of strong ground motions in Indonesia will be better characterized. Consequently, analogue regions of similar attenuation could be identified from which to base ShakeMap ground-shaking hazard. In time, data collected by BMKG's strong-motion networks could yield GMPEs developed specifically for Indonesia from Indonesian data, and these should be used for real-time ShakeMaps.

Site amplification factors for ground-motion estimates are based on the method of Wald & Allen (2007), which uses topographical gradient as a proxy for determining time-averaged shear-wave velocity in the upper 30 m of the crust (V_{S30}).

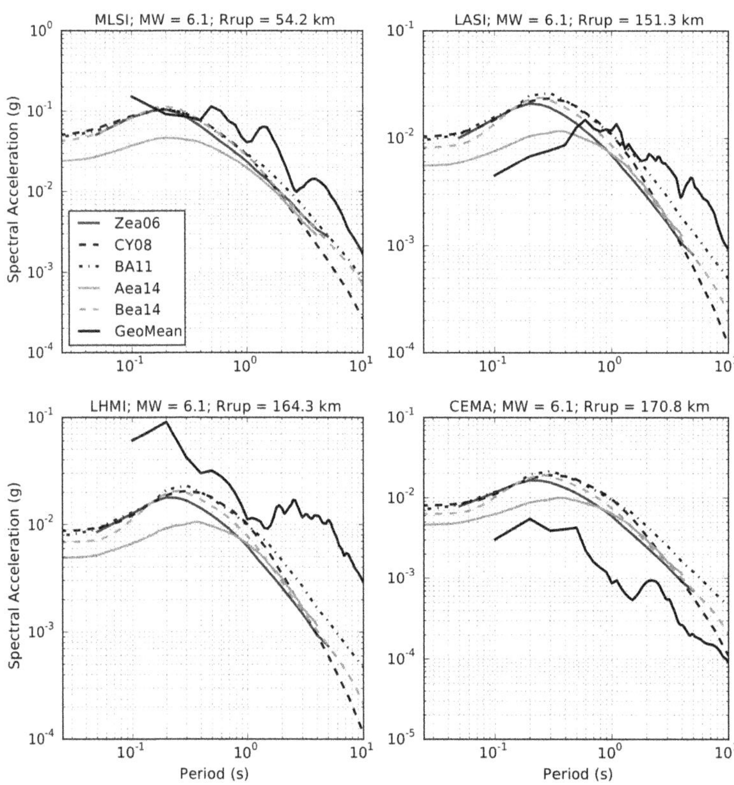

Fig. 5. Comparison of the geometric mean of 5% damped pseudo-spectral acceleration of the horizontal components from the 2 July 2013 M_w 6.1 Bener Meriah, Sumatra earthquake against several commonly used ground-motion prediction equations: Zhao *et al.* (2006: Zea06), Chiou & Youngs (2008: CY08), Atkinson & Boore (2011: BA11), Akkar *et al.* (2014: Aea14) and Boore *et al.* (2014: Bea14). The ground motions were recorded on the accelerographs located at Meulaboh (MLSI), Langsa Aceh (LASI), Lhouksemawe (LHMI) and Matai Aceh (CEMA). Ground-motion predictions assume a V_{S30} of 560 m/s (i.e. NEHRP site class C) at the specified rupture distance, R_{rup}.

Despite the method's limitations (Wald & Allen 2007; Allen & Wald 2009), this method has been demonstrated to provide robust estimates of seismic site conditions in the absence of detailed geotechnical information (e.g. Harmandar *et al.* 2007; Wills & Gutierrez 2008; Silva *et al.* 2011; Thompson *et al.* 2011). For the application of the Wald & Allen (2007) method, we use the Shuttle Radar Topography Mission (Farr & Kobrick 2000) 30 arc-second coverage for Indonesia.

The 2 July 2013 M_w 6.1 Bener Meriah, Sumatra Earthquake

A deadly magnitude 6.1 earthquake occurred in Aceh province (4.698° N, 96.687° E), near Bener Meriah, Sumatra at 14:37 WIB (7:37 UTC) on 2 July 2013. Thirty-five people perished in the event, with over 2400 reported injuries (BNPB 2014). The Aceh earthquake was recorded on BMKG's strong-motion network. These data were used in near real-time to improve ShakeMap ground shaking estimates (Fig. 4a). Following the location of the earthquake and the computation of the response spectral parameters from the real-time data, the first version of the ShakeMap was issued at 14:44 WIB (local time), 7 min after the earthquake origin. This information was immediately transferred to Indonesia's National Emergency Management Agency (BNPB), whereupon the potential impact was assessed and emergency response actions were initiated.

Example response spectra, as calculated by *scwfparam* are shown for four sites within BMKG's network (Fig. 5). In total, seven strong-motion instruments within 200 km recorded the earthquake. The attenuation of the ground motion acceleration is plotted against distance for 0.2 and 1.0 s spectral acceleration (Fig. 6). We have assumed that the typical site condition for the BMKG strong-motion network is approximately equivalent to NEHRP site class C (Building Seismic Safety Council 2004) with a V_{S30} of 560 m s^{-1}. Overall, ground motions recorded from this earthquake are approximately consistent with ground motions estimated from commonly used active crustal GMPEs (e.g. Zhao *et al.* 2006; Chiou & Youngs 2008; Atkinson & Boore 2011; Akkar *et al.* 2014; Boore *et al.* 2014). Table 1 lists peak ground accelerations from the BMKG strong-motion network for the Bener Meriah earthquake.

The 22 October M_w 5.4 Aceh Besar, Sumatra Earthquake

On 22 October 2013 at 12:40 WIB (05:40 UTC), a moderate magnitude M_w 5.4 earthquake occurred near Aceh Besar, Sumatra (5.103° N, 95.971° E). The earthquake had few consequences and occurred near the junction of the Aceh and Seulimeum segments of the Sumatran Fault System (SFS: Sieh & Natawidjaja 2000). Nevertheless, the earthquake was well recorded across BMKG's strong-motion network, which was used to calibrate the event's

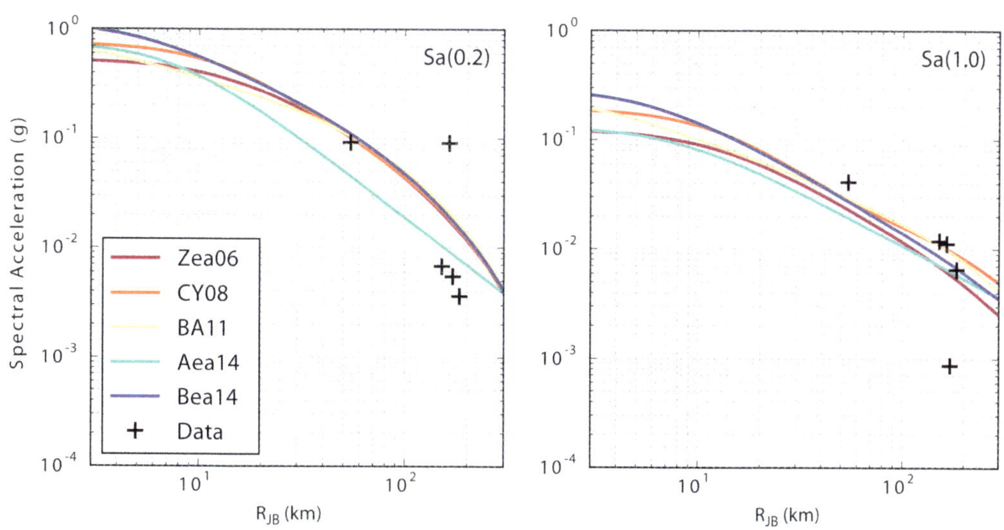

Fig. 6. Attenuation of the geometric mean of 5% damped pseudo-spectral acceleration with Joyner–Boore distance (R_{JB}) of the horizontal components from the 2 July 2013 M_w 6.1 Bener Meriah, Sumatra, earthquake plotted against common GMPEs (see the Fig. 5 caption for details). Ground-motion predictions assume a V_{S30} of 560 m s^{-1}.

Table 1. *Peak ground accelerations recorded from the 2 July 2013 M_w 6.1 Bener Meriah, Sumatra, earthquake*

Station	Station code	Latitude (°N)	Longitude (°E)	PGA (cm s^{-2})			R_{hyp} (km)
				Vertical	North	East	
Meulaboh	MLSI	4.27	96.40	19.893	45.890	48.792	54.16
Lhouksemawe	LHMI	4.46	97.97	18.159	30.730	20.622	164.33
Sta Klim Indrapuri Aceh	CERI	5.40	95.46	4.928	10.566	9.327	149.63
Sta Met Blang Bintang Aceh	CEBI	5.52	95.42	3.540	8.960	9.738	161.22
Langsa Aceh	LASI	4.47	97.95	4.488	8.172	8.202	151.28
Sta Geof Matai Aceh	CEMA	5.50	95.30	4.039	7.227	7.241	170.82
Sta Mar Belawan Medan	MEBA	3.80	98.70	–	4.110	5.162	252.78
Sta Klim Sampali Medan	MESA	3.62	98.79	1.641	5.146	−4.865	270.53
Sta Met Polonia Medan	MEPO	3.56	98.68	–	3.469	−2.532	262.05
Kutacane	KCSI	3.52	97.77	1.650	3.093	2.067	183.94
Sinabang	SNSI	2.41	96.33	1.345	2.292	2.062	256.43
Tuntungan	TSI	3.50	98.56	1.093	2.089	2.209	254.70
Mandailing Natal	MNSI	0.80	99.58	0.118	0.337	0.403	544.58
Sta Met Binaka Gn. Sitoli	NIBI	1.16	97.70	0.300	0.735	0.895	410.97
Sta Geof Parapat Medan	MEPA	2.70	98.92	0.133	0.226	0.174	339.92
Kerinci	KRJI	−2.09	101.46	0.148	0.225	0.187	926.50
Gunung Sitoli	GSI	1.30	97.58	0.144	0.195	0.202	391.93
Aek Godang	SBSI	1.55	98.89	0.131	0.413	0.302	431.32
Mukomuko	MKBI	−2.45	101.24	0.123	0.085	0.113	945.29
Saibi Mentawai Sumbar	SISI	−1.33	99.09	0.081	0.116	0.130	723.49
Pulau Pagai Mentawai	PPSI	−2.76	100.01	0.022	0.043	–	910.37
Sungai Dareh Sumbar	SDSI	−0.93	101.43	0.020	0.035	0.034	822.75

R_{hyp}, hypocentral distance.

ShakeMap (Fig. 4b). In Figure 7, example response spectra are shown for four sites within BMKG's network. In total, four strong-motion instruments within approximately 100 km recorded the earthquake. In Figure 8, the attenuation of the ground-motion acceleration is plotted against distance for 0.2 and 1.0 s spectral acceleration. Again, we observe that ground motions recorded from this earthquake are generally consistent with those estimated from commonly used ground-motion models. Table 2 lists peak ground accelerations from the BMKG strong-motion network for the Aceh Besar earthquake.

Future directions

Herein, we have provided a summary of the advances being made towards strong-motion monitoring in Indonesia and its subsequent benefits in generating ShakeMaps, which provide a rapid indication of spatial distribution of strong ground shaking from earthquakes relative to major population centres. The logical next progression is to provide rapid estimates of earthquake losses (e.g. human casualties and economic loss). Global systems, such as the USGS Prompt Assessment of Global Earthquakes for Response (PAGER: Earle *et al.* 2009), already provide estimates of loss based on regional fatality and loss models (Jaiswal & Wald 2010, 2013; Jaiswal *et al.* 2011). However, global systems, such as PAGER, are unlikely to fully capture the spatial distribution of a national population and building inventory. Consequently, there is scope for an Indonesian-based solution to provide more accurate details of population distribution and other elements at risk. Tools such as the Indonesia Scenario Assessment for Emergencies (InaSAFE: 2013) now allow for a variety of users – from scientists to local and national governments and non-profit agencies – to evaluate the potential impacts from natural disasters in the region of their choice. The InaSAFE programme has been endorsed by BNPB, and efforts are underway to map inventories of people and buildings for the greater Jakarta region and other major centres in Indonesia. Together with Indonesian-specific casualty and building fragility models (e.g. Sengara *et al.* 2013), the inclusion of locally derived inventory models will eventually yield more reliable estimates of earthquake impact when combined with data-driven ShakeMaps.

Looking further into the future, the densification of strong-motion networks in Indonesia will yield even more gains in rapidly assessing emergency situations and responding accordingly. Thereby,

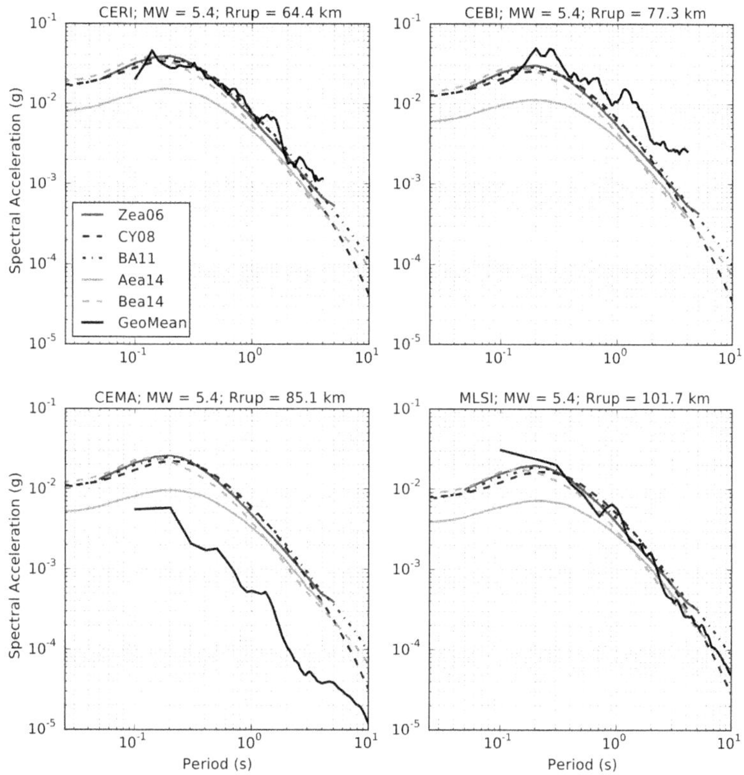

Fig. 7. Comparison of the geometric mean of 5% damped pseudo-spectral acceleration of the horizontal components from the 10 October 2013 M_w 5.4 Aceh Besar, Sumatra earthquake against several commonly used ground-motion prediction equations (see the Fig. 5 caption for details). The ground motions were recorded on the accelerographs located at Klim Indrapuri Aceh (CERI), Met Blang Bintang Aceh (CEBI), Matai Aceh (CEMA) and Meulaboh (MLSI). Ground-motion predictions assume a V_{S30} of 560 m/s (i.e. NEHRP site class C) at the specified rupture distance, R_{rup}.

reducing response times and potentially saving lives. Furthermore, data recorded by these networks will inform national hazard assessments that underpin Indonesia's building regulations and standards.

Concluding remarks

Since the great 2004 M_w 9.1 Sumatra–Andaman Islands earthquake and tsunami, the Government of Indonesia has invested significantly towards the development of a national strong-motion network of some 500 stations (Prakoso & Sukanta 2012). Once complete, the data from these stations, coupled with the data from additional donor-funded networks, have the potential to provide critical near-real-time information on the level of ground shaking and potential impact from Indonesian earthquakes near communities using the USGS ShakeMap software (Wald *et al.* 1999).

In developing this nascent capability, several challenges had to be overcome. First, different networks were being used to capture strong-motion data in Indonesia. Each network had varying levels of documentation, and data from each network was being stored on different servers. In some cases, these data were not being archived. Through this activity, we have developed standard operating procedures for the management of Indonesia's strong-motion network and data archives. Communications between sites and BMKG's national observatory in Jakarta have been significantly enhanced, allowing access to strong-motion data streams in near real-time. These data are now consolidated onto a single SeisComP3 data server, ensuring a consistent data format.

The addition of the *scwfparam* module, which interfaces between SeisComP3 and ShakeMap, now allows the use of real-time strong-motion data to calibrate ground-shaking estimates with ground-truthed observations. Being able to better

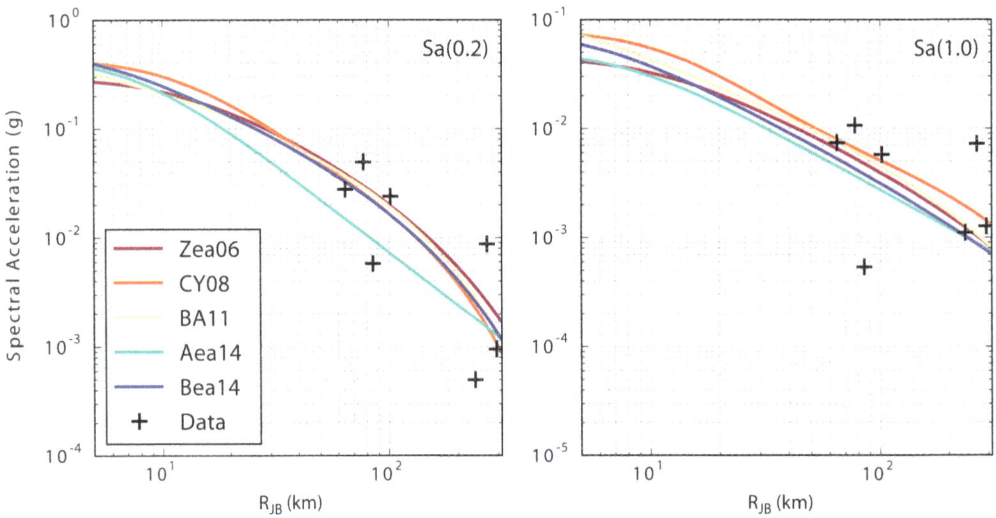

Fig. 8. Attenuation of the geometric mean of 5% damped pseudo-spectral acceleration with Joyner–Boore distance (R_{JB}) of the horizontal components from the 10 October 2013 M_w 5.4 Aceh Besar, Sumatra, earthquake plotted against common GMPEs (see the Fig. 5 caption for details). Ground-motion predictions assume a V_{S30} of 560 m s^{-1}.

characterize ground shaking following earthquakes will improve estimates of potential impact on Indonesian communities. This capability will lead to improved decision-making for emergency management and response, and therefore has the potential to save many lives following earthquake disasters in Indonesia.

Finally, the routine archiving of instrument-calibrated strong-motion data recorded on Indonesia's strong-motion networks will benefit future Indonesian and global earthquake hazard research. Data analysed from the two events discussed herein suggest that earthquakes in the northern Sumatra region yield ground motions that are generally consistent with those expected from common active crustal GMPEs for a given magnitude event. However, it is difficult to determine the most appropriate model to use in the Sumatra region from these two events alone. More rigorous analyses of data from these and additional earthquakes in the future will enable the development of Indonesian-specific GMPEs that will improve real-time situational

Table 2. *Peak ground accelerations recorded from the 10 October 2013 M_w 5.4 Aceh Besar, Sumatra, earthquake*

Station	Station code	Latitude (°N)	Longitude (°E)	PGA (cm s^{-2})			R_{hyp} (km)
				Vertical	North	East	
Sta Met Blang Bintang Aceh	CEBI	5.522	95.416	4.557	13.358	15.232	77.3
Sta Geof Matai Aceh	CEMA	5.496	95.296	4.987	7.072	7.210	85.1
Sta Klim Indrapuri Aceh	CERI	5.404	95.464	6.043	9.326	12.924	64.4
Kutacane	KCSI	3.522	97.772	0.253	0.585	0.370	267.2
Langsa Aceh	LASI	4.474	97.951	0.467	0.712	0.701	236.9
Sta Mar Belawan Medan	MEBA	3.8	98.7	–	0.327	0.327	340.2
Sta Met Polonia Medan	MEPO	3.563	98.677	–	0.343	0.350	349.2
Sta Klim Sampali Medan	MESA	3.62	98.793	–	0.357	0.445	357.9
Meulaboh	MLSI	4.267	96.404	4.554	8.688	10.568	101.7
Sta Met Binaka Gn. Sitoli	NIBI	1.164	97.704	0.051	0.086	0.112	473.1
Aek Godang	SBSI	1.554	98.889	0.015	0.046	0.034	508.9
Sinabang	SNSI	2.409	96.327	0.375	0.554	0.262	294.0
Tuntungan	TSI	3.5012	98.5645	0.120	0.250	0.283	341.5

R_{hyp}, hypocentral distance.

awareness and earthquake hazard assessments in Indonesia.

We acknowledge the Australian Agency for International Development (AusAID) through the Australia–Indonesia Facility for Disaster Reduction and BNPB, who contributed towards funding this work. In particular, the authors would like to acknowledge assistance from Trevor Dhu, Ole Neilson and Radhietya Hadikusuma. We thank Bruce Worden for provision of support in the installation of ShakeMap at BMKG and his review of this manuscript. We also thank John Clinton, Carlo Cauzzi and Jan Becker for their efforts in leading the development of the *scwfparam* module. Comparisons of Indonesian ground-motion data with modern GMPEs use codes supplied through the Openquake Hazard Library (hazardlib: Pagani *et al.* 2014). We thank John Cassidy for providing an internal review of this manuscript. Finally, Geoscience Australia authors publish with the permission of the Chief Executive Officer.

References

Akkar, S., Sandikkaya, M.A. & Bommer, J.J. 2014. Empirical ground-motion models for point- and extended-source crustal earthquake scenarios in Europe and the Middle East. *Bulletin of Earthquake Engineering*, **12**, 359–387, https://doi.org/10.1007/s10518-013-9461-4

Allen, T.I. & Wald, D.J. 2009. On the use of high-resolution topographic data as a proxy for seismic site conditions (V_{S30}). *Bulletin of the Seismological Society of America*, **99**, 935–943, https://doi.org/10.1785/0120080255

Allen, T.I., Marano, K.D., Earle, P.S. & Wald, D.J. 2009. PAGER-CAT: a composite earthquake catalog for calibrating global fatality models. *Seismological Research Letters*, **80**, 57–62, https://doi.org/10.1785/gssrl.80.1.57

Atkinson, G.M. & Boore, D.M. 2011. Modifications to existing ground-motion prediction equations in light of new data. *Bulletin of the Seismological Society of America*, **101**, 1121–1135, https://doi.org/10.1785/0120100270

BNPB. 2014. *Data and Informasi Bencana Indonesia.* Badan Nasional Penanggulangan Bencana (Indonesian National Disaster Management Agency), Jakata, http://dibi.bnpb.go.id/DesInventar/simple_data.jsp

Boore, D.M., Stewart, J.P., Seyhan, E. & Atkinson, G.M. 2014. NGA-West 2 equations for predicting PGA, PGV, and 5%-damped PSA for shallow crustal earthquakes. *Earthquake Spectra*, **30**, 1057–1085, https://doi.org/10.1193/070113EQS184M

BUILDING SEISMIC SAFETY COUNCIL. 2004. *NEHRP Recommended Provisions for Seismic Regulations for New Buildings and other Structures, 2003 Edition.* **FEMA 450**. Building Seismic safety Council, Washington, DC.

Casey, R. 2012. *Portable Data Collection Center (PDCC) v3.8 User Manual.* Incorporated Research Institutions for Seismology (IRIS), Washington, DC.

Cauzzi, C., Clinton, J., Becker, J. & Kästli, P. 2013. scwfparam: a tool for rapid parameterisation of ground motions and input to ShakeMap in SeisComP3. *Seismological Research Letters*, **84**, 356, https://doi.org/10.1785/0220130011

Cauzzi, C., Edwards, B. *et al.* 2015. New predictive equations and site amplification estimates for the next-generation Swiss ShakeMaps. *Geophysical Journal International*, **200**, 421–438, https://doi.org/10.1093/gji/ggu404

Chiou, B.S.-J. & Youngs, R.R. 2008. An NGA model for the average horizontal component of peak ground motion and response spectra. *Earthquake Spectra*, **24**, 173–215, https://doi.org/10.1193/1.2894832

Clinton, J.F., Cua, G., Huérfano, V., von Hillebrandt-Andrade, C.G. & Martínez-Cruzado, J. 2006. The current state of seismic monitoring in Puerto Rico. *Seismological Research Letters*, **77**, 532–543.

Cosgrave, J. 2007. *Synthesis Report: Expanded Summary. Joint Evaluation of the International Response to the Indian Ocean Tsunami.* Tsunami Evaluation Coalition, London.

Earle, P.S., Wald, D.J. *et al.* 2009. *Prompt Assessment of Global Earthquakes for Response (PAGER): A System for Rapidly Determining the Impact of Earthquakes Worldwide.* United States Geological Survey, Open-File Report, **2009-1131**.

Earthquake Engineering Research Institute 2006. *Learning from Earthquakes: The Mw 6.3 Java, Indonesia, Earthquake of May 27, 2006.* Earthquake Engineering Research Institute, Oakland, CA.

Farr, T.G. & Kobrick, M. 2000. Shuttle Radar Topography Mission produces a wealth of data. *Eos, Transactions of the American Geophysical Union*, **81**, 583–585.

Harmandar, E., Oye, V., Lindholm, C. & Bungum, H. 2007. *Soil Condition Maps Based on Topographic Slope.* Network of Earthquake Research Institutes for Earthquake Seismology (NERIES) JRA3 Report.

Ionescu, C., Danet, A., Sorensen, M.B., Zaharia, B., Stromeyer, D. & Grunthal, G. 2007. Implementation of the near real time ShakeMap system in Romania. *In*: *AGU Fall Meeting 2007*, San Francisco, CA, 10–14 December 2007, Abstracts. American Geophysical Union, Washington, DC, Abstract S51A-0215.

Jaiswal, K. & Wald, D. 2010. An empirical model for global earthquake fatality estimation. *Earthquake Spectra*, **26**, 1017–1037, https://doi.org/10.1193/1.3480331

Jaiswal, K. & Wald, D.J. 2013. Estimating economic losses from earthquakes using an empirical approach. *Earthquake Spectra*, **29**, 309–324, https://doi.org/10.1193/1.4000104

Jaiswal, K., Wald, D. & D'Ayala, D. 2011. Developing empirical collapse fragility functions for global building types. *Earthquake Spectra*, **27**, 775–795, https://doi.org/10.1193/1.3606398

Macintyre, A.G., Barbera, J.A. & Smith, E.R. 2006. Surviving collapsed structure entrapment after earthquakes: a 'time-to-rescue' analysis. *Prehospital Disaster Medicine*, **21**, 4–19.

Michelini, A., Faenza, L., Lauciani, V. & Malagnini, L. 2008. ShakeMap implementation in Italy.

Seismological Research Letters, **79**, 688–697, https://doi.org/10.1785/gssrl.79.5.688

MÜNCH, U., RUDLOFF, A. & LAUTERJUNG, J. 2011. The GITEWS Project – results, summary and outlook. *Natural Hazards and Earth Systems Science*, **11**, 765–769, https://doi.org/10.5194/nhess-11-765-2011

PAGANI, M., MONELLI, D. ET AL. 2014. OpenQuake engine: an open hazard (and risk) software for the Global Earthquake Model. *Seismological Research Letters*, **85**, 692–702, https://doi.org/10.1785/0220130087

PRAKOSO, W.A. & SUKANTA, I.N. 2012. Benchmark for country-level earthquake strong-motion instrumentation program. *Makara Journal of Technology*, **16**, 129–134.

PRIBADI, K.S., KUSUMASTUTI, D. & RILDOVA,. 2008. Learning from recent Indonesian earthquakes: an overview to improve structural performance. *In*: *Proceedings of the 14th World Conference on Earthquake Engineering*. 12–17 October 2008, Beijing, China.

RUDLOFF, A., LAUTERJUNG, J., MÜNCH, U. & TINTI, S. 2009. The GITEWS Project (German–Indonesian Tsunami Early Warning System). *Natural Hazards and Earth Systems Science*, **9**, 1381–1382, https://doi.org/10.5194/nhess-9-1381-2009

RUDYANTO, A. 2013. *Development of Strong-Motion Database for the Sumatra-Java region*. Master of Philosophy, Australian National University.

SENGARA, I.W., SUARJANA, M. ET AL. 2010. *The 30th September West Sumatra Earthquake, Padang Region Damage Survey*. Geoscience Australia, Record, **2010/44**.

SENGARA, I.W., SUARJANA, M., PRIBADI, K.S., ADIPUTRA, I. & SAGALA, S. 2013. Development of empirical earthquake fatality model for Indonesia. *In*: *15th World Conference on Earthquake Engineering*, 24–28 September 2012, Lisbon, Portugal.

SIEH, K. & NATAWIDJAJA, D. 2000. Neotectonics of the Sumatran fault, Indonesia. *Journal of Geophysical Research*, **105**, 28,295–28,326.

SILVA, W., THOMPSON, E., MAGISTRALE, H. & WILLS, C. 2011. *Development of a V_S (30 m) Map for Central and Eastern North America (CENA): Median Estimates and Uncertainties*. PEER NGA-East (Pacific Earthquake Engineering Research Center), University of California, Berkeley, CA.

SO, E., SPENCE, R., KHAN, A. & LINDAWATI, T. 2008. Building damage and casualties in recent earthquakes and tsunamis in Asia: a cross-event survey of survivors. *In*: *Proceedings of the 14th World Conference on Earthquake Engineering*. 12–17 October 2008, Beijing, China, Paper, **S31–039**.

THOMPSON, E.M., BAISE, L.G., KAYEN, R.E., MORGAN, E.C. & KAKLAMANOS, J. 2011. Multiscale site-response mapping: a case study of Parkfield, California. *Bulletin of the Seismological Society of America*, **101**, 1081–1100, https://doi.org/10.1785/0120100211

WALD, D.J. & ALLEN, T.I. 2007. Topographic slope as a proxy for seismic site conditions and amplification. *Bulletin of the Seismological Society of America*, **97**, 1379–1395, https://doi.org/10.1785/0120060267

WALD, D.J., QUITORIANO, V., HEATON, T.H., KANAMORI, H., SCRIVNER, C.W. & WORDEN, B.C. 1999. TriNet 'ShakeMaps': rapid generation of peak ground-motion and intensity maps for earthquakes in southern California. *Earthquake Spectra*, **15**, 537–556.

WALD, D.J., JAISWAL, K., MARANO, K., EARLE, P. & ALLEN, T. 2011. Advancements in casualty modeling facilitated by the USGS Prompt Assessment of Global Earthquakes for Response (PAGER) System. *In*: SPENCE, R., SO, E. & SCAWTHORN, C. (eds) *Human Casualties in Earthquakes*. Advances in Natural and Technological Hazards Research, **29**. Springer, Dordrecht, 221–230.

WEBER, B., BECKER, J. ET AL. 2007. SeisComP3 – automatic and interactive real time data processing. *Geophysical Research Abstracts*, **9**, 09219, http://meetings.copernicus.org/www.cosis.net/abstracts/EGU2007/09219/EGU2007-J-09219.pdf

WILLS, C. & GUTIERREZ, C. 2008. *Investigation of Geographic Rules for Improving Site-Conditions Mapping*. California Geological Survey, Final Technical Report, Award Number 07HQGR0061, Sacramento, CA.

WORDEN, C.B., WALD, D.J., ALLEN, T.I., LIN, K., GARCIA, D. & CUA, G. 2010. A revised ground-motion and intensity interpolation scheme for ShakeMap. *Bulletin of the Seismological Society of America*, **100**, 3083–3096, https://doi.org/10.1785/0120100101

ZHAO, J.X., ZHANG, J. ET AL. 2006. Attenuation relations of strong ground motion in Japan using site classification based on predominant period. *Bulletin of the Seismological Society of America*, **96**, 898–913, https://doi.org/10.1785/0120050122

Empirical fatality model for Indonesia based on a Bayesian approach

I. WAYAN SENGARA[1]*, MADE SUARJANA[1], M. ADDIFA YULMAN[1], HADI GASHEMI[2] & HYEUK RYU[2]

[1]*Civil and Environmental Engineering Faculty, Institut Teknologi Bandung, Lebak Siliwangi, Coblong, Bandung City, West Java 40132, Indonesia*

[2]*Geoscience Australia, GPO Box 378, Canberra ACT 2601, Australia*

**Correspondence: iws@geotech.pauir.itb.ac.id*

Abstract: An empirical fatality model for Indonesia has been developed by relating the macroseismic intensity to the fatality rate using compiled sub-district level fatality rate data and the numerically simulated ground-shaking intensity for four recent damaging events. The fatality rate data were compiled by collecting population and fatality statistics of the regions affected by the selected events. The ground-shaking intensity was numerically estimated by incorporating a finite-fault model of each event and local site conditions approximated by topographically based site amplifications. The macroseismic intensity distribution of each event was generated using ShakeMap software, combining a selected pair of ground motion prediction equations and ground motion to intensity conversion equations. The developed fatality model is a Bayesian generalized linear model in which the fatality rate is assumed to follow a mixture of Bernoulli and gamma distributions. The model was validated by calculating the fatalities in past events from the EXPO-CAT catalogue and comparing the estimates with the EXPO-CAT fatality records. Although the model can provide an estimate of the range of fatalities for future events, it needs ongoing refinement by the incorporation of additional fatality rate data from past and future events.

The risk of natural disasters is high in Indonesia as a result of the high potential for natural hazard events and the extensive building stock, coupled with a large human population. This risk has been manifested through several severe events with devastating consequences. Recent earthquakes in Indonesia, such as the Nias (28 March 2005), Yogyakarta (26 May 2006), West Java (2 September 2009) and West Sumatra (30 September 2009) earthquakes, have caused a large number of fatalities and widespread damage to buildings. Recent seismic hazard mapping for Indonesia has confirmed that many cities and urban areas are located in high-hazard regions with the potential to cause many human fatalities (Irsyam *et al.* 2010). Consequently, there is a constant need to develop and enhance fatality models to better estimate human fatalities in future earthquakes as part of earthquake impact and risk assessments.

Several researchers have developed fatality models based on the intensity of ground-shaking that are either global or country-specific. Allen *et al.* (2009) developed a global fatality model with an exponential function using the EXPO-CAT catalogue to rank regions in terms of earthquake vulnerability. Jaiswal & Wald (2010) developed country-specific fatality models with a log-normal cumulative distribution function for the Prompt Assessment of Global Earthquakes for Response (PAGER) system of the United States Geological Survey (USGS) using the EXPO-CAT catalogue. Sengara *et al.* (2012) developed two fatality models for Indonesia, one with an exponential function and one with a log-normal cumulative distribution function, using the data collected from four recent damaging events. The model with an exponential function was implemented as one of the impact functions available in the InaSAFE software (Inasafe.org 2015). A limitation of this model is that it can only estimate the mean (or expected) number of fatalities.

We applied a Bayesian generalized linear modelling technique to develop a fatality model for Indonesia to estimate uncertainty in addition to the expected number of fatalities. This paper presents the development of the model from data collection on fatalities and ground motion field generation to the application of the Bayesian method. To validate the developed model, we estimated the fatalities in past events using exposure data from EXPO-CAT and compared the estimates with the fatality data from EXPO-CAT. The limitations and directions for future development are discussed.

Collection of data on fatalities

The performance of an empirically derived fatality model relies heavily on the accuracy of the collected

From: CUMMINS, P. R. & MEILANO, I. (eds) 2017. *Geohazards in Indonesia: Earth Science for Disaster Risk Reduction*. Geological Society, London, Special Publications, **441**, 179–187.
First published online April 24, 2017, https://doi.org/10.1144/SP441.9

Table 1. *Selected historical earthquake events in Indonesia*

Event	Date	No. of fatalities
Nias	28 March 2005	850
Yogyakarta	26 May 2006	5167
West Java	2 September 2009	11
West Sumatra	30 September 2009	783

population and fatality data. Hence considerable effort was put into the collection of reliable data. Population information and the number of fatalities in the regions affected by the selected events were collected with the assistance of the relevant government agencies in Indonesia, including Badan Pusat Statistik (the Central Agency on Statistics of Indonesia) and local officials. This yielded population and fatality data at the sub-district level for four major recent events in Indonesia associated with a significant number of fatalities. Table 1 summarizes the reported number of fatalities in the four events. The fatality numbers reflect the number of fatalities due to ground-shaking and not those due to secondary earthquake effects such as landslides or tsunamis.

Figure 1 shows the sub-districts close to the epicentre of the selected events. In this figure, the sub-districts with reported fatalities are displayed as coloured polygons and other sub-districts as white patches. There were many sub-districts with no reported fatality data, even in the vicinity of the epicentre.

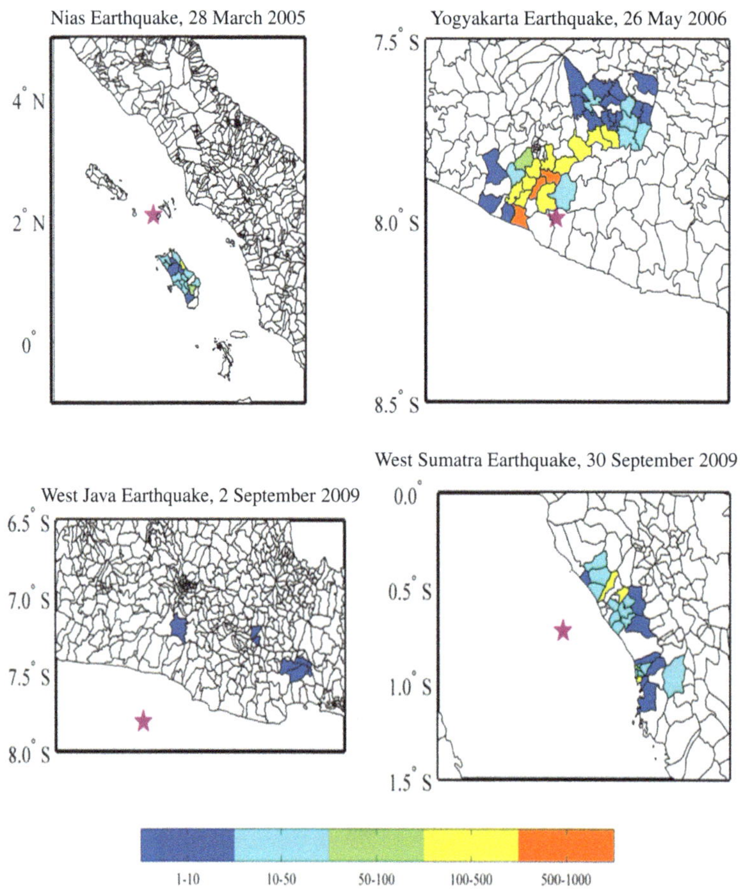

Fig. 1. Spatial distribution of the sub-districts located in the vicinity of the selected earthquakes. The earthquake epicentres are shown as a star and the sub-districts with reported fatalities are displayed as polygons coloured according to the number of fatalities.

Table 2. *GMPE and GMICE combination by tectonic setting*

	GMPE			GMICE
	Active shallow crust	Subduction interface	Subduction intra-slab	
BMKG	Zhao *et al.* (2006)	Zhao *et al.* (2006)	Zhao *et al.* (2006)	Worden *et al.* (2012)
EXPO-CAT	Boore *et al.* (1997)	Youngs *et al.* (1997)	Youngs *et al.* (1997)	Wald *et al.* (1999)

Ground motion field generation

The modified Mercalli intensity (MMI) was used as an explanatory variable to link ground-shaking to the fatality rate. The level of ground-shaking caused by the selected earthquakes was numerically estimated using the USGS ShakeMap software (Wald *et al.* 2006).

ShakeMap uses ground motion prediction equations (GMPE) in combination with ground motion to intensity conversion equations (GMICE) to estimate the MMI for a particular earthquake over a spatial grid. We choose the same GMPE and GMICE combination adopted by Badan Meteorologi, Klimatologi, dan Geofisika (BMKG), the Indonesian Agency for Meteorology, Climatology and Geophysics, because the developed fatality model will be applied to ShakeMaps generated by BMKG, the government agency responsible for generating ShakeMaps for earthquakes in Indonesia. Table 2 summarizes BMKG's preferred GMPEs by tectonic setting. The GMPE and GMICE combination used in the EXPO-CAT was selected to validate the model against the EXPO-CAT fatality records.

Finite-fault models, local site conditions and MMI observations can also be incorporated to enhance the quality of the MMI distribution maps. A finite-fault model was adopted from a corresponding study for each of the selected events in this study (Table 3). Local site effects were also included using the topographic slope–V_{S30} correlations of Allen & Wald (2009). Among the selected events (Table 3), observed MMI values were only available for the 2006 Yogyakarta earthquake and were used to generate a data-informed ShakeMap for this event.

Figure 2 is a map of the estimated V_{S30} and MMI values for the 2006 Yogyakarta earthquake. The level of ground-shaking decreased as the source to site distance increased, as expected. The MMI level varied with the local site conditions – that is, for the same source to site distance, the MMI level was higher at the sites on soft soil than at the sites on firm soil. The ShakeMaps are sensitive to the selection of GMPEs and GMICEs. The estimated MMI values using the GMICE of Wald *et al.* (1999) were generally higher than those using the model of Worden *et al.* (2012) at closer distances to the fault. There was a reasonable correlation between the estimated ShakeMaps and the location of the sub-districts with reported fatalities. To establish a hazard magnitude for each sub-district, an MMI value at the centre of each sub-district was estimated by two-dimensional linear interpolation.

Development of fatality model

The fatality rate was calculated for each sub-district using the recorded number of fatalities and population information. It was then linked with the estimated MMI value based on the choice of GMPE

Table 3. *Seismological parameters of the selected earthquakes in Indonesia*

Event	Moment magnitude	Longitude of epicentre (°)	Latitude of epicentre (°)	Depth of epicentre (km)	Tectonic setting	Finite-fault model
Nias	8.7	97.013	2.0764	30	Subduction Interface	Konca *et al.* (2007)
Yogyakarta	6.3	110.43	−7.995	17	Active Shallow Crust	Tsuji *et al.* (2009)
West Java	7.0	107.259	−7.8088	46	Subduction Intra-slab	Madlazim *et al.* (2010)
West Sumatra	7.6	99.856	−0.7254	81	Subduction Intra-slab	Sladen (2009)

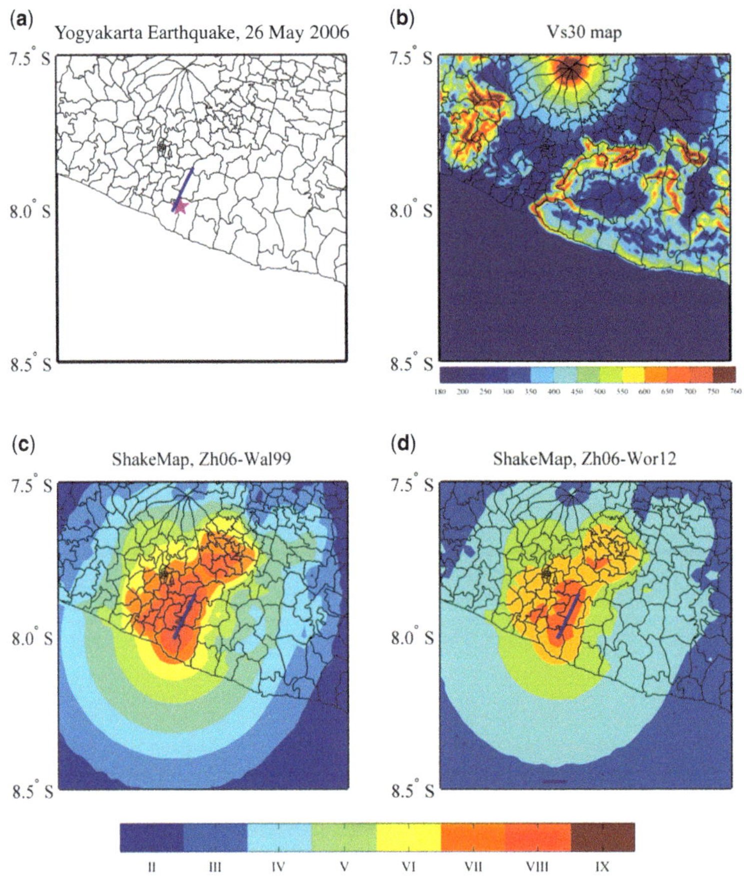

Fig. 2. (**a**) Sub-districts located in the vicinity of the Yogyakarta earthquake. The epicentre and finite-fault model are shown as a star and line, respectively. (**b**) Map of the average shear wave velocity between 0 and 30 m depth (V_{S30}, m s^{-1}). (**c**) ShakeMap of the Yogyakarta earthquake using the Zhao *et al.* (2006)–Wald *et al.* (1999) GMPE–GMICE pair and (**d**) the Zhao *et al.* (2006)–Worden *et al.* (2012) GMPE–GMICE pair.

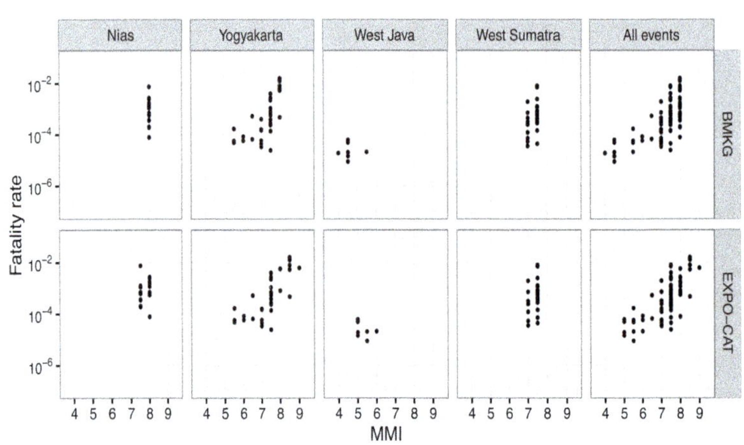

Fig. 3. Calculated fatality rates v. modified Mercalli intensity (MMI) for selected events. The MMI values are based on the selected GMPE–GMICE pairs listed in Table 2.

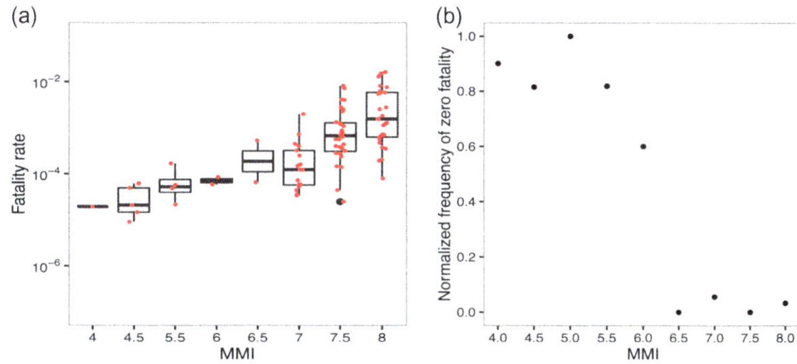

Fig. 4. (**a**) Boxplot of Fatality Data – BMKG on a logarithmic scale with jittered data points and (**b**) scatter plot of normalized frequency of zero fatality over modified Mercalli intensity (MMI).

and GMICE combination. The population information was based on the Indonesian 2010 census data. Figure 3 shows that the relationship between shaking intensity and fatality rate varied across the events. The MMI values were rounded to the nearest 0.5 MMI (e.g. an MMI between 7.75 and 8.25 was rounded to MMI 8.0). For the purpose of fatality model development, the fatality data from the four events were merged into databases and denoted as Fatality Data – BMKG or Fatality Data – EXPO-CAT based on the choice of GMPE and GMICE combination. The fatality model was accordingly denoted as Fatality Model – BMKG or Fatality Model – EXPO-CAT based on the fatality database.

Figure 4a shows a box plot of the Fatality Data – BMKG on a logarithmic scale. The fatality rate on average increased with the increase in intensity, suggesting a linear relationship between the log of the fatality rate and the MMI. The fatality rate at each MMI intensity showed an approximately symmetrical distribution around the median on a logarithmic scale, which suggests a log-normal distribution for the fatality rate. Figure 4b shows the normalized frequency of zero fatality along the range of MMI, which suggests that the zero fatality rate decreased on average with an increase in the intensity.

We analysed the fatality rate data with a Bayesian generalized linear model. Bayesian analysis yielded complete distributional information with respect to the parameter values in the regression model. In the generalized linear model, the fatality rate λ was described as a random variable, the distribution of which was a mixture of a Bernoulli and a log-normal distribution:

$$\lambda \sim \begin{cases} p, & \lambda = 0 \\ (1-p) \times \text{log-normal}\ (\mu, \sigma), & \lambda > 0 \end{cases} \quad (1)$$

where p is the probability of zero fatality rate at intensity s, and μ and σ are the mean and standard deviation of the log-normal distribution at intensity s.

The Bernoulli distribution was used to account for the zero fatality data, whereas the log-normal distribution was used to fit the skewed distribution of the non-zero fatality data. This approach has also been used to fit the fatality rate data from the 2005 Pakistan earthquake (Noh *et al.* 2013).

The mean log of the non-zero fatality rate μ was related to a linear function of shaking intensity s, as can be inferred from Figure 4a, whereas the probability of zero fatality rate p was related to a linear function of shaking intensity s with the logit link function as follows:

$$\begin{aligned} \mu &= a + b \times s \\ \text{logit}(p) &= c + d \times s \end{aligned} \quad (2)$$

where a, b, c and d are the parameters of the linear function of shaking intensity s.

The posterior distributions of the five parameters were calculated using the Hamilton Monte Carlo algorithm implemented in the package RStan (Stan Development Team 2014), with modifications from the version of the programs written by Kruschke (2014). A Cauchy distribution with centre 0 and scale 2.5 was used for the prior distribution for a, b, c and d. A half-Cauchy distribution with centre 0 and scale 2.5 was used for the prior distribution of σ. These choices imply weakly informative priors (Gelman *et al.* 2008). Four chains were used with burn-in steps of 1000 and a total of 8000 steps per chain were saved.

Figure 5 shows the marginal distributions of the five parameters of the model based on the Fatality Data – BMKG and Table 4 summarizes the statistics for the posterior distribution of parameters based on either fatality database. As expected, the parameter b, the slope of the linear function related to the mean log of the non-zero fatality rate, was estimated to be positive, whereas the parameter d,

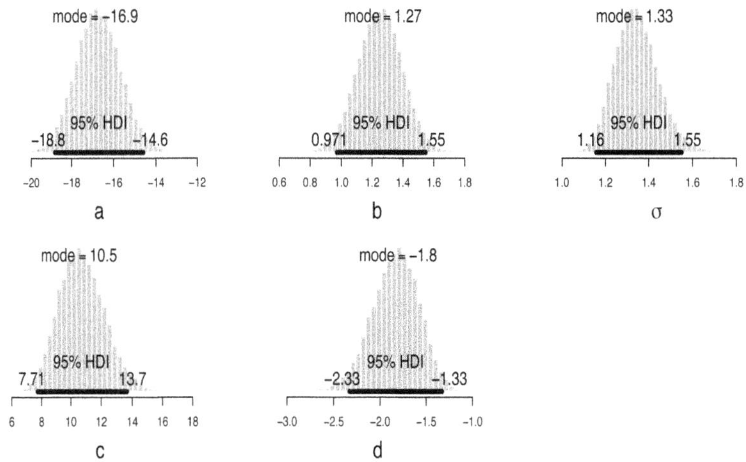

Fig. 5. Marginal distribution of the five parameters of the model based on the Fatality Data – BMKG.

the slope of the linear function related to the probability of zero fatality rate, was estimated to be negative.

Figure 6 shows the posterior predictions of fatality rate by the Fatality Model – BMKG. The predicted fatality rate shows comparable results against the observed data.

Fatality estimation

The number of fatalities for an earthquake event can be calculated by the sum of the products of fatality rate and population over a range of shaking intensities as follows:

$$N = \sum_{s} \lambda(s) \times P(s) \tag{3}$$

where N is the estimated number of fatalities, λ is the fatality rate at the shaking intensity s and P is the population exposed to the shaking intensity s.

With each set of sampled values of the posterior distribution of the parameters a, b, c, d and σ, the probability of zero fatality p and the two parameters of log-normal were calculated using equation (1). The fatality rate was randomly sampled from a Bernoulli log-normal distribution with the computed values of the parameters μ, σ and p. With each of the sampled values of λ, the number of fatalities was calculated using equation (3).

Figure 7a shows histograms of the sampled fatalities for the four events, where the dots represent the reported fatalities. The probability of fatalities being in a specific range can be easily calculated by counting. Figure 7b shows the probability distributions of

Table 4. *Statistical summary of posterior distribution of the parameters*

Parameter	Mean	Median	Mode	Effective sample size	Lower limit of 95% Highest Density Interval	Upper limit of 95% Highest Density Interval
Fatality Model – BMKG						
a	−16.736	−16.744	−16.867	10 083.7	−18.809	−14.603
b	1.251	1.253	1.271	10 112.2	0.971	1.549
σ	1.345	1.340	1.335	13 465.4	1.157	1.551
c	10.727	10.658	10.542	10 502.2	7.709	13.668
d	−1.833	−1.820	−1.799	10 517.2	−2.331	−1.328
Fatality Model – EXPO-CAT						
a	−18.764	−18.765	−18.807	9382.9	−21.018	−16.577
b	1.516	1.516	1.495	9385.7	1.210	1.815
σ	1.260	1.4	1.243	12 245.0	1.083	1.451
c	13.137	13.025	12.662	8994.3	9.487	16.931
d	−2.100	−2.081	−2.040	8956.6	−2.693	−1.522

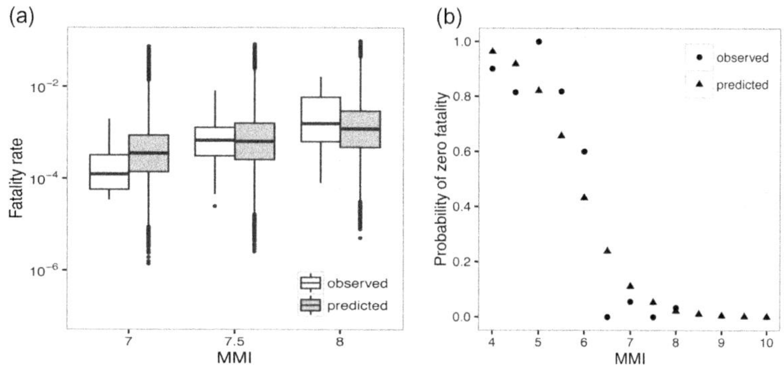

Fig. 6. Comparison of fatality rate between that predicted by Fatality Model – BMKG and Fatality Data – BMKG. (**a**) Boxplot of non-zero fatality rate on a logarithmic scale for selected modified Mercalli intensity (MMI) intensities and (**b**) scatter plot of probability of zero fatality over MMI range.

fatalities for each order of magnitude, together with the reported fatalities presented as dots. Although the model predicts the highest probability as being of the same order of magnitude as the reported fatalities, there is still a large uncertainty in the fatality estimate of up to three orders of magnitude.

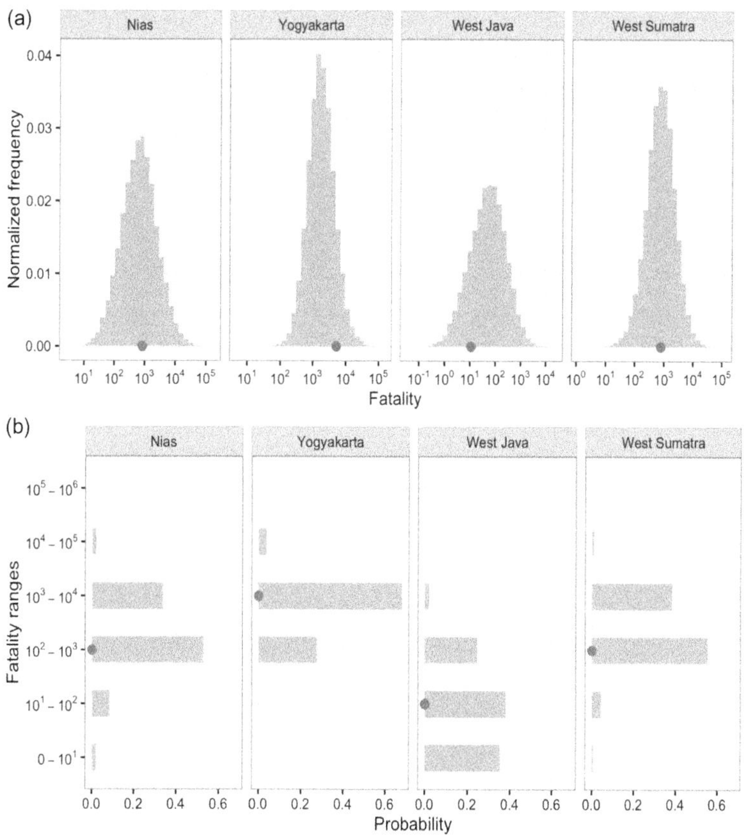

Fig. 7. Distributions of sampled fatalities for the four events using the Fatality Model – BMKG (**a**) Histogram of estimated fatalities and (**b**) probability of fatalities being in each order of magnitude.

Table 5. *Summary of performance of Fatality Model – EXPO-CAT against reported fatalities*

Number of events for which the reported number of fatalities corresponds to the estimated fatality range with:	Fatality range				
	$0-10^1$	10^1-10^2	10^2-10^3	10^3-10^4	Total
Highest probability	17(11)	6(7)	4(5)	1(1)	28(24)
Second highest probability	5(7)	8(4)	1(0)	0(0)	14(11)
Remaining	23(27)	3(6)	0(0)	1(1)	27(34)
Total	45	17	5	2	69

Data in parentheses are the estimated fatalities using the Fatality Model – BMKG.

Fatality model validation with EXPO-CAT data

To assess the performance of the developed model, we computed the fatalities of 69 historical Indonesian earthquakes for the period 1975–2007. The earthquakes were selected from the EXPO-CAT catalogue (Allen *et al.* 2009), which compiles the population estimates at each macroseismic intensity level and the total shaking-related fatalities. The four events used for model development were excluded from the selection.

We computed the fatalities of the selected earthquakes using the Fatality Model – EXPO-CAT and then compared the computed fatalities with the fatality records from the catalogue. We also computed the fatalities using the Fatality Model – BMKG to assess the effect of inconsistency of the GMPE–GMICE combination between the fatality model and the population data on the estimation of fatalities.

Table 5 summarizes the number of events for which the reported fatalities corresponded to the estimated fatality range with the first and second highest probability. Both models performed better for the events with high fatalities than for those with low fatalities, which can be attributed to the fact that the input data used for the model development were largely from events with high fatalities. Not surprisingly, the Fatality Model – EXPO-CAT performed better than the Fatality Model – BMKG, which highlights the importance of the consistency of the GMPE–GMICE combination between the fatality model and the population data.

Conclusions

We applied a Bayesian technique to develop an earthquake fatality model for Indonesia using collected fatality data and the simulated macroseismic intensity for four recent damaging events. The developed model differs from the model developed by Sengara *et al.* (2012) in that the parameters of the developed model have distributions rather than single values, which enabled us to estimate the uncertainty in the prediction in addition to an expected value. The provision of a range of fatalities with an associated probability (e.g. the probability of fatality being between 100 and 1000) should be more effective than a single point estimate when it comes to communication with the general public and government emergency management agencies.

The fatality database compiled in this study is unique in the sense that the fatality rate is directly related to the ground-shaking intensity and the recorded fatalities for a given sub-district, whereas EXPO-CAT only provides fatality records aggregated by event. To complement the current fatality database, systematic post-disaster surveys need to be carried out for future events in Indonesia. Although the model has only been developed for fatalities in this study, a similar approach could be applied for the development of a non-fatal injury model to estimate the number of people with non-fatal injuries for future events and also the number of people who need to be evacuated.

The comparison between the predicted fatalities using the developed model and the recorded fatalities highlights a need for continuing enhancement with additional fatality data. It also suggests a need for the development of a fatality model based on building damage (e.g. the PAGER analytical fatality model) given the fact that human casualties are usually caused by building collapse.

The developed fatality model for Indonesia will be implemented in the InaSAFE tool, which will provide fatality estimations for earthquake scenarios, including real-time earthquake impact assessment. The GMPE–GMICE consistency must be met in applying a GMPE–GMICE dependent fatality model for fatality prediction. The Fatality Model – BMKG should therefore be combined with a ShakeMap generated by BMKG. In a real-time or near-real-time application in which there might be no finite-fault or felt-report information, the ShakeMap generated would be considered as the best representation of ground-shaking. The initial assessment will be revised once additional information is available to update the ShakeMap.

The authors appreciate input provided by Trevor Allen through information and the facilitation of the ShakeMap program. The authors also thank Krishna S. Pribadi, Saut Sagala and Irwan Meilano for providing various inputs to this study. Support in fatality data collection by research assistants at the Research Center for Disaster Mitigation-Institut Teknologi Bandung – Indradi Adisaputra, Aria Mariany, Inln Wahdiny, Anin Utami and Bayu Novianto–is highly appreciated. The authors thank the Australia–Indonesia Facility for Disaster Reduction and the Indonesian Disaster Management Agency for financial support of the study. This paper is published with permission of the CEO, Geoscience Australia.

References

ALLEN, T.I. & WALD, D.J. 2009. On the use of high-resolution topographic data as a proxy for seismic site conditions (VS30). *Bulletin of the Seismological Society of America*, **99**, 935–943.

ALLEN, T.I., WALD, D.J., EARLE, P.S., MARANO, K.D., HOTOVEC, A.J., LIN, K. & HEARNE, M.G. 2009. An atlas of ShakeMaps and population exposure catalog for earthquake loss modeling. *Bulletin of Earthquake Engineering*, **7**, 701–718.

BOORE, D.M., JOYNER, W.B. & FUMAL, T.E. 1997. Equations for estimating horizontal response spectra and peak acceleration from western North American earthquakes: a summary of recent work. *Seismological Research Letters*, **68**, 128–153.

GELMAN, A., JAKULIN, A., PITTAU, M.G. & SU, Y. 2008. A weakly informative default prior distribution for logistic and other regression models. *The Annals of Applied Statistics*, **2**, 1360–1383.

INASAFE.ORG 2015. The InaSAFE Project, http://inasafe.org

IRSYAM, M., ASRURIFAK, M., HENDRIYAWAN, BUDIONO, B., TRIYOSO, W. & FIRMANTI, A. 2010. Development of spectral hazard maps for proposed revision of Indonesia seismic building code. *Geomechanics and Geoengineering an International Journal*, **5**, 35–47.

JAISWAL, K. & WALD, D. 2010. An empirical model for global earthquake fatality estimation. *Earthquake Spectra*, **26**, 1017–1037.

KONCA, A.O., HJORLEIFSDOTTIR, V. *ET AL*. 2007. Rupture kinematics of the 2005 Mw 8.6 Nias-Simeulue earthquake from the joint inversion of seismic and geodetic data. *Bulletin of the Seismological Society of America*, **97**, S307–S322, https://doi.org/10.1785/0120050632

KRUSCHKE, J.K. 2014. *Doing Bayesian Data Analysis. A Tutorial with R, JAGS, and Stan*. 2nd ed. Academic Press, London.

MADLAZIM, M., SANTOSA, B.J. & UTAMA, W. 2010. Three components seismogram analysis of Tasikmalaya, Indonesia on September 2nd 2009 earthquake to investigate the Coulomb stress change and seismicity rate change. *Indonesian Journal of Physics*, **21**, 71–76.

NOH, H., KIREMIDJIAN, A.S., JAISWAL, D. & SO, E. 2013. Bayesian analysis of earthquake fatality rates. *In*: *Proceedings of the 11th International Conference on Structural Safety and Reliability*, June 16–20, New York, NY, USA.

SENGARA, I.W., SUARJANA, M., PRIBADI, K.S., ADIPUTRA, I. & SAGALA, S. 2012. Development of empirical earthquake fatality model for Indonesia. *In*: *Proceedings of the 15th World Conference of Earthquake Engineering*, 24–28 June, Lisbon, Portugal.

SLADEN, A. 2009. *Preliminary result (Mw 7.6)*, Padang, Indonesia, http://www.tectonics.caltech.edu/slip_history/2009_padang/padang.html

STAN DEVELOPMENT TEAM 2014. RStan: the R interface to Stan. Version 2.5, http://mc-stan.org/rstan.html

TSUJI, T., YAMAMOTO, K. *ET AL*. 2009. Earthquake fault of the 26 May 2006 Yogyakarta earthquake observed by SAR interferometry. *Earth, Planets and Space*, **61**, e29–e32.

WALD, D.J., QUITORIANO, V., HEATON, T.H. & KANAMORI, H. 1999. Relationships between peak ground acceleration, peak ground velocity, and modified Mercalli intensity in California. *Earthquake Spectra*, **15**, 557–564.

WALD, D.J., WORDEN, C.B., QUITORIANO, V. & PANKOW, K.L. 2006. ShakeMap® manual, technical manual, users guide, and software guide, http://pubs.usgs.gov/tm/2005/12A01/pdf/508TM12-A1.pdf

WORDEN, C.B., GERSTENBERGER, M.C., RHOADES, D.A. & WALD, D.J. 2012. Probabilistic relationships between ground motion parameters and modified Mercalli intensity in California. *Bulletin of the Seismological Society of America*, **102**, 204–221.

YOUNGS, R.R., CHIOU, S.J., SILVA, W.J. & HUMPHREY, J.R. 1997. Strong ground motion attenuation relationships for subduction zone earthquakes. *Seismological Research Letters*, **68**, 58–73.

ZHAO, J.X., ZHANG, J., *ET AL*. 2006. Attenuation relations of strong ground motion in Japan using site classification based on predominant period. *Bulletin of the Seismological Society of America*, **96**, 898–913.

Evaluating a volcanic ash ground-loading hazard at Gunung Ciremai, West Java, Indonesia using PF3D

A. N. BEAR-CROZIER[1]*, N. KARTADINATA[2], A. HERIWASESO[2] & O. NIELSEN[3]

[1]*Geoscience Australia, GPO Box 378, Canberra, ACT 2601, Australia*

[2]*Pusat Vulkanologi dan Mitigasi Bencana Geologi, Jalan Diponegoro No. 57, Bandung, Indonesia*

[3]*Australia–Indonesia Facility for Disaster Reduction, Menara Thamrin Building Suite 1505, Jalan M.H. Thamrin Kav. 3, Jakarta 10250 Indonesia*

**Correspondence: Adele.Bear-Crozier@ga.gov.au*

Abstract: Volcanic ash represents a serious hazard to communities living in the vicinity of active volcanoes in developing countries such as Indonesia. Geoscience Australia (GA), the Australia–Indonesia Facility for Disaster Reduction (AIFDR) and Badan Geologi (Indonesia's Geological Agency) have adapted an existing open-source volcanic ash dispersion model for use in Indonesia. An application example is presented here for Gunung Ciremai in West Java, Indonesia. A stochastic set of eruption events was simulated using eruptive parameters within the acceptable range of possible future events for this volcano: granulometry and a meteorological dataset that represents the complete range of possible wind conditions expected during seasonal wind conditions for the region. Implications for nearby communities of dry v. rainy season conditions on volcanic ash hazard were investigated. Communities on the western side of Gunung Ciremai were highly susceptible to volcanic ash ground loading regardless of the season: communities on the eastern side, however, were more susceptible during the rainy season months. This is attributed to prevailing wind conditions during the rainy season that include a strong easterly component. Disaster risk reduction workers can use hazard maps like those produced here to inform decision-making, and to focus mitigation efforts on communities most at risk.

In recent times, volcanic hazards have featured a greater potential to affect growing populations in developing countries across the Asia-Pacific region. In addition to its continental mainland, the Asia-Pacific region encompasses a vast oceanic component throughout which thousands of islands are dispersed. These islands, which represent a number of developing countries in the region, are inherently more vulnerable to volcanic hazards due to their tectonic origins along plate boundaries associated with the Pacific Ring of Fire, meaning they experience high levels of volcanic activity (Terry & Goff 2012). Of these countries, the densely populated islands of Indonesia are the most exposed, where volcanic ash, the most widespread of all volcanic hazards, has the potential to affect hundreds of thousands, or even millions, of people. Indonesia has approximately 127 active volcanoes distributed across the 13 000 islands that make up the archipelago. According to statistical analyses undertaken on the relationship between eruption magnitude and frequency of volcanoes across the Asia-Pacific region by Simpson *et al.* (2011), Indonesia has the highest frequency of Volcanic Explosivity Index (VEI) 4 or greater eruptions, and over 100 000 lives have been lost to volcanic disasters in the last 200 years. There is limited information available for this region on the hazard posed by volcanic ash, particularly from volcanoes that have not erupted in recent times. Therefore, computational models capable of accurately predicting volcanic ash dispersal at ground level, coupled with field observations of historical or on-going eruptive activity, are required to inform and focus the preparedness efforts of emergency managers mandated with assessing communities at risk.

This paper will present modelled volcanic ash dispersal from an active volcanic centre, Gunung Ciremai in West Java, Indonesia. This case study will be based on an eruptive scenario within the acceptable range of possible future eruptive events for this volcano (i.e. taking into consideration what is known about eruptions historically at Ciremai combined with the historical behaviour of analogue volcanoes from the surrounding region), granulometry (grain-size characteristics) as determined through field studies, and a meteorological dataset that represents the range of possible wind conditions for the region. This work builds on previous modelling for Gunung Gede in West Java

From: Cummins, P. R. & Meilano, I. (eds) 2017. *Geohazards in Indonesia: Earth Science for Disaster Risk Reduction*. Geological Society, London, Special Publications, **441**, 189–198.
First published online March 24, 2016, https://doi.org/10.1144/SP441.1

(Bear-Crozier *et al.* 2012), increasing the knowledge base for the volcanic hazards in this region, and providing higher levels of certainty for provincial and Indonesian disaster managers.

The role of advection–diffusion–sedimentation (ADS) models

Historically, volcanic ash hazard maps in Indonesia have been constructed by investigating the eruptive history of a volcano through field-based approaches and delineating the hazard zone accordingly. Whilst understanding the historical behaviour of an active volcano is important, it cannot provide insights into what may happen if: (1) the volcano erupts differently to previous events; or (2) volcanoes that have not erupted in historical times become eruptive in the future. Ideally, a probabilistic approach should be taken, whereby an eruptive scenario is defined and then simulated using a tephra dispersal computational model (Cioni *et al.* 2003).

The role of advection–diffusion–sedimentation (ADS) models in assessing volcanic ash hazard has significantly increased in recent decades for both ground deposition and atmospheric concentration (Barberi *et al.* 1990; Hurst 1994; Searcy *et al.* 1998; Hurst & Turner 1999; Folch & Felpeto 2005; Macedonio *et al.* 2005, 2008; Costa *et al.* 2006, 2009; Barsotti *et al.* 2008; Folch *et al.* 2008, 2009; Folch & Sulpizio 2010; Sulpizio *et al.* 2012; Bonadonna & Costa 2013). Numerical simulations of volcanic ash fallout hazard generally involve running an eruptive scenario that represents the most likely event (based on historical investigation and/or modern analogues) over a period of time sufficiently large as to capture all possible meteorological conditions (Folch & Sulpizio 2010). Ideally, each volcanological parameter should be derived from empirical data collected at the relevant volcano, and sampled using an associated probability density function (PDF) that takes into consideration natural variability and uncertainty (Cioni *et al.* 2003). However, in the absence of such information, data from analogue volcanoes should be used. When combined with traditional field-based approaches, these models have the potential to greatly enhance the volcanic ash hazard information currently available for Indonesia.

FALL3D and the development of PF3D

FALL3D is a 3D time-dependant Eulerian model used for simulating the dispersal of volcanic ash, including both particle concentration in the atmosphere and particle loading at ground level (Costa *et al.* 2006; Folch *et al.* 2008, 2009; Macedonio *et al.* 2008). A python-scripted user interface called python-FALL3D (PF3D) was developed jointly between Geoscience Australia (GA), the Australia–Indonesia Facility for Disaster Reduction (AIFDR) and Badan Geologi (Indonesia's Geological Agency) to modify the modelling procedure of FALL3D in line with the needs of government agencies and emergency managers in the Asia-Pacific region. The authors have previously presented a detailed account of the modelling methodology and development of PF3D using the aforementioned case study for Gunung Gede (Bear-Crozier *et al.* 2012).

PF3D introduces a number of enhancements to the modelling procedure for FALL3D that do not change the operation or functionality of the core model but greatly simplify its overall use. The python scripts wrap around the existing model, providing a buffer between the user and FALL3D. These enhancements streamline the installation process, automating and simplifying both the pre-processing of input meteorological datasets and the configuration and execution of each utility program into a single-step process, thus reducing the likelihood of errors. PF3D produces geospatially referenced hazard maps for time-dependant ash load (kg m^{-2}), time-dependant ash thickness (mm, cm or m) and time-dependant ash concentration in the atmosphere (kg m^{-3}). These hazard maps can be integrated directly with exposure datasets, including population density, exposure of the built environment and crop extents, in order to assess potential impacts. They are intended for use by government agencies (or other users) to assess the risk and potential impact of volcanic ash on communities.

PF3D can be used to undertake probabilistic volcanic ash hazard assessments. This involves running one eruption scenario against thousands of potential wind fields for a particular region of interest. A stochastic catalogue of eruption events is generated and a script for computing the probability of ground ash-load exceedance (%) is used to merge these events into a single probabilistic ground ash-load hazard map for an ash-load threshold of interest (e.g. an ash-load threshold of 10 kg m^{-2} of could be used to assess the potential impact of ash on agricultural crops).

Gunung Ciremai, West Java, Indonesia

Tectonic setting

The evolution of the Indonesian archipelago since the late Palaeozoic has been dominated by a developing system of island arcs, with the present-day arc becoming a single entity in the mid-Tertiary (Katili 1975; Hamilton 1979; Widiyantoro & Van Der Hilst 1997; Soeria-Atmadja & Noeradi 2005; Sendjaja *et al.* 2009). The Indonesian island arcs are underlain by the Australian and Eurasian plates, characterized by a crustal thickness of approximately 20 km. The Australian Plate is being subducted beneath the Eurasian Plate at a convergence rate of 6–7 cm a^{-1} (Widiyantoro & Van Der Hilst 1997). Seismicity associated with the convergent boundary can be traced down to a depth of 600 km (Ghose *et al.* 1990; Puspito & Shimazaki 1995; Widiyantoro & Van Der Hilst 1996; Widiyantoro & Van Der Hilst 1997; Sendjaja *et al.* 2009). The Indonesian volcanic arc stretches 6000 km from northern Sumatra east to the Molluca Sea, and is typically subdivided from west to east into the Sumatra cordilleran sector, the Sunda arc (the focus of this study), the Banda arc and the Molluca Sea collision complex (Fig. 1) (Hamilton 1979; Sendjaja *et al.* 2009).

The Sunda Arc

This study focuses on the Sunda Arc. The Sunda Arc comprises islands from Java to Flores and has existed in its current configuration for 5 myr (Katili 1975; Hamilton 1979; Sendjaja *et al.* 2009). It is likely, however, that a volcanic arc has been present here since the early Eocene. Quaternary volcanics overlie late Miocene–Pliocene volcanic rocks and sediments; the densest clustering of currently active volcanic centres occurs on the island of Java. Most Quaternary volcanic centres are calc-alkaline (pyroxene andesite) in composition, with minor dacites also represented. The Sunda Arc is divided into two parallel volcanic lineaments. The southern volcanic range, named the volcanic front (VF), and the northern volcanic range, named the rear-arc (RA) (Sendjaja *et al.* 2009). The VF includes the active volcanic centres of Papandayan, Cikuray, Guntur and Galunggung (Fig. 1). The RA includes Tangkuban Parahu, Tamporanas and Ciremai (Fig. 1).

Eruption history

Gunung Ciremai is a stratovolcano with an elevation of 3072 m above sea level and a diameter of approximately 10 km. Ciremai is located on the boundary between Majalengka and Kunugan regency approximately 20 km south of the city of Cirebon in West Java (Fig. 1). Little is known about the pre-historic eruption history of Ciremai. However, Situmorang (1995) produced a geological map that delineates the distribution of volcanic facies from recent eruptions. According to Situmorang (1995), Ciremai has experienced five eruptions during historical times (since AD 1600). These occurred in 1698, 1772, 1775, 1805 and 1937, at intervals of 3–112 years. Eruptive activity has been characteristically both effusive and explosive in nature, producing lava flows, pyroclastic fall, pyroclastic flows and breccias of predominantly andesite composition (Situmorang 1995; Sendjaja *et al.* 2009).

Field observations and quantitative grain-size analysis

A field study was undertaken at Gunung Ciremai in order to ground truth assumptions made about the granulometry of pyroclastic fall deposits associated with historical eruptions. Pyroclastic fall deposits were mapped radially around the central vent and 11 type localities were identified (Fig. 1). Detailed qualitative descriptions of pyroclastic fall deposits at each locality were made and recorded in stratigraphic logs (Fig. 2).

Descriptions at the outcrop scale include bedding, depositional structures, stratigraphic boundaries and palaeosoils, as well as characterization of the componentry (i.e. five largest pumice clasts, five largest lithic clasts, crystal content and so on: Fig. 2). Where possible, an attempt was made to correlate facies laterally. Where preservation of the pyroclastic fall deposits was sufficiently high (the tropical climate is not conducive to ash preservation), samples were taken for semi-quantitative grain-size analysis (grain-size sieving).

Grain-size sieving was undertaken between 4ø (31–62.5 μm) and −4ø (16–32 mm) at 1ø intervals in order to derive weight percentage (wt%) v. grain size. For componentry finer than 4ø, volume percentage (vol%) v. grain size was derived using laser diffraction at grain sizes between 0.020 and 2000 μm. Density measurements of the componentry (juvenile, lithics, crystals) were also analysed. All sampled sites were located proximal to the source (within 10–15 km) and it is acknowledged that the resulting total grain-size distribution is fines-depleted as a result of this. The fines-depleted nature of the field data was considered acceptable, as the objective was to address proximal hazard assessments using PF3D.

Definition of the simulated eruptive scenario

Source term

The source term describes how volcanic ash particles are distributed throughout the eruptive column with time. A procedure, first coined by Suzuki (1983) for the distribution of mass that considers particle turbulent diffusion and wind advection, was used to simulate the source term for an eruption column height of 15 km (Suzuki 1983; Folch *et al.* 2009; Folch & Costa 2010). The assignment of eruption column height was based on historical eruptions as an indication of the acceptable range of possible

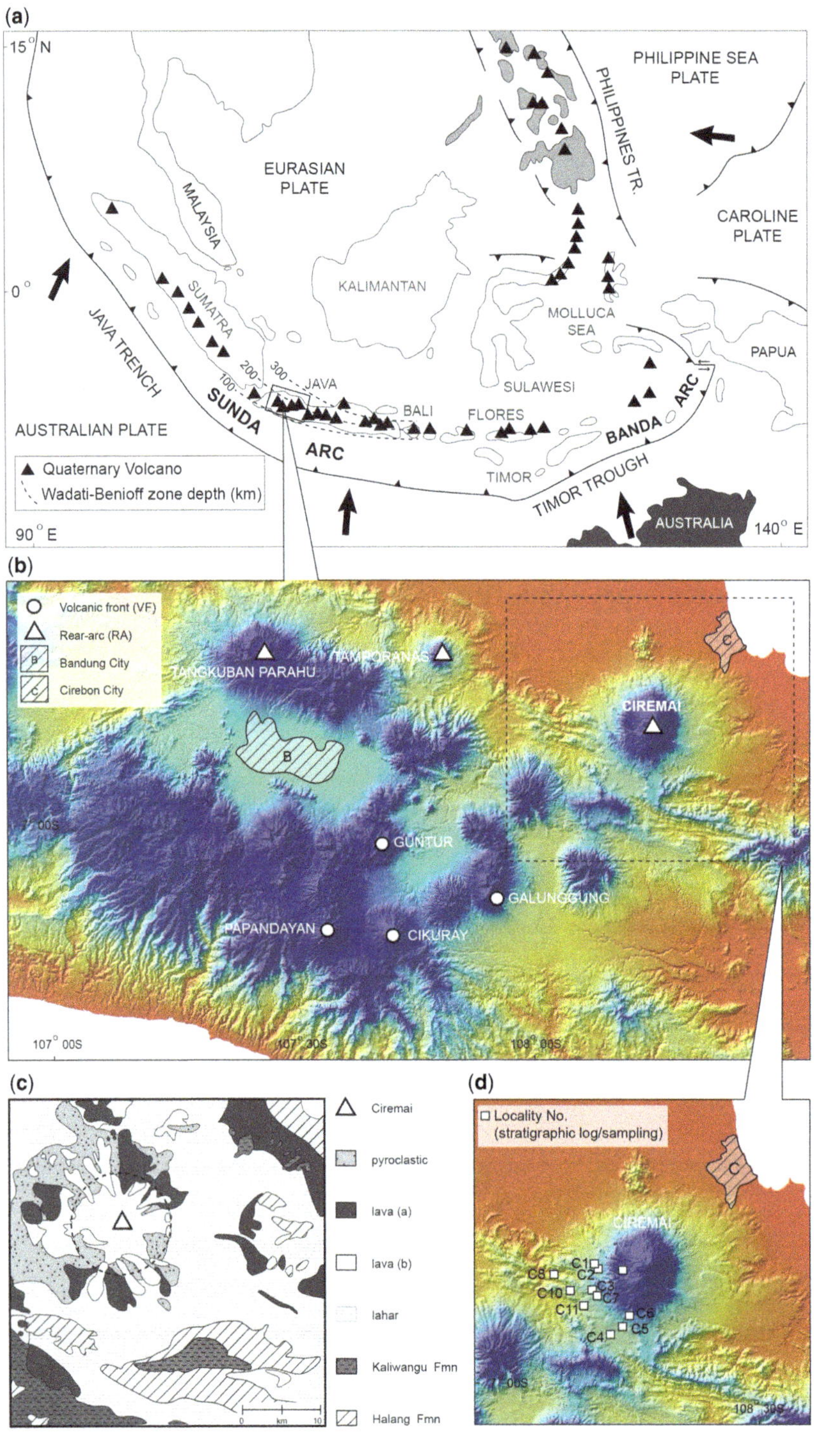

Fig. 1. (**a**). Tectonic setting of Indonesia showing the Sunda arc and the location of Java modified after Hamilton (1979) and Sendjaja *et al.* (2009). (**b**) The two volcanic ranges of the Sunda arc – the volcanic front (VF) and the rear arc (RA) – modified after Sendjaja *et al.* (2009). (**c**) Geological map of Gunung Ciremai modified after Situmorang (1995). (**d**) Field localities identified during the present study at Gunung Ciremai.

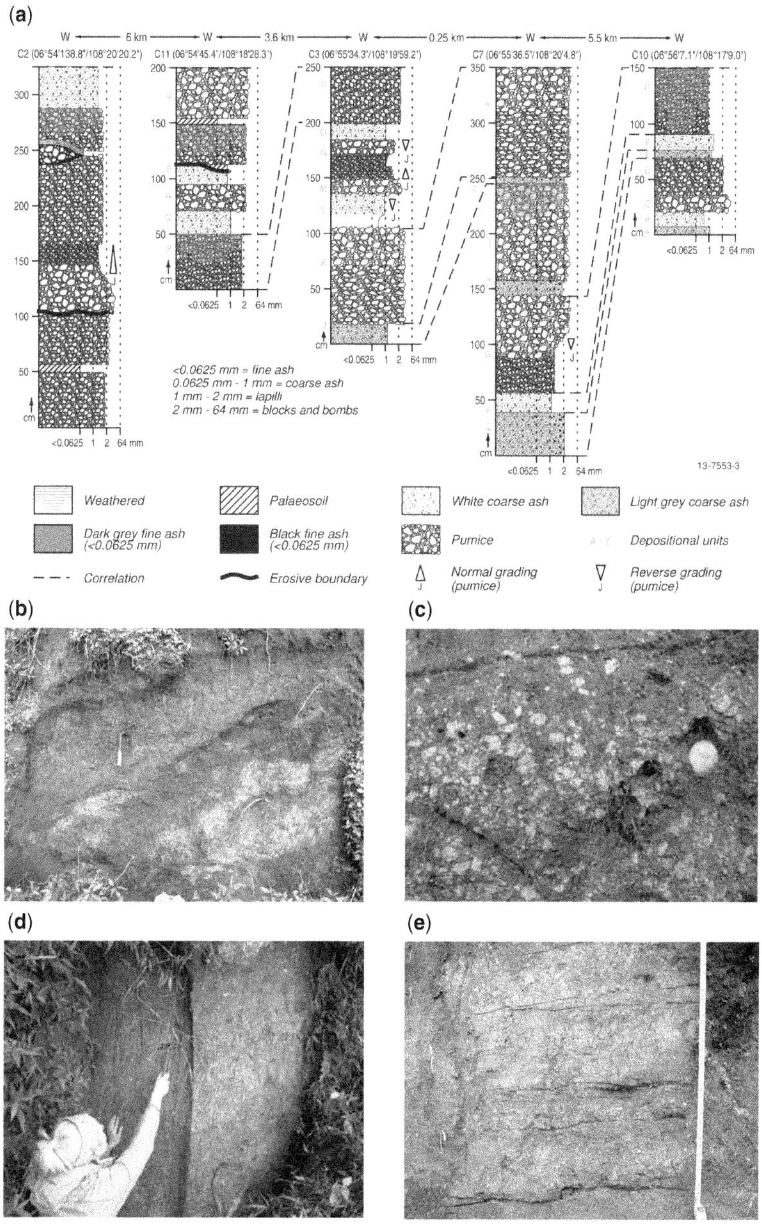

Fig. 2. (**a**) Stratigraphic logs depicting grain-size characteristics of pyroclastic fall deposits associated with historical eruptions at Ciremai correlated across multiple type locations. (**b**) Field photograph featuring a well-sorted pyroclastic fall deposit at Ciremai. (**c**) Field photograph of the juvenile clasts and lithic clasts associated with a pyroclastic fall deposit at Ciremai. (**d**) Field photograph of a massive, lithied pyroclastic fall deposit at Ciremai. (**e**) Field photograph of multiple pyroclastic fall units at Ciremai.

future events at Ciremai. Mass flow rate was estimated by PF3D based on empirical fits of magma discharge rate v. eruption column height. An eruption duration of 6 h was simulated allowing for an additional 3 h of post-eruptive particle settling time (a total simulation duration of 9 h). The modelled parameters for the simulated eruption of Ciremai are summarized in Table 1.

Table 1. *Summary of input parameters used to simulate volcanic ash fallout at Ciremai using PF3D*

Input parameter	Value
Computational domain	111877E 9136544S (lower-left corner)
Cell size	1342 m
Vertical discretization	*c.* 1 km
Eruption column height	15 km
Mass flow rate	3.45×10^6 kg s^{-1} (estimate by FALL3D using empirical relationships)
Total erupted mass	7.4×10^{10} kg
Duration of eruption	6 h
Duration of PES*	3 h
Initial grain size	Eight classes −4ø to 4ø (16–0.0625 mm)

*Post-eruptive setting of particles.

Granulometry

Granulometry parameters (minimum, maximum and average grain size) were derived through averaging across the sampled locations. For modelling purposes, the total grain-size distribution was grouped into eight classes at 1ø intervals, with the end members fixed at −4ø (16–32 mm in diameter) and 4ø (31–62.5 μm in diameter). For modelling purposes, a density range between 570 and 1300 kg m^{-3} was set for juvenile clasts, and a constant value of 2300 kg m^{-3} was set for lithic and crystal fragments.

Meteorological data

Meteorological data (http://www.esrl.noaa.gov/psd/data/gridded/data.ncep.reanalysis.html) was extracted for a 3 year period between 1 January 2007 and 31 December 2009 from the National Centre for Environmental Protection/National Centre for Atmospheric Research (NCEP/NCAR) in order to provide a representative suite of seasonal wind conditions for the West Java region. Air temperature (in K), geo-potential height (in metres) and the vector components of wind (*u* and *v* in m s^{-1}) were extracted and used to generate 4045 vertical wind profiles at the vent location, which represent changing wind conditions with increasing altitude at intervals of 6 h. The 4045 representative wind profiles for the region were divided into dry season (April–September) and rainy season (October–March) in order to assess the impact of seasonal wind conditions on the distribution of volcanic

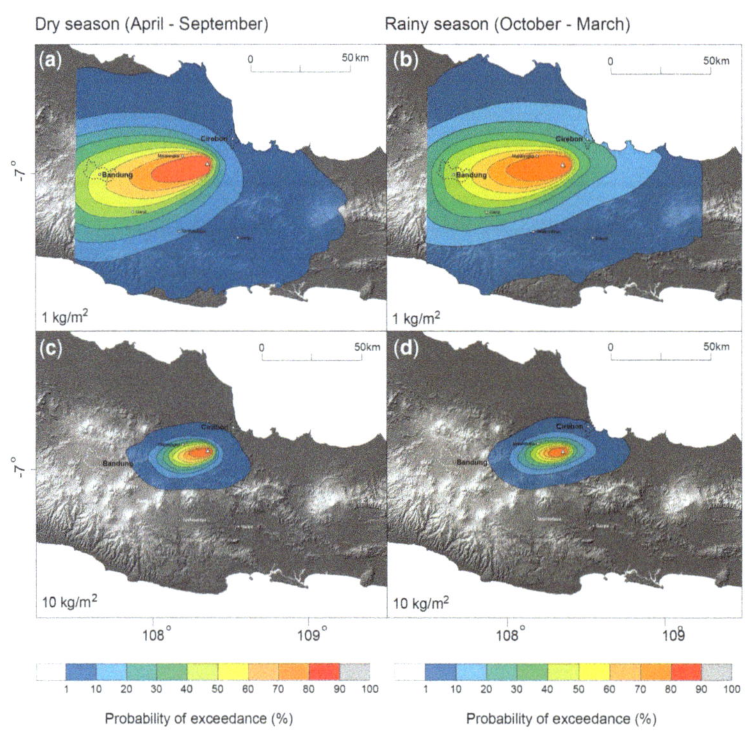

Fig. 3. Ground-load probability maps showing the probability (in %) that the deposit load exceeds 1 kg m^{-2} (top) and 10 kg m^{-2} (bottom) for dry season (left) and rainy season (right) wind conditions.

ash ground loading. The eruptive scenario defined above was simulated for each wind profile, generating a stochastic catalogue of 2184 dry season events and 1856 rainy season events.

Results

Ground-loading probability maps are presented here for ash-load thresholds of 1 and 10 kg m^{-2} for dry and rainy season meteorological conditions. These load thresholds represent conservative estimates of agricultural crop damage and surficial damage to buildings based on the observed impact of volcanic ash during historical eruptions in the Asia-Pacific region and analogous studies worldwide (Blong 2003; Folch & Sulpizio 2010). Each hazard map depicts the percentage probability of exceeding the designated ash-load threshold given the thousands of potential wind fields. Dry and rainy season maps are considered to account for seasonal variations in wind speed, direction and temperature with altitude within the region. It is important to note that ‘rainy season’ is a term used here to describe a time period (i.e. October–March) and not ‘wet’ ash fall. Rainy season events do not include the effects of rainfall on the aggregation of falling ash particles or compaction of the resulting deposit as a result of post-depositional rainfall events.

The percentage of scenarios at a given location where volcanic ash deposit load exceeds 1 and 10 kg m^{-2} (*c.* 1 mm and 1 cm, respectively) for dry season v. rainy season meteorological conditions is shown in Figure 3. Accumulations of volcanic ash load exceeding 1 kg m^{-2} (*c.* 1 mm equivalent) are distributed to a distance of 110 km from the summit region under dry season conditions, predominantly in a westerly direction, although a marked elongation to the east is also observed. This observation is highlighted in Figure 3a, where the percentage probability of exceeding 1 kg m^{-2} during dry season conditions in the city of Bandung, located west of Gunung Ciremai, is 50–60% and the percentage probability for the city of Cirebon, located east of Gunung Ciremai, is a lower 1–10%.

Accumulations of ash load exceeding 1 kg m^{-2} during rainy season conditions are similar to those presented for dry season conditions. The inundation footprint extends up to 100 km in a westerly direction: however, it features a pronounced easterly elongation to a distance of 110 km from the summit region. This is highlighted in Figure 3b, where the

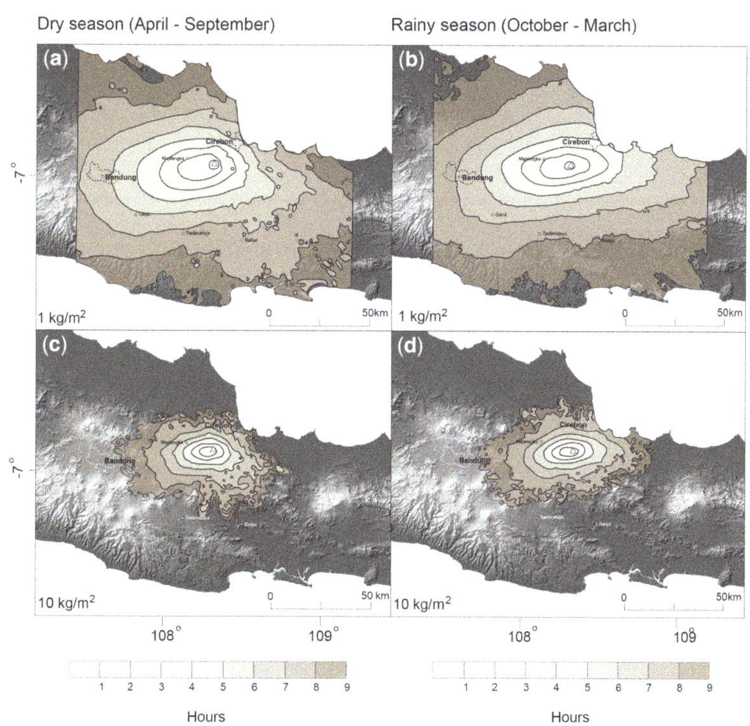

Fig. 4. Isochron maps showing the averaged accumulation time (in h) for the volcanic ash-load threshold values of 1 kg m^{-2} (top) and 10 kg m^{-2} (bottom) for dry season (left) and rainy season (right) conditions.

percentage probability of exceeding 1 kg m^{-2} during rainy season conditions in Bandung is 40–50% and 20% for Cirebon.

Ash-load deposits exceeding 10 kg m^{-2} (*c.* 1 cm equivalent) during dry season conditions are also concentrated to the west (up to 60 km), but are also similarly elongated to the east under rainy season conditions, where accumulations exceed 10 kg m^{-2} up to 50 km from the summit region. The implications for ash load exceeding 10 kg m^{-2} during dry and rainy season conditions in Bandung and Cirebon are highlighted in Figure 3c, d. In Bandung, there is a <1% likelihood of exceeding 10 kg m^{-2} during both dry and rainy conditions: in Cirebon, however, there is a <1% likelihood during the dry season, but a 1–10% probability during rainy season conditions of exceeding this ash-load threshold.

Isochron maps are presented here for the ash-load thresholds and meteorological conditions delineated above. Isochron maps depict the accumulation time in hours for volcanic ash load to exceed the designated threshold value. These maps have important implications for the temporal aspects of volcanic ash hazard (i.e. how long will it takes for ash to exceed a threshold at a site of interest, such as a hospital or school).

The averaged accumulation times (in hours) for the two threshold values of ground load (1 and 10 kg m^{-2}) and for both dry and rainy season meteorological conditions are presented as isochron contours in 1 h increments in Figure 4. Accumulation rates for ash loads exceeding 1 kg m^{-2} during dry season conditions are typically higher at locations west of the active vent compared with locations to the east. In rainy season conditions, accumulation rates are more evenly spread to the east and west of locations, exceeding 1 kg m^{-2}. For ash loads exceeding 10 kg m^{-2}, the pattern of accumulation is similar for dry season conditions, with a slight bias to the west, and for rainy season conditions where accumulation rates are more evenly distributed to the east and west. Typical accumulation times for ground loading at medial to distal localities was 8–9 h for dry season conditions and <8 h for rainy season conditions.

Discussion

The dispersal of volcanic ash through the atmosphere and subsequent deposition at ground level is heavily influenced by the prevailing wind conditions for the region, and the affects of seasonal variations in wind speed, direction and temperature are reflected in the distribution footprint. These seasonal variations have important implications for impact on crop damage/destruction, building damage/collapse and transportation accessibility associated with surrounding communities at risk. During the dry season (April–September), a high proportion of volcanic ash is likely to accumulate to the west of Ciremai. This suggests that damage and/or destruction of crops (1 kg m^{-2}) and surficial damage to buildings (10 kg m^{-2}) in and around the proximal communities of the western flanks of Ciremai is likely to occur during the dry season wind conditions. The city of Bandung to the west is likely to be significantly impacted by volcanic ash from Ciremai during dry season wind conditions. Crop damage or destruction is highly likely (>50%): however, surficial damage to buildings is less likely (<1%) owing to the distance from the vent. Although, even a small degree of inundation will have important implications on the disruption of transportation pathways during an eruption (e.g. airport closure, major road closures). Comparatively less ash was simulated on the eastern slopes of Ciremai across the city of Cirebon during dry season wind conditions, which is likely to be the result of strong prevailing westerly winds at this time of year. The implications are that communities located on the eastern flanks of Gunung Ciremai, including Cirebon, are comparatively less susceptible, but not completely immune, to volcanic ash ground loading during dry season meteorological conditions (1–10% likelihood of crop damage/destruction) than those located on the western side.

During rainy season meteorological conditions, the accumulation of ground load follows a similar pattern of distribution towards the west (up to 100 km) as during the dry season. However, a marked elongation of the footprint to the east has important implications for Cirebon. A higher proportion of crops could be adversely impacted by volcanic ash ground loading (20% likelihood) and, owing to its proximity to Ciremai, there is a 1–10% probability that buildings across Cirebon would experience surficial damage related to volcanic ash ground loading during an eruption that occurred in rainy season wind conditions.

The averaged accumulation times (or number of hours before ash load exceeds 1 and 10 kg m^{-2}) at a given location to the west of Ciremai (e.g. western Bandung) during dry season meteorological conditions (6–7 h) v. rainy season meteorological conditions (7–8 h) supports the presence of a strong westerly wind influence during the dry season that distributes ground load further from the vent source.

Conclusions

Volcanic ash represents a serious hazard to communities living in the vicinity of active volcanoes in developing countries such as Indonesia. An application example for the volcanic ash dispersion model

PF3D is presented here for Gunung Ciremai in West Java, Indonesia. A stochastic set of eruption events was simulated using eruptive parameters within the acceptable range of possible future events for this volcano: granulometry, as determined through field studies, and a meteorological dataset that represents the complete range of possible wind conditions expected during the dry and rainy seasons for the region. Communities on the western flanks of Ciremai were found to be highly susceptible to volcanic ash ground loading regardless of the season: whereas, communities on the eastern side were found to be more susceptible during the rainy season months. The city of Bandung to the west was found to be highly susceptible (50–60%) to ash loads exceeding 1 kg m^{-2} during dry season wind conditions, which has important implications for crop damage/destruction and the disruption of critical transportation pathways during an eruption. The city of Cirebon to the east was found to be more susceptible to surficial building damage (1–10%) during rainy season wind conditions than Bandung. This is attributed to prevailing wind conditions during the rainy season that include a strong easterly component and the more proximal location of Cirebon to the active vent. Emergency managers and disaster risk reduction workers can use volcanic ash hazard maps, such as those produced from this study, to inform decision-making and to focus mitigation efforts on communities most at risk.

This paper presents research undertaken in collaboration between Geoscience Australia, Badan Geologi and the Australia–Indonesia Facility for Disaster Reduction. We greatly acknowledge A. Folch and A. Costa for their technical assistance and numerous scientific discussions throughout the development of PF3D. The authors also wish to thank T. Kuske, V. Miller and A. Jones for their constructive reviews that greatly improved the quality of this manuscript.

References

BARBERI, F., MACEDONIO, G., PARESHCI, M.T. & SANTACROCE, R. 1990. Mapping tephra fallout risk: an example from Vesuvius, Italy. *Nature*, **344**, 142–144.

BARSOTTI, S., NERI, A. & SCIRE, J.S. 2008. The VOLCALPUFF model for atmospheric ash dispersal: 1. Approach and physical formulation. *Journal of Geophysical Research-Solid Earth*, **113**, B03208, https://doi.org/10.1029/2006JB004623

BEAR-CROZIER, A.N., KARTADINATA, N., HERIWASESO, A. & NIELSEN, O. 2012. Development of python-FALL3D: a modified procedure for modelling volcanic ash dispersal in the Asia-Pacific region. *Natural Hazards*, **64**, 821–838.

BLONG, R. 2003. Building damage in Rabaul, Papua New Guinea, 1994. *Bulletin of Volcanology*, **65**, 43–54.

BONADONNA, C. & COSTA, A. 2013. Modeling of tephra sedimentation from volcanic plumes. *In*: FAGENTS, S.A., GREGG, T.K.P. & LOPES, R.M.C. (eds) *Modeling Volcanic Processes. The Physics and Mathematics of Volcanism.* Cambridge University Press, Cambridge, 173–202.

CIONI, R., LONGO, A., MACEDONIO, G., SANTACROCE, R., SBRANA, A., SULPIZIO, R. & ANDRONICO, D. 2003. Assessing pyroclastic fall hazard through field data and numerical simulations: example from Vesuvius. *Journal of Geophysical Research: Solid Earth*, **108**, 2063.

COSTA, A., MACEDONIO, G. & FOLCH, A. 2006. A three-dimensional Eulerian model for transport and deposition of volcanic ashes. *Earth and Planetary Science Letters*, **241**, 634–647.

COSTA, A., DELL'ERBA, F., DI VITO, M.A., ISAIA, R., MACEDONIO, G., ORSI, G. & PFEIFFER, T. 2009. Tephra fallout hazard assessment at Campi Flegrei caldera (Italy). *Bulletin of Volcanology*, **71**, 259–273.

FOLCH, A. & COSTA, A. 2010. *FALL3D-6.2 User Guide.* Barcelona Supercomputing Center – Centro Nacional de Supercomputacion, Barcelona.

FOLCH, A. & FELPETO, A. 2005. A coupled model for dispersal of tephra during sustained explosive eruptions. *Journal of Volcanology and Geothermal Research*, **145**, 337–349.

FOLCH, A. & SULPIZIO, R. 2010. Evaluating long-range volcanic ash hazard using supercomputing facilities: application to Somma–Vesuvius (Itay), and consequences for civil aviation over the Central Mediterranean Area. *Bulletin of Volcanology*, **72**, 1039–1059.

FOLCH, A., JORBA, O. & VIRAMONTE, J. 2008. Volcanic ash forecast – application to the May 2008 Chaiten eruption. *Natural Hazards and Earth System Sciences*, **8**, 927–940.

FOLCH, A., COSTA, A. & MACEDONIO, G. 2009. FALL3D: a computational model for transport and deposition of volcanic ash. *Computers and Geosciences*, **35**, 1334–1342.

GHOSE, R., YOSHIOKA, S. & OIKE, K. 1990. Three dimensional numerical simulation of the subduction dynamics in the Sunda arc region, southeast Asia. *Tectonophysics*, **181**, 223–255.

HAMILTON, W. 1979. *Tectonics of Indonesia Region.* United States Geological Survey, Professional Paper, **1078**.

HURST, A.W. 1994. *ASHFALL – A Computer Program for Estimating Volcanic Ash Fallout (Report and User Guide).* Institute of Geological & Nuclear Sciences Science Report, **94**.

HURST, A.W. & TURNER, R. 1999. Performance of the program ASHFALL for forecasting ashfall during the 1995 and 1996 eruptions of Ruapehu volcano. *New Zealand Journal of Geology and Geophysics*, **42**, 615–622.

KATILI, J.A. 1975. Volcanism and plate tectonics in the Indonesian island arcs. *Tectonophysics*, **26**, 165–188.

MACEDONIO, G., COSTA, A. & LONGO, A. 2005. A computer model for volcanic ash fallout and assessment of subsquent hazard. *Computers and Geosciences*, **31**, 837–845.

MACEDONIO, G., COSTA, A. & FOLCH, A. 2008. Ash fallout scenarios at Vesuvius: numerical simulations and implications for hazard assessment. *Journal of Volcanology and Geothermal Research*, **178**, 366–377.

PUSPITO, N.T. & SHIMAZAKI, K. 1995. Mantle structure and seismotectonics of the Sunda and Banda arcs. *Tectonophysics*, **251**, 215–228.

SEARCY, C., DEAN, K. & STRINGER, W. 1998. PUFF: a high-resolution volcanic ash tracking model. *Journal of Volcanology and Geothermal Research*, **80**, 1–16.

SENDJAJA, Y.A., KIMURA, J.I. & SUNARDI, E. 2009. Across-arc geochemical variation of Quaternary lavas in West Java, Indonesia: mass-balance elucidation using arc basalt simulator model. *Island Arc*, **18**, 201–224.

SIMPSON, A., JOHNSON, R.W. & CUMMINS, P. 2011. Volcanic threat in developing countries of the Asia-Pacific region: probabilistic hazard assessment, population risks, and information gaps. *Natural Hazards*, **57**, 151–165.

SITUMORANG 1995. *Geological Map of Mount Ciremai.* Directorate of Volcanology, Indonesia [in Indonesian].

SOERIA-ATMADJA, R. & NOERADI, D. 2005. Distribution of early tertiary volcanic rocks in south Sumatra and west Java. *Island Arc*, **14**, 679–686.

SULPIZIO, R., FOLCH, A., COSTA, A., SCAINI, C. & DELLINO, P. 2012. Hazard assessment of far-range volcanic ash dispersal from a violent Strombolian eruption at Somma-Vesuvius volcano, Naples, Italy: implications on civil aviation. *Bulletin of Volcanology*, **74**, 2205–2218.

SUZUKI, T. 1983. A theoretical model for dispersion of tephra. *In*: SHIMOZURU, D. & YOKOYAMA, I. (eds) *Arc Volcanism: Physics and Tectonics*. Terra Scientific Publishing, Tokyo, 95–116.

TERRY, J.P. & GOFF, J.R. 2012. The special vulnerability of Asia-Pacific islands to natural hazards. *In*: TERRY, J.P. & GOFF, J.R. (eds) *Natural Hazards in the Asia-Pacific Region: Recent Advances and Emerging Concepts*. Geological Society, London, Special Publications, **361**, 3–5, https://doi.org/10.1144/SP361.2

WIDIYANTORO, S. & VAN DER HILST, R. 1996. Structure and evolution of the lithospheric slab beneath the Sunda arc, Indonesia. *Science*, **271**, 1566–1570.

WIDIYANTORO, S. & VAN DER HILST, R. 1997. Mantle structure beneath Indonesia inferred from high-reolution tomographic imaging. *Geophysical Journal International*, **130**, 167–182.

Understanding the trigger for the LUSI mud volcano eruption from ground deformation signatures

HERI ANDREAS[1]*, HASANUDDIN Z. ABIDIN[1], TEGUH P. SIDIQ[1], IRWAN GUMILAR[1], YOSUKE AOKI[2], AGUS L. HAKIM[3] & PRIHADI SUMINTADIREDJA[4]

[1]*Geodesy Research Division, Institute of Technology Bandung, Indonesia*

[2]*Earthquake Research Institute, Tokyo University, Tokyo, Japan*

[3]*PT, Asaba, Jakarta, Indonesia*

[4]*Applied Geology Research Group, Institute of Technology Bandung, Indonesia*

**Correspondence: heriandreas49@gmail.com*

Abstract: The LUSI mud volcano in the sub-district of Porong, Sidoarjo, East Java, Indonesia started to erupt on 29 May 2006. An almost continuous eruption of a mixture of mud, water and gas has occurred around this area since this date. The eruption triggered vertical and horizontal ground deformation. From June 2006 to December 2010, 14 global positioning system campaigns were conducted to observe the ground deformation using *c.* 50 stations sparsely located up to 10 km from the eruption centre. Field observations of cracks, terrestrial laser scanning and geo-electrical measurements have also been used to infer the ground deformation signature around the LUSI mud volcano. More than 150 pairs of interferograms generated from 66 ALOS PALSAR images from June 2006 to December 2009 have also been used to study the ground deformation caused by the LUSI mud volcano. The LUSI mud eruption began only 200 m from where the Lapindo Inc. oil company was drilling for oil and gas. The drilling may have pierced a deeper high-pressure zone, causing an underground blow-out of the drillhole into a hydrofracture. Alternatively, the magnitude 6.3 Yogyakarta earthquake, which was located *c.* 275 km from the eruption site and occurred two days before the LUSI eruption, may have shaken the area sufficiently to cause the eruption by reactivating a fault in the region and liquefying the mud. These two hypotheses for triggering the mud volcano have been argued vehemently and still remain controversial. The ground deformation signatures provide important clues to understanding the trigger for the eruption and to solve this controversy. Co-seismic fault reactivation has its own typical ground deformation signature. This study used global positioning system and InSAR techniques, as well as field observations of cracks, terrestrial laser scanning and geo-electrical measurements, to determine the signature of ground deformation around the LUSI mud volcano and to explain the triggering mechanism.

A mud volcano started to form in the sub-district of Porong, Sidoarjo, East Java, Indonesia on 29 May 2006 (Davies *et al.* 2007; Istadi 2007; Mazzini *et al.* 2007). It is referred to here as the LUSI mud volcano (LUSI = Lumpur Sidoarjo; *lumpur* means mud in the Indonesian language). A massive extrusion of mud, water and gas inundated an area of several square kilometres and caused significant damage to the livelihoods of local people, the environment and the local infrastructure, which continues to this day.

The volumes of mud erupted increased from an initial 5000 m^3 day^{-1} in the early stages to 120 000 m^3 day^{-1} in August 2006. Peaks of 160 000 and 170 000 m^3 day^{-1} of erupted material followed earthquakes swarms during September 2006; in December 2006 the flux reached a maximum of 180 000 m^3 day^{-1} and in June 2007 the mud volcano was still expelling >110 000 m^3 day^{-1} (Mazzini *et al.* 2007). By four years after the eruption, the volume of mud erupted each day had significantly decreased.

Ground deformation occurred immediately after the eruption as a result of the effects of mud loading, collapse of the overburden due to the removal of mud from the subsurface, and land settlement caused by ground works (e.g. the construction of dykes) around the LUSI mud volcano. At the start of the eruption (2006–07), global positioning system (GPS) surveys carried out by Abidin *et al.* (2008) and InSAR analysis by Fukushima *et al.* (2009) clearly explained the ground deformation around the LUSI mud volcano. Vertical and horizontal deformation rates of up to *c.* 4 and 1 cm day^{-1}, respectively, were observed in this period.

The LUSI mud eruption began only 200 m from where the oil company Lapindo Inc. was drilling for

From: Cummins, P. R. & Meilano, I. (eds) 2017. *Geohazards in Indonesia: Earth Science for Disaster Risk Reduction*. Geological Society, London, Special Publications, **441**, 199–212.
First published online July 10, 2017, https://doi.org/10.1144/SP441.10

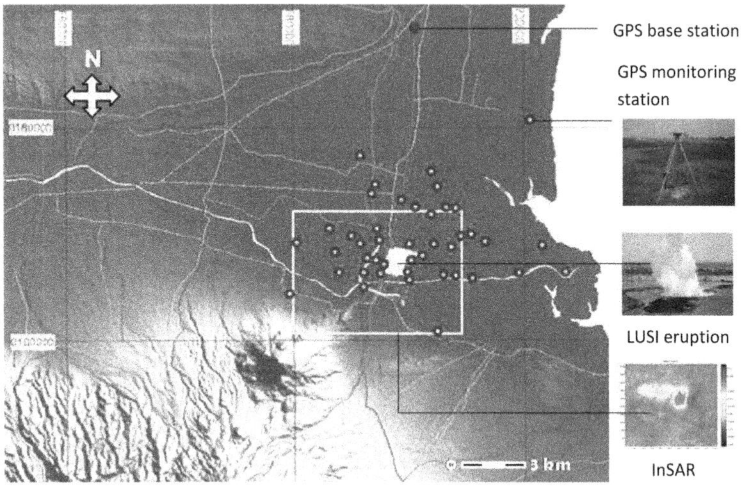

Fig. 1. Map of the LUSI eruption area, together with the GPS base and monitoring stations and the InSAR study area used in the ground deformation investigations.

oil and gas. Drilling may have pierced a deeper high-pressure zone, causing an underground blow-out of the drillhole into a hydrofracture. Alternatively, the magnitude 6.3 Yogyakarta earthquake, located *c.* 275 km from the eruption and occurring two days before the LUSI eruption, may have shaken the area sufficiently to cause the eruption by reactivating a fault in the region and liquefying the mud. These two alternative hypotheses for the trigger have been argued vehemently and remain controversial.

To solve this controversy, we show here that an analysis of ground deformation is the key to understanding the triggering of the LUSI eruption. Co-seismic fault reactivation has its own typical ground deformation signature. This paper uses GPS and InSAR techniques, as well as field observations of ground surface cracks, terrestrial laser scanning and geo-electrical measurements to determine the signature of ground deformation around the LUSI mud volcano and to elucidate the triggering mechanism.

Ground deformation at LUSI monitored by GPS and InSAR

GPS observations, in both campaign and continuous modes, were conducted to study the ground deformation following the birth and development of the LUSI mud volcano. Fourteen GPS campaigns were conducted between June 2006 and December 2010. GPS surveys were performed with up to 50 stations deployed over an area within 10 km of the centre of eruption using dual-frequency geodetic-type receivers, with observation session lengths of about 5–10 h. Continuous subsidence monitoring was conducted by GPS on some stations from 22 September 2006 to early 2007. Figure 1 shows the locations of the GPS monitoring stations on a map of the LUSI area. As a result of the change in the coverage area of the mud, the number of observed GPS stations was different from one survey to the next and the observed stations were not always the same. The locations of the GPS stations were also restricted by the mud coverage and its progression.

Data processing of the GPS survey was conducted using the GPS processing software Bernese 5.0 (Beutler *et al.* 2007). In general, the standard deviations of the estimated coordinates are of the order of several millimetres in both the horizontal and vertical components. We applied simple differencing to the coordinates processed for each period of the GPS surveys to derive the ground deformation value.

Using the InSAR measurements, >150 pairs of interferograms were generated from 66 ALOS (Advanced Land Observing Satellite) PALSAR images from June 2006 to December 2009. A reference satellite image from 19 May 2006 (before the eruption started) was used to generate interferograms for the periods between this date and subsequent dates. The temporal resolution of the images varied from 46 to 138 days. The interferometric pairs were selected with a maximum perpendicular baseline of 2 km, which allowed us to resolve the signals associated with the activity of the LUSI mud volcano. Interferograms from both ascending and descending orbits were used to allow us to constrain the direction of deformation. Figure 1 shows

Table 1. *Magnitude of subsidence one, two and four years after the LUSI mud volcano eruption*

Point	Distance from LUSI eruption centre (km)	Magnitude of subsidence (m)		
		First year after eruption	Two years since eruption	Four years since eruption
Inner caldera (0–1 km from LUSI eruption)				
RIG1	0.10	−8.716	No observations	
JBSR	0.37	−6.868		
PBRK	0.54	−4.06		
TOLL	0.47	−3.638		
SIRN	0.46	−3.544		
JTRJ	0.95	−1.732		
PSKO	0.76	−1.564		
Rim of caldera (1–2 km from LUSI eruption)				
JLSR	1.00	−0.696	−0.973	−1.250
RMKN	1.08	−0.576	−0.783	−0.949
SNGN	1.24	−0.542	−0.788	−0.989
PRTL	1.32	−0.723	−0.963	−1.153
GLAG	1.40	−0.341	−0.520	−0.679
BPN1	1.62	−0.322	−0.454	−0.583
1210	1.74	−0.060	−0.090	−0.130
PJKN	1.78	−0.126	−0.156	−0.196
GMPL	1.84	−0.310	−0.400	−0.490
SWH1	1.91	−0.171	−0.241	−0.301
Outer rim of caldera (>2 km from LUSI eruption)				
JBPR	2.32	−0.090	−0.120	−0.170
PGN1	2.32	−0.090	−0.099	−0.159
BMT2	2.33	−0.125	−0.211	−0.243
BT16	2.48	−0.060	−0.100	−0.137
BMT4	2.53	−0.069	−0.114	−0.149
KCMT	3.21	−0.050	−0.078	−0.105
CNDI	3.21	−0.250	−0.391	−0.565
KLDN	3.40	−0.149	−0.190	−0.222
GPOL	3.88	−0.105	−0.145	−0.163
BM08	3.99	−0.080	−0.125	−0.143
ORFF	4.41	−0.075	−0.083	−0.110

the InSAR investigation area and the GPS monitoring stations and topographic map.

The interferograms were interpreted as a time-dependent range using a least-squares inversion and a linear inversion methodology. Quality control of the data was achieved by choosing only those pixels whose coherence exceeded a threshold of 0.4. This low value of coherence was applied because the coherence is generally low in the study area.

GPS results

Tables 1 and 2 present the magnitude and rates of subsidence for one, two and four years since the LUSI eruption started, with the GPS monitoring stations arranged in order of distance from the eruption so that we could carry out spatial analyses to determine how far the eruption affected the deformation and how the magnitude was related to the distance from the eruption. Other researchers have mainly focused on the vertical component of deformation or subsidence (e.g. Abidin *et al.* 2008; Fukushima *et al.* 2009; Rudolph *et al.* 2013), whereas we consider here both the vertical and horizontal components. Table 3 shows the results for the horizontal components.

To interpret the ground deformation around the LUSI mud volcano, we divided the measurements into spatial domains corresponding to inside the caldera, near the rim of the caldera and outside the caldera (Davies *et al.* 2007, 2010). Our data distinguished the different characteristics of deformation among these spatial domains. Deformation inside the caldera is associated with caldera formation processes and is characterized by rapid subsidence, while the rim and outside of the caldera have normal

Table 2. *Rates of subsidence one, two and four years years after the LUSI eruption*

Point	Distance from LUSI eruption (km)	Rate of subsidence at year 1 (m a^{-1})	Rate of subsidence at year 2 (m a^{-1})	Rate of subsidence at year 4 (m a^{-1})
Inner caldera (0–1 km from LUSI eruption)				
RIG1	0.10	−8.716	No observations	
JBSR	0.37	−6.868		
SIRN	0.46	−3.544		
TOLL	0.47	−3.638		
PBRK	0.54	−4.060		
BND2	0.76	−1.995		
JTRJ	0.95	−1.732		
Rim of caldera (1–2 km from LUSI eruption)				
JLSR	1.00	−0.696	−0.277	−0.093
RMKN	1.08	−0.576	−0.207	−0.068
SNGN	1.24	−0.542	−0.246	−0.068
PRTL	1.32	−0.723	−0.240	−0.060
GLAG	1.40	−0.341	−0.179	−0.054
BPN184	1.62	−0.322	−0.132	−0.044
1210	1.74	−0.060	−0.030	−0.020
PJKN	1.78	−0.126	−0.030	−0.020
GMPL	1.84	−0.310	−0.090	−0.030
SWH1	1.91	−0.171	−0.070	−0.020
Outer of Caldera (>2 km from LUSI eruption)				
JBPR	2.32	−0.090	−0.030	−0.020
PGN1	2.32	−0.090	−0.009	−0.020
BMT2	2.33	−0.125	−0.086	−0.021
BT16	2.48	−0.060	−0.040	−0.010
BMT4	2.53	−0.069	−0.044	−0.012
KCMT	3.21	−0.050	−0.028	−0.009
CNDI	3.21	−0.250	−0.142	−0.056
KLDN	3.40	−0.149	−0.041	−0.014
GPOL	3.88	−0.105	−0.040	−0.008
BM08	3.99	−0.080	−0.045	−0.008
ORFF	4.41	−0.075	−0.008	−0.007

rates of subsidence (Andreas *et al.* 2010), which are most probably (Abidin *et al.* 2008) associated with the effects of mud loading and ground relaxation due to the outflow of mud.

Note that ground deformation (e.g. subsidence) around the inside of the caldera was not observable after one year post-eruption, which is probably due to the catastrophic nature of the collapse processes during caldera formation. In addition, the expansion of mud coverage and its loading effect has made observing, or even the extrapolation, of ground deformation nearly impossible. For this reason, we did not include information on such deformation inside the caldera after the first year of eruption (Tables 1 & 2, Figs 2–4).

For one year after the LUSI eruption, the magnitudes of subsidence inside the caldera varied between −1 and −8 m, whereas values between −1 and −7 dm were recorded around the caldera rim. Outside the caldera we observed only a few centimetres of subsidence. Four years after the eruption, −0.3 to −1.2 m of subsidence was recorded around the caldera rim, whereas several decimetres of subsidence were recorded outside the caldera. The inside of the caldera is assumed to be 0–1 km, the rim of the caldera is assumed to be 1–2 km and the outside of the caldera is assumed to be >2 km from the LUSI eruption.

Figure 2a shows the contours of the magnitude of subsidence for two years after the start of the LUSI eruption and Figure 2b describes the same for four years after the start of the eruption. The typical cone-shaped caldera of subsidence created by the LUSI mud volcano is clearly seen in both images. Figures 3 and 4 each show profiles of the magnitude and rate of subsidence for one, two and four years after the LUSI eruption. It can be seen that subsidence mostly affected an area within 1 km of the eruption centre (i.e. inside the caldera) and the rates clearly decrease after four years of eruption.

For the horizontal components (Figs 5 & 6), we can see that the displacement vectors consistently

Table 3. *Vectors of horizontal ground displacement from random epoch of LUSI mud volcano system*

Point name	Vectors of horizontal ground displacement from random epoch	
	East (m)	North (m)
RIG1	−4.66	−8.716
JBSR	−2.29	−6.868
SIRN	−1.85	−3.544
TOLL	−1.92	−3.638
PBRK	−2.15	−4.060
PSKO	−0.84	−1.564
PS02	−0.52	−1.564
BND2	−0.89	−1.995
JTRJ	−0.92	−1.732
JBSR	−0.29	−0.696
RMKN	−0.20	−0.576
SNGN	−0.27	−0.542
PRTL	−0.34	−0.723
GLAG	−0.17	−0.341
BPN184	−0.11	−0.322
1210	0.00	−0.060
PJKN	−0.05	−0.126
GMPL	−0.15	−0.310
SWH1	−0.09	−0.171
JBPR	−0.02	−0.090
PGN1	−0.03	−0.090
BMT2	−0.03	−0.125

point towards the centre of the LUSI eruption. This orientation of vectors is consistent with the cone-shaped pattern of subsidence expected when subsurface material is extruded to the surface (e.g. a mud volcano eruption). It is difficult to explain this pattern by fault motion because this should involve a discontinuity of either the horizontal (strike-slip faulting) or vertical (thrust or normal faulting) motion along the trace of the fault. Thus a combined analysis of both the horizontal and vertical signatures of deformation is crucial in investigating the trigger of the LUSI eruption.

InSAR results

After processing the InSAR data, the ground deformation was generally well resolved in the LUSI area, with a cone-shaped ellipsoid pattern clearly visible from the interferograms (Fig. 7, area A). This type of ellipsoid pattern represents a typical caldera formation, as suggested by the GPS results. West of the LUSI area (Fig. 7, area B), we can see an additional signal of ground deformation, which may correspond to oil and gas extraction because many wells are established in this area. Fukushima *et al.* (2009) and Abidin *et al.* (2008) reached the same conclusion. Further analysis is recommended to better understand the uplift signal on the east side of the LUSI eruption; this signal is of unknown origin and

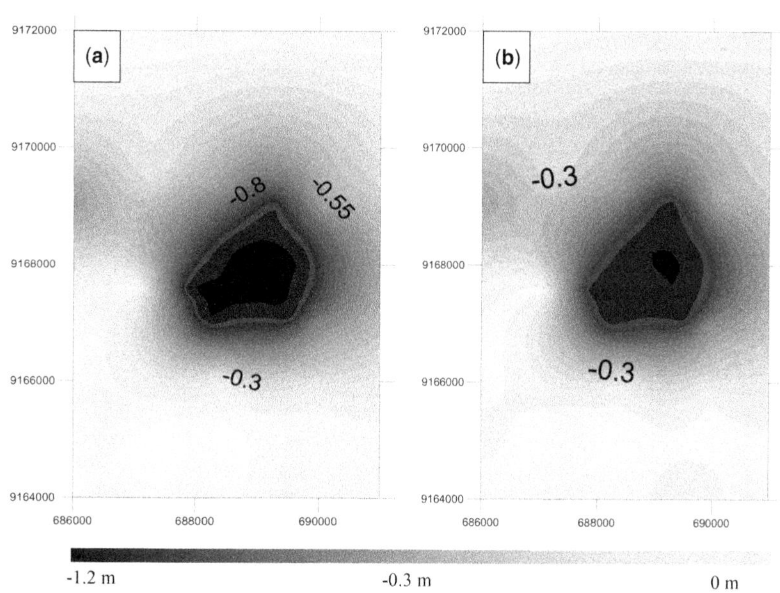

Fig. 2. Filled contour graph of subsidence magnitude for (**a**) two and (**b**) four years after the LUSI mud volcano eruption. The typical cone-shaped subsidence caldera can be clearly seen.

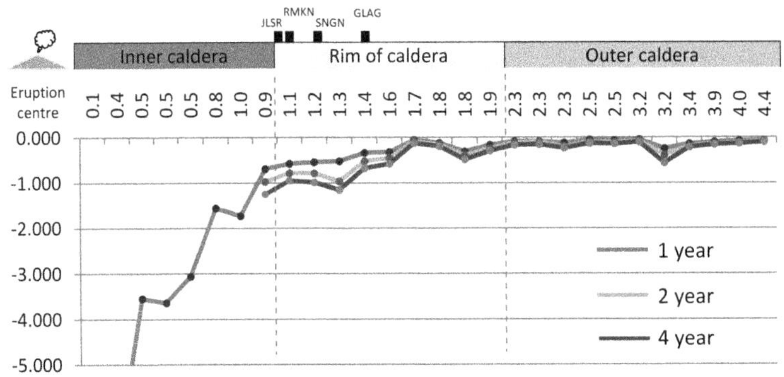

Fig. 3. Subsidence magnitude profiles one, two and four years after the LUSI mud volcano eruption; the horizontal axis is the distance from the eruption centre (in kilometres) and the vertical axis is the magnitude of subsidence (in metres). Each year is represent by different colour (please see online version for colour reference). The positions of some of the principal GPS monitoring stations (e.g. JLSR, RMKN, SNGN and GLAG) are also plotted.

as yet we can only suggest that it may be an atmospheric effect or some other signal processing error. InSAR also shows a decreased rate of subsidence. Table 4 and the graph of the InSAR control points (e.g. CP01 and CP02) (see Figure 9) represent the rate of subsidence.

As a result of low coherence inside and adjacent to the edifice of LUSI (inside the caldera), the ground deformation there was poorly resolved. In these low coherence areas, the InSAR result cannot be shown to be consistent with the ground deformation measured by GPS. Nevertheless, very good agreement between the GPS and InSAR results can be seen everywhere else. A detailed comparison between the GPS and InSAR results is beyond the scope of this paper, but we suggest it should be conducted in the future to better constrain the ground deformation. Some differences are likely to occur due to the different technologies and analyses used. It is important to note that the InSAR results show the line-of-sight deformation, not the true vertical or horizontal deformation.

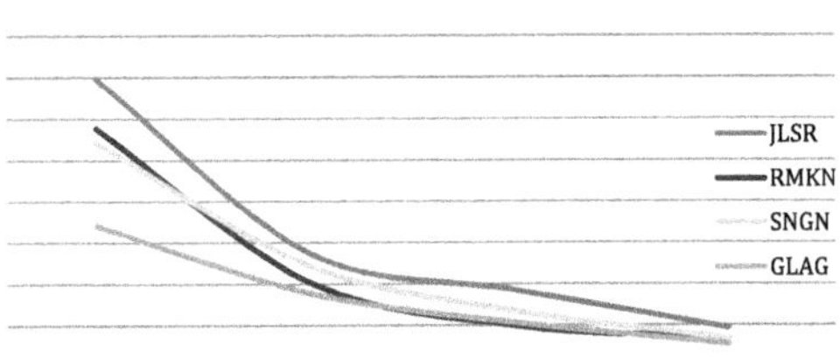

Fig. 4. Subsidence histories measured yearly at several GPS monitoring stations (e.g. JLRS, RMKN, SNGN and GLAG), each located on the caldera rim. y-axis = rate (m a^{-1}), x-axis = year. Each monitoring station shows a decreasing rate from one year to the next, especially after four years.

Figure 8 shows the magnitude of subsidence derived from InSAR processing from May 2006 to the months of April 2007, July 2007, October 2007, January 2008, April 2008, July 2008, October 2008, January 2009 and April 2009. As a result of the low coherence inside and adjacent to the edifice of LUSI (the inner caldera), any large surface deformation there is only poorly revealed. In this case, the InSAR result cannot resolve the surface deformation revealed by the GPS results closest to the caldera.

Table 4 and Figure 9 show the InSAR check points (CP01 and CP02) prepared for spatial and temporal analysis, as well as a general comparison with the GPS results. The maximum subsidence recorded by InSAR (around the rim of caldera) was −12 cm in the year after eruption and −51 cm three years after the eruption. A decrease in the rate of subsidence can be seen after one year and later from both CP01 and CP02 (−0.33 to −0.06 m for CP01 and −0.16 to −0.10 m for CP02, each for years 1 and 3). The positive agreement between the GPS and InSAR results is clear.

There has been a decrease in the rate of subsidence since the co-eruption deformation, shown as an exponential decay in our GPS and InSAR results. This decrease in the deformation rate may indicate that the forces that initially drove the eruption have been spent, perhaps signalling the demise of the LUSI mud volcano eruption. Rudolph *et al.* (2013) reached the same conclusion and were confident that the lifetime of the eruption will probably be shorter than other workers have predicted. This is good news for mitigation and adaptation in the area surrounding LUSI. Nevertheless, without further monitoring and analysis we cannot know whether the eruption will resume in the future.

Fig. 5. Horizontal ground displacement vectors around the LUSI mud volcano overlain on a Google Earth graphic. (Google Earth images from Landsat/Copernicus.)

Ground deformation at LUSI from crack observations and geophysical measurements

In addition to the GPS observations (both the campaign and the continuous measurements) and the InSAR measurements, which both clearly show ground deformation around the LUSI mud volcano, ground deformation can also be observed directly in the field as cracks in the ground surface. Mazzini *et al.* (2007) used these observations as evidence of fault scarp formation indicating fault reactivation. Figure 10 shows a map of our crack observations, together with photographs. As it was important in this study to understand the direction and dip of the cracks, geo-electrical measurements were conducted around the LUSI area. Figure 11 shows these geo-electrical measurements together with the results of modelling.

From the crack observations and geo-electrical measurements, we found interesting signatures that could explain the physical nature of the LUSI eruption, including what might have been its

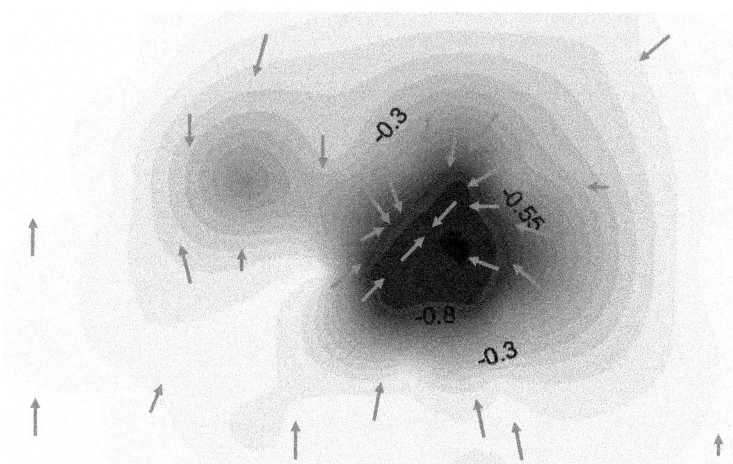

Fig. 6. Horizontal ground displacement pattern around the LUSI mud volcano overlain with the subsidence plotted as filled contours.

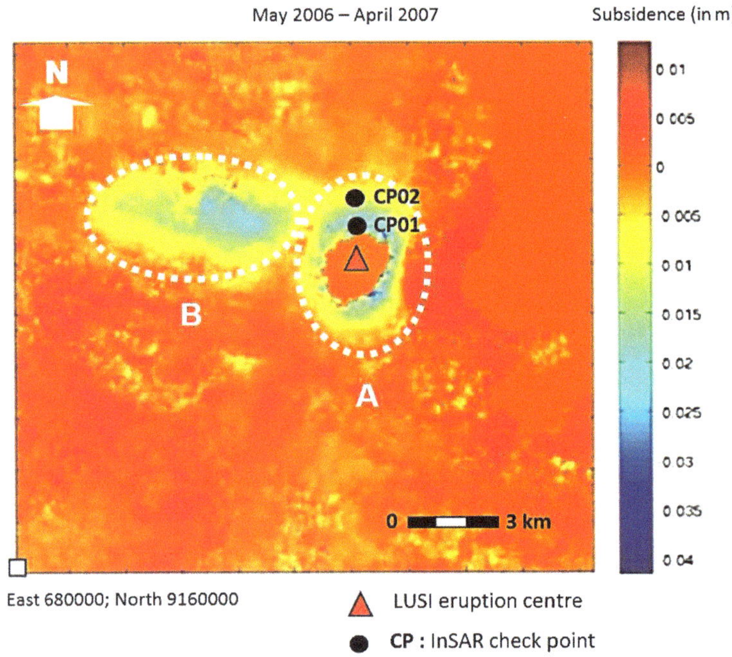

Fig. 7. Example of subsidence image from InSAR (May 2006–April 2007). An ellipsoid pattern is clearly visible from the interferograms (area A). We also can see a signal of subsidence in the western part (area B). Also shown are the mark of the LUSI centre eruption site and the CP01 and CP02 InSAR checkpoints.

triggering mechanism. We observed cracks surrounding the centre of the LUSI eruption with dips directed inwards, which clearly suggests caldera formation rather than scarp formation on reactivated faults. Thus the suggestion that fault reactivation by the Yogyakarta earthquake was a major factor in the LUSI eruption seems tenuous.

Understanding the LUSI trigger from ground deformation signatures

Base on earthquake-induced hydrological effects and a reactivated the fault system, Mazzini *et al.* (2007, 2009) concluded that the Yogyakarta earthquake triggered the LUSI eruption. Davies *et al.* (2010) compared the distance and magnitude of the earthquake with the relationship between the distance and magnitude of historical earthquakes that have caused sediment liquefaction, triggered the eruption of mud volcanoes or caused other hydrological responses. Using this comparison, they found that triggering of the LUSI eruption by the Yogyakarta earthquake was unlikely. Lupi *et al.* (2013) proposed that a concave rock layer capping the mud reservoir could have focused and amplified incoming seismic waves from the Yogyakarta earthquake and could thus have triggered the LUSI mud eruption.

Our approach, taking into account some of the details of ground deformation associated with the LUSI eruption, can explain whether there is a

Table 4. *Magnitude and rate of subsidence (yearly) at InSAR check points*

Point	Magnitude of subsidence (m)					Rate of subsidence at year 2 (m a^{-1})	Rate of subsidence at year 3 (m a^{-1})
	Year 1	Year 1.5	Year 2	Year 2.5	Year 3		
CP01	−0.120	−0.300	−0.450	−0.500	−0.510	−0.330	−0.060
CP02	−0.040	−0.120	−0.220	−0.270	−0.300	−0.160	−0.100

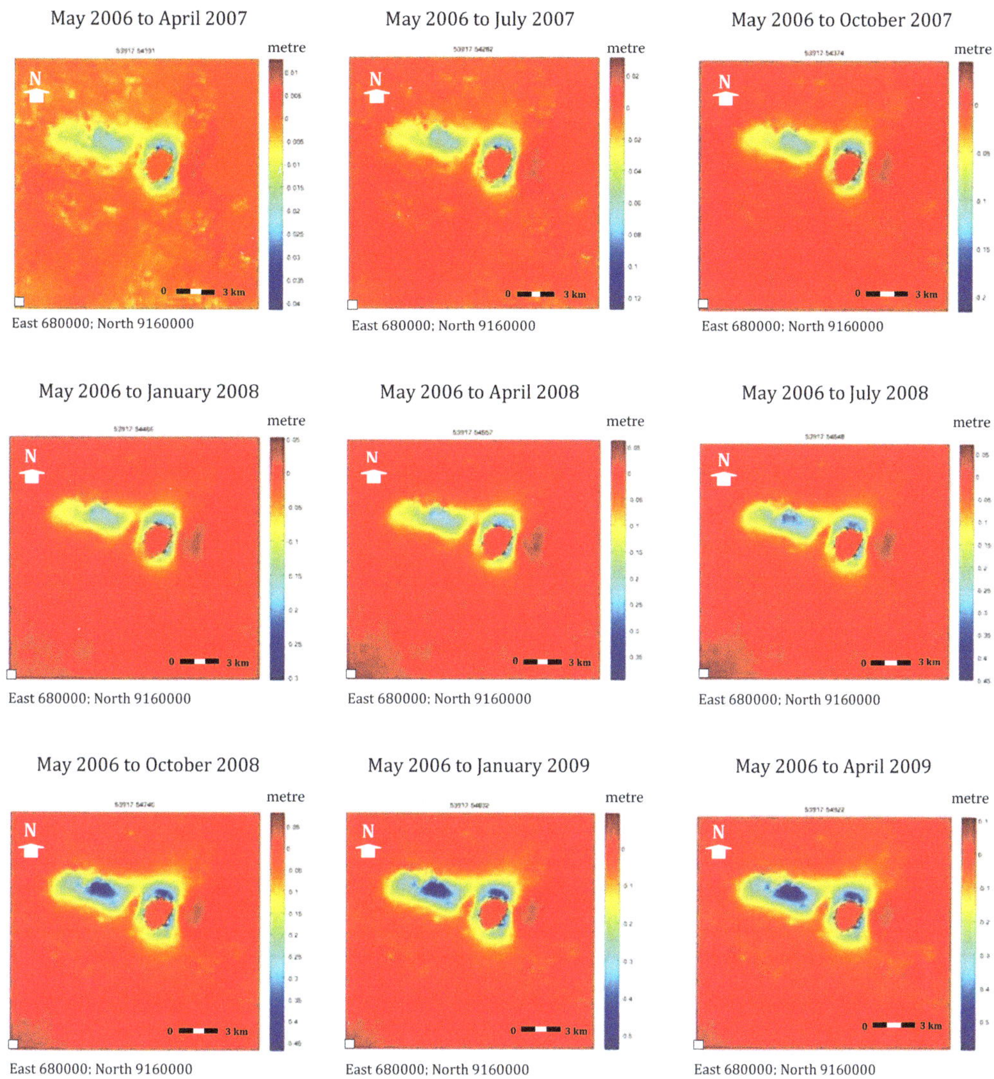

Fig. 8. Images of subsidence (in metres) derived from InSAR processing from May 2006 to the subsequent months of April 2007, July 2007, October 2007, January 2008, April 2008, July 2008, October 2008, January 2009 and April 2009. The maximum deformations recorded are: April 2007, 0.04 m; July 2007, 0.12 m; October 2007, 0.20 m; January 2008, 0.30 m; April 2008, 0.35 m; July 2008, 0.45 m; October 2008, 0.45 m; January 2009, 0.50 m; and April 2009, 0.51 m. As there was poor coherence near the mud eruption, the large deformation signal (as in the GPS results) cannot be resolved by InSAR.

connection between the eruption and the 2006 Yogyakarta earthquake and, in particular, whether the eruption was triggered through the reactivation of faults in the LUSI area. Our approach is supported by data from GPS, InSAR, crack surveys, geo-electrical measurements and terrestrial laser scanning. A synthetic interferogram calculated from a model of co-seismic deformation from the 2010 Haiti earthquake, overlaid with GPS measured co-seismic displacement vectors (courtesy of UNAVCO) were also used in our explanation (Fig. 12).

Using a long-range MAPTEC terrestrial laser scanner, we obtained a highly accurate and detailed map of the morphology of the LUSI caldera and its surroundings. More than 15 scan positions were used in December 2015 to generate a cloud of several million data points (Fig. 13). Registering all

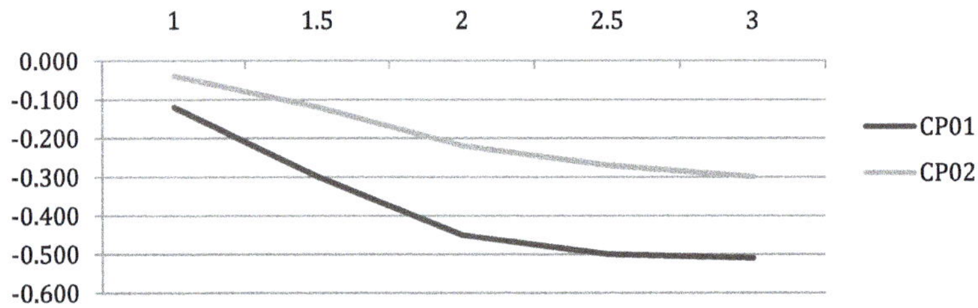

Fig. 9. Graph of subsidence on the InSAR checkpoints (CP01 and CP02), each located on the rim of the caldera. The vertical axis shows the subsidence in metres and the horizontal axis shows the number of years since the start of the eruption. Each checkpoint shows decreasing subsidence from one year to the next.

the points using the cloud-to-cloud technique, as well as constraints from high-accuracy GPS control points, resulted in accuracies of few millimetres on the registered points cloud. After filtering and filling of the points cloud, we generated a model of the morphology of the LUSI caldera and surroundings. With this millimetre-accurate model, we should be able to identify any existing fault offset (Fig. 14).

Figure 12b shows an example from the 2010 Haiti earthquake of a typical co-sesmic ground deformation signal due to fault rupture. The GPS horizontal displament vectors abruptly change direction as the fault is crossed from north to south and the InSAR deformation signal is elongated along the direction of the fault. Figure 10a,d show the ground deformation following the LUSI eruption, which, in contrast to Figure 10b, shows a compact, cone-shaped area of subsidence, such as might be expected when subsurface material is extruded onto the ground surface (e.g. in a mud volcano eruption). In particular, there is no discernible elongation of the InSAR pattern along any particular direction, nor is there any abrupt change in direction of the horizontal displacement vectors. Figure 10c shows a Google Earth image on which the LUSI mud volcano is visible and the Watukosek Fault trace is overlain; as can be seen from the figure, apart from position there is no obvious correlation in deformation patterns with the fault orientation.We therefore suggest that our ground deformation data do not support the hypothesis that the LUSI eruption was triggered by fault reactivation.

We can identify any existing fault line with the millimetre-accuracy model of the morphology of the LUSI caldera and its surroundings created by

Fig. 10. Map of crack observations and photographs. We can see clearly the cracks surrounding the centre of the LUSI eruption, which is indicative of caldera formation rather than fault reactivation.

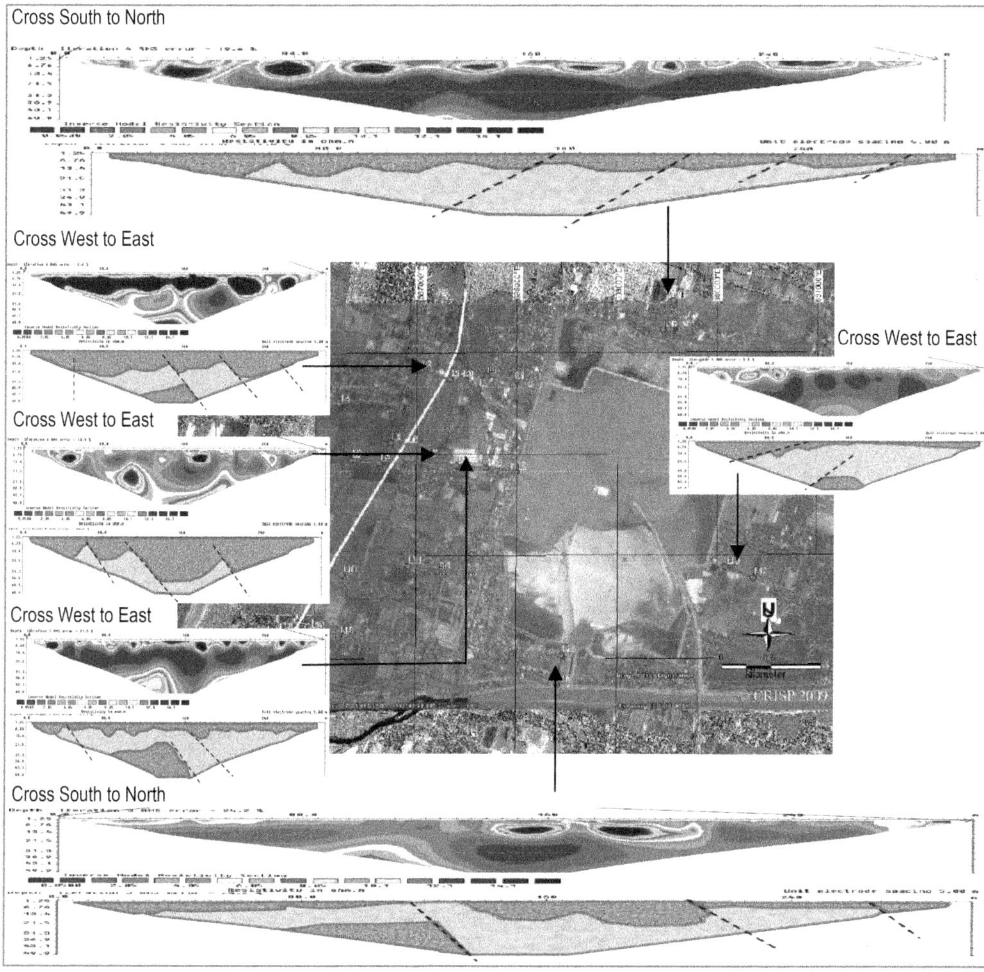

Fig. 11. Map showing locations of geo-electrical measurements along with modelling results. We can see an inwards-dipping crack pattern rounding the centre of the LUSI eruption, suggesting that the cracks are associated with caldera formation rather than fault scarp reactivation.

terrestrial laser scanning. By overlaying the Watukosek interpreted fault line with the morphology of the LUSI caldera and its surroundings determined from terrestrial laser scanning (Fig. 14), we concluded that there is no fault along the proposed Watukosek fault line. The model only shows the circle of caldera formation; if we zoom in on the image, we can see the crop circle on the rim of the caldera.

From observations of cracks and our geo-electrical measurements, we can see the cracks surrounding the centre of the LUSI eruption with an inward-dipping crack pattern that corresponds to caldera formation, not a scarp due to fault reactivation. Again, the connection to the trigerring of the LUSI eruption by fault reactivation due to the Yogyakarta earthquake is not supported by these data.

Discussion and conclusions

Previously published studies of the LUSI mud volcano eruption hypothesize that it was triggered by the reactivation of a local fault due to seismic waves from the Yogyakarta earthquake. Mazzini *et al.* (2007, 2009) concluded that the Yogyakarta earthquake triggered the LUSI eruption due to earthquake-induced hydrological effects that reactivated a fault in the LUSI area. Lupi *et al.* (2013) proposed that a concave rock layer capping the mud reservoir could have focused and amplified incoming seismic waves from the Yogyakarta earthquake and could thus have reactivated the fault and triggered the LUSI mud eruption. Nevertheless, the ground deformation signature

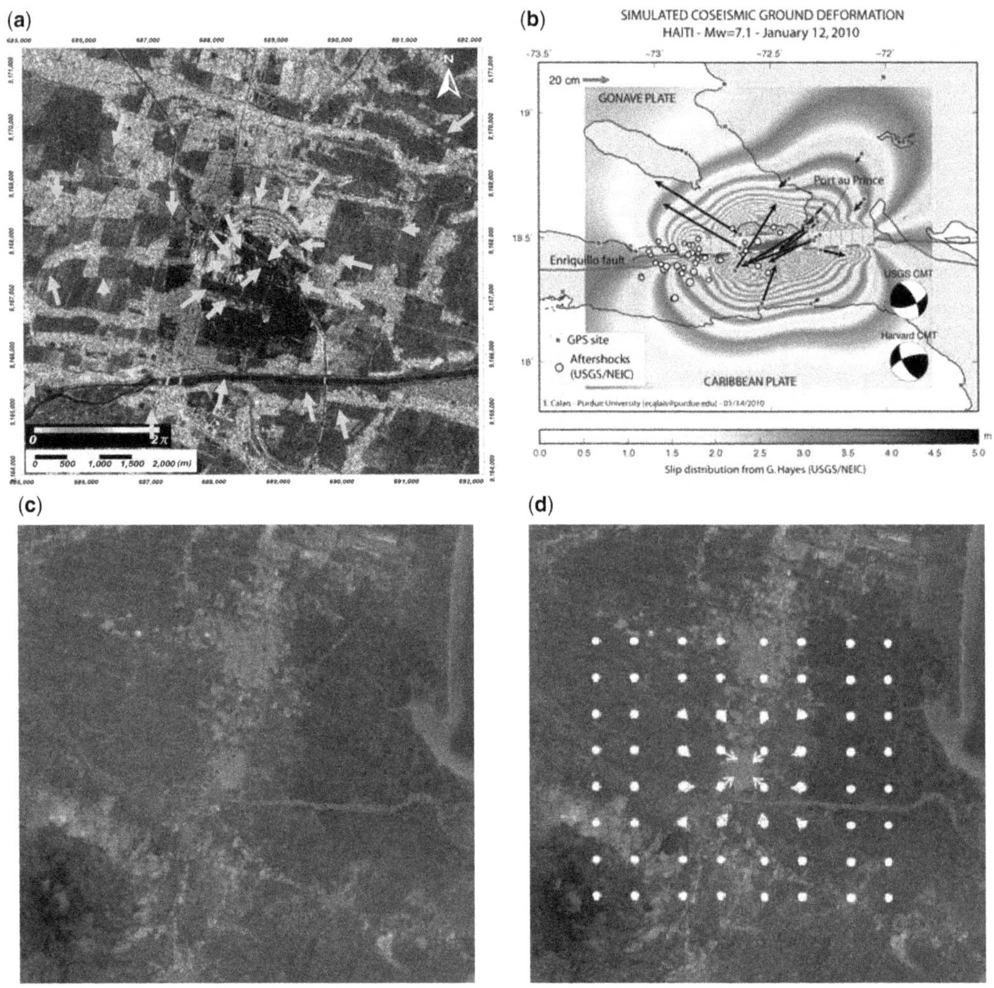

Fig. 12. (**a**) Horizontal vectors from GPS (not to scale), overlain with InSAR interferogram result. (**b**) Synthetic interferogram calculated from a model of co-seismic deformation from the 2010 Haiti earthquake, overlain with GPS measured co-seismic displacement vectors (courtesy of UNAVCO Science Highlights (2010) Event Response – $M_w = 7$ Haiti Earthquake 1 December 2010). (**c**) Google Earth image showing the extent of the LUSI mud volcano overlain with the Watukosek Fault. (**d**) Google Earth image overlain with horizontal model vectors (not to scale) from interpolated GPS and InSAR measurements. (Google Earth images from Landsat/Copernicus.)

following the LUSI eruption is of cone-shaped subsidence, which would be expected if subsurface material was extruded onto the surface (e.g. in a mud volcano eruption). It is difficult to conclude that the observed deformation pattern is associated with fault reactivation. The morphology of the LUSI mud volcano determined using a high-accuracy terrestrial laser scanning survey does not support reactivation of the Watukosek Fault. Although it is difficult to rule out the seismic focusing argument of Lupi *et al.* (2013) based on our data alone, the connection of the LUSI mud volcano with the Yogyakarta earthquake seems tenuous at best.

Supporting this conclusion, Davies *et al.* (2007) compared the distance and magnitude of the earthquake with the relationship between the distance and magnitude of historical earthquakes that have caused sediment liquefaction, triggered the eruption of mud volcanoes or caused other hydrological responses. They found that, by this comparison, an earthquake trigger is unlikely.

Detailed consideration of the ground deformation signatures at the start of the eruption

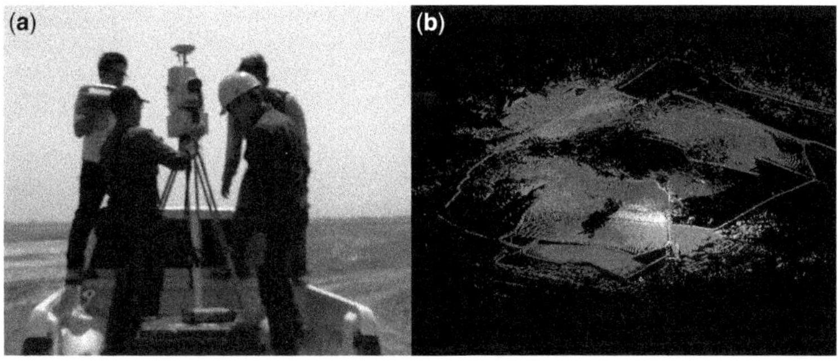

Fig. 13. (**a**) Long-range MAPTEC terrestrial laser scanner survey used to create high-accuracy model of morphology of LUSI caldera and its surroundings. (**b**) More than 15 scanned positions were taken and these generated millions of cloud points.

(2006–07), GPS surveys and InSAR measurements by Abidin *et al.* (2008) and InSAR measurements by Fukushima *et al.* (2009) have all clearly explained how ground deformation occurred around the LUSI mud volcano. Vertical and horizontal rates up to about 4 and 1 cm day^{-1}, respectively, were observed in this time period. We can also see a linear trend of deformation from their results. Abidin *et al.* (2008) and Davies *et al.* (2010) made a projection of 100 m of subsidence during the following decade. Such a large ground deformation signal would be extraordinary.

At present, more than four years after the eruption, the observed ground deformation is showing an exponential decay pattern, with rates having slowed to only several cm a^{-1} four years after the eruption. Following the large deformation rates associated with the intial eruption, the exponential

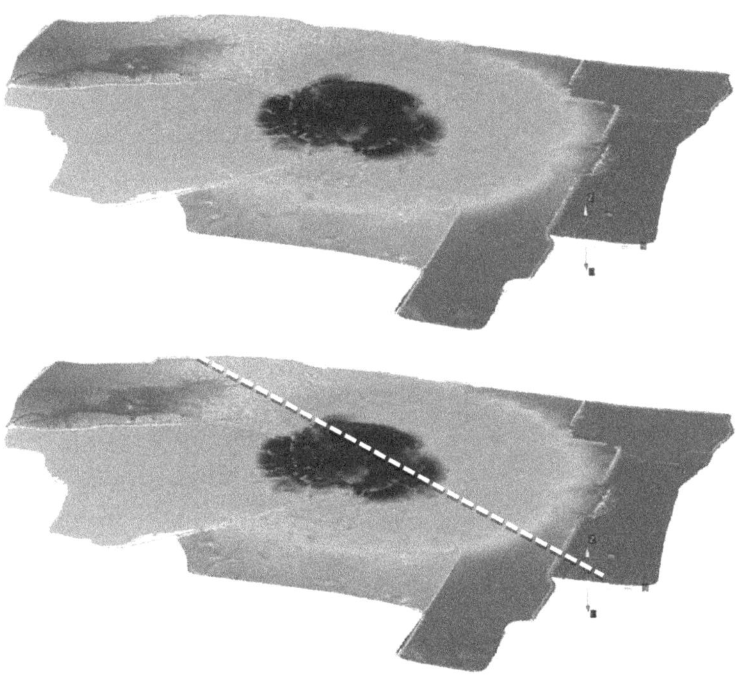

Fig. 14. High-accuracy model of morphology of LUSI caldera and surroundings created from terrestrial laser scanning survey. The Watukosek interpreted fault line is overlain on the morphology of the LUSI caldera and its surroundings.

decay in these results might be indicating the end of the LUSI mud volcano eruption. A decrease in the deformation rate may indicate that the forces driving the eruption have been spent, so that the mud eruption is finally coming to an end.

References

ABIDIN, H.Z., DAVIES, R.J., KUSUMA, M.A. & ANDREAS, H. 2008. Subsidence and uplift of Sidoarjo (East Java) due to the eruption of the LUSI mud volcano (2006–Present). *Environmental Geology*, **57**, 845, https://doi.org/10.1007/s00254-008-1363-4

ANDREAS, H., ABIDIN, H.Z., KUSUMA, M.A., SUMINTADIREJA, P. & GUMILAR, I. 2010. Ground displacement around LUSI mud volcano Indonesia as inferred from GPS surveys. Paper presented at the *FIG Congress 2010 (Facing the Challenge-Building Capacity)*, 11–16 April 2010, Sydney Australia, https://www.fig.net/resources/proceedings/fig_proceedings/fig2010/papers/ts02d/ts02d_andreas_abidin_et_al_3875.pdf

BEUTLER, G., BOCK, H. ET AL. 2007. Bernese GSP Software version 5.0. *In*: DACH, R., HUGENTOBLER, U., FRIDEZ, P. & MEINDL, M. (eds) *Bernese GSP Software version 5.0. Documentation*. Astronomical Institute, University of Bern, 1–612.

DAVIES, R.J., SWARBRICK, R.E., EVANS, R.J. & HUUSE, M. 2007. Birth of a mud volcano: east Java, 29 May 2006. *GSA Today*, **17**, 4–9, https://doi.org/10.1130/GSAT01702A.1

DAVIES, R.J., MANGA, M., TINGAY, M., LUSIANGA, S. & SWARBRICK, R. 2010. Sawolo *et al.* (2009) the LUSI mud volcano controversy: was it caused by drilling? *Marine and Petroleum Geology*, **27**, 1651–1657.

FUKUSHIMA, Y. & MORI, J., HASHIMOTO, M., KANO, Y. 2009. Subsidence associated with the LUSI mud eruption, East Java, investigated by SAR interferometry. *Marine and Petroleum Geology*, **26**, 1740–1750.

ISTADI, B. 2007. LUSI – Birth of a Mud Volcano (History and Impact of Mud Volcano Disaster in East Java). PT Lapindo Brantas presentation, Jakarta, June 2007.

LUPI, M., SAENGER, E.H., FCHS, F. & MILLER, S.A. 2013. Lusi mud eruption trigger by geometric focusing of seismic waves. *Nature Geoscience Letters*, **6**, 642–646.

MAZZINI, A., SVENSEN, H., AKHMANOV, G.G., ISTADI, B. & PLANKE, S. 2007. Pulsating and quasi-hydrothermal mud volcanism at LUSI, Indonesia. *Geophysical Research Abstracts*, **9**, 09677.

MAZZINI, A., NERMOEN, A., KROTKIEWSKI, M., PODLADCHIKOV, Y., PLANKE, S. & SVENSEN, H. 2009. Strike slip faulting as a trigger mechanism for overpressure release through piercement structure. Implication for the Lusi mud volcano, Indonesia *Marine and Petroleum Geology*, **26**, 1751–1765.

RUDOLPH, M.L., SHIRZAE, M., MANGA, M. & FUKUSHIMA, Y. 2013. Evolution and future of the LUSI mud eruption inferred from ground deformation. *Geophysical Research Letters*, **40**, 1089–1092.

Index

Page numbers in *italics* refer to Figures. Page numbers in **bold** refer to Tables.